INSTRUCTOR'S MANUAL
TO ACCOMPANY

Algebra and Trigonometry

HERBERT A. HOLLISTER

Harper
& Row

Instructor's Manual to accompany

ALGEBRA AND TRIGONOMETRY

Instructor's Manual to accompany

ALGEBRA AND TRIGONOMETRY

Herbert A. Hollister

Bowling Green State University

Harper & Row, Publishers

New York Hagerstown San Francisco London

Instructor's Manual to accompany ALGEBRA AND TRIGONOMETRY

For information, address Harper & Row, Publishers, Inc., 10 East 53rd Street, New York, N. Y. 10022.

ISBN 0-06-362972-0

CHAPTER 1

Section 3

4. (a) false (b) true
   (c) true (d) true
   (e) false (f) false

6. $|1/a| > |1/b|$

Section 4

2. 2, 3, 5, 7, 11, 13, 17, 19, 23, 29, 31, 37, 41, 43, 47, 53, 59, 61, 67, 71, 73, 79, 83, 89, 91, 97

6. 13

Section 5

2. (a) 9 (b) 1/9
   (c) no solution (d) 729
   (e) 9 (f) 140.29
   (g) 729 (h) 729

4. (a) 1/6 (b) $6^{1/10}$
   (c) 6 (d) $6^{11/6}$
   (e) $6^{5.83}$ (f) $6^{.42}$
   (g) 1 (h) $6^{11/16}$

6. (a) $2\sqrt{14}$ (b) $5\sqrt{6}$
   (c) $2x^2y^2\sqrt{3x}$ (d) $2x^3y^3\sqrt{10x}$
   (e) $3x^8y^{10}\sqrt{5xy}$ (f) $4a^6b^2c^5\sqrt{3a}$
   (g) $6x^2y^3$ (h) $x^6y^7\sqrt{238xy}$

8. (a) $-2\sqrt[3]{5}$ (b) $2xy^2\sqrt{6x^2y}$
   (c) $.1\sqrt[5]{.001}$ (d) 8
   (e) $\sqrt[3]{-19}$ (f) $\sqrt[3]{14}/2$

(g) $\sqrt[4]{63}/3$ (h) $\sqrt[3]{3x^2}/x$

(i) $3a^3b^2\sqrt[3]{2a^2bc}$ (j) $2x^4y^7z\sqrt[7]{2x^4y^5z^3}$

10. (a) $3\sqrt{2}$ (b) $2\sqrt[3]{9}$

(c) 10 (d) 70

(e) .2 (f) $\sqrt{3}$

(g) 10 (h) $2\sqrt[5]{3}$

12. (a) $\sqrt{15}/3$ (b) $\sqrt{14}/2$

(c) $\sqrt{2}/2$ (d) $\sqrt[3]{63}/3$

(e) $\sqrt{xy}/y$ (f) $\sqrt{xy^2}/y$

(g) $\sqrt[4]{12}/2$ (h) $\sqrt[3]{.06}/.1$

(i) $\sqrt{3}\ \sqrt[3]{4}/2$ (j) $\sqrt{2}\ \sqrt[3]{6}/2$

14. (a) $5a^3b^{3/2}$ (b) $3\sqrt{2}x^{9/2}y^3$

(c) $2x^{11/3}y^3$ (d) $243a^{10}b^5$

(e) $\sqrt{3}\sqrt[3]{5}a^{7/2}b^{17/6}$ (f) $(\sqrt[3]{6}x^{1/3}y^{5/3})/\sqrt[3]{5}$

(g) $(6a^{4/5}b^{-1/10})/\sqrt[3]{5}$ (h) $x^{14/5}y^{23/10}$

Section 6

2. (a) $6x^2 + 17x - 45$ (b) $20x^2 + 23x - 21$

(c) $6y^2 + 23y - 55$ (d) $14z^2 - z - 30$

(e) $24b^2 - 14ab - 55a^2$ (f) $18x^2 - 5xy - 33y^2$

(g) $15t^2 - 37tz - 76z^2$ (h) $12x^2 + 52xy + 51y^2$

4. (a) $x^4 + 4x^3 - 3x^2 + 20x - 10$

(b) $x^3 - y^3$

(c) $5x^4 + 11x^3 - 14x^2 + 13x - 3$

(d) $x^4 + x^3y + 4x^2y^2 - 3xy^3 + 5y^4$

(e) $x^5 - 4x^4 - 12x^3 - 21x^2 + 6x + 8$

(f) $a^6 - 2a^5b - 35a^4b^2 + 47a^3b^3 - 62a^2b^4 + 21ab^5 + 18b^6$

6. (a) $(a + 2b)^2$

(b) $(x + 2y)(x - 2y)(x^2 - 2xy - 4)(x^2 + 2xy + 4)$

(c) $x^2(x + 6y)(x - y)$

(d) $3a^2(a + 4b)(a - 3b)$

(e) $(x^2 + 9)(x + 3)(x - 3)$

(f) $5x(x - 3)(x + 1)$

(g) $(a - 1)(a^2 + a + 1)$

(h) $(y - 3)(y^2 + 3y + 9)$

(i) $(2a + 1)(4a^2 - 2a + 1)$

(j) $(4x^2 + 1)(2x + 1)(2x - 1)$

(k) $(y - \sqrt[3]{3})(y^2 + \sqrt[3]{3}y + \sqrt[3]{9})$

(l) $(x - 1)(x^4 + x^3 + x^2 + x + 1)$

(m) $(b - \sqrt[5]{11})(b^4 + \sqrt[5]{11}b^3 + \sqrt[5]{121}b^2 + \sqrt[5]{1331}b + \sqrt[5]{1464})$

(n) $(y - 2)(y^2 + 2y^3 + 4y^2 + 8y + 16)$

(o) $(x + 2)(x^6 - 2x^5 + 4x^4 - 8x^3 + 16x^2 - 32x + 64)$

8. (a) $(5x - 7y)(2x + 3y)$ (b) $(8x - 7y)(2x + y)$

(c) $2(5x + 3)(2x - 1)$ (d) $(4x + 3)(5x - 3)$

(e) not factorable (f) $(6x - 5)(x + 4)$

(g) $2x(3x - 5)(3x - 4)$ (h) $3y^2(2x^2 + 15x - 24)$

(i) $5x^3y^3(x - y)(x^2 + xy + y^2)$ (j) $(x^3 - 4)(x^3 + 3)$

(k) $(x + \sqrt{10})(x - \sqrt{10})$ (l) $(y + \sqrt{11})(y - \sqrt{11})$

(m) $(x - 2)(x + \frac{1}{5})$ (n) $(x + \frac{1}{2})^2$

10. (a) $\dfrac{5(\sqrt{7} + \sqrt{3})}{4}$ (b) $\dfrac{\sqrt{13} - \sqrt{5}}{8}$

(c) $-(6 + \sqrt{35})$ (d) $\dfrac{15 - 2\sqrt{26}}{11}$

(e) $\dfrac{x - 2\sqrt{xy}\ y}{x - y}$ (f) $\dfrac{\sqrt{35} + \sqrt{21}}{2}$

(g) $\dfrac{\sqrt[3]{49} + \sqrt[3]{35} + \sqrt[3]{25}}{2}$ (h) $\dfrac{\sqrt[3]{a^2} - \sqrt[3]{ab} + \sqrt[3]{b^2}}{a - b}$

12. (a) $(x + 3)(x - 5)(x + 4)$ (b) $(x - 2)(x^2 + 5x + 7)$

(c) $(x + 1)(x^3 - 4x^2) - (3x + 10)$

(d) $(x - 1)(x + 1)(x - 5)(x + 4)$

(e) $(x - 1)(x - 2)(x + 6)(x + 2)$

Section 7

2. (a) $\frac{x + y}{x - y}$ (b) $\frac{x + 3}{x + 2}$

(c) $\frac{x + 3}{x + 2}$ (d) $\frac{x + 2y}{5(x - 2y)}$

(e) $\frac{y^4 + y^3 + y^2 + y + 1}{(y^2 + 1)(y + 1)}$ (f) $\frac{x + 5}{x - 2}$

(g) $\frac{2x + 5y}{3x - y}$ (h) $\frac{a^2 + ab + b^2}{a + b}$

4. (a) $\frac{13y^2 - 4y}{(2y + 1)(y - 1)}$ (b) $\frac{3(3x - 2)}{x^2 - 4}$

(c) $\frac{x + 4y}{x - 4y}$ (d) $\frac{x - 1}{2(x + 1)}$

(e) $\frac{13 - 2x}{12y}$ (f) $\frac{(x + 3)((x - 3)^2 + 1)}{x - 3}$

(g) $\frac{4(y + 3)}{3y(y - 2)}$ (h) $\frac{7(y^2 + 2y + 4)}{5(y + 2)}$

6. (a) $\frac{x^2 + 27x + 9}{(x - 5)(x + 2)(x + 4)}$ (b) $\frac{13x^2 + 66x + 66}{x(x + 2)(x + 3)}$

(c) $\frac{3x(x - 55)}{(x + 5)^2(x - 5)^2}$ (d) $\frac{x^3 - 2x^2 + 16x - 15}{(x - 3)(x - 6)(x + 3)}$

(e) $\frac{5y^3 + 6y^2 - 12y + 5}{(y + 1)(y - 1)^2}$ (f) $\frac{-2(2x^2 - 11x + 9)}{(x + 3)(x - 3)}$

(g) $\frac{-5(2x^2 + 7x + 4)}{(x - 4)(x + 3)(x - 5)}$ (h) $\frac{-(y^4 - 5y^3 + 4y^2 + 3y - 16)}{(y + 2)(y - 2)(y^2 - 2y + 4)}$

8. (a) $\frac{(3x + 1)(x - 4)}{(x + 3)(5x + 2)}$ (b) $\frac{(4x - 1)(x - 3)}{(x + 10)(x + 4)}$

(c) $\frac{(x - 2)(x + 3)(x - 3)}{(x + 6)(x + 1)(x - 4)}$ (d) $\frac{(x - 2)(x + 1)}{(x - 1)(x + 2)}$

(e) $\frac{a^2 + 2}{a^2 - 3}$ (f) $\frac{b}{a + b}$

(g) $\frac{(y + 1)^2(y - 1)(y^2 - y + 1)}{(y + 2)^2(y - 2)(y^2 - 2y + 4)}$

(h) $\frac{(x + 1)^5}{(x + 2)^2(x^2 + 2x + 4)}$

CHAPTER 2

Section 1

2. (a) $x = \frac{100}{3}$ (b) $y = \frac{45}{68}$

(c) $x = 1$ (d) $x = -\frac{295}{53}$

(e) $-1.03/(\sqrt[3]{9} - \sqrt[3]{4})$

4. (a) $x = 3, -5$ (b) $x = 9, -5$

(c) $x = -4, -10$ (d) $x = 4, -5$

(e) $x = \frac{16}{5}, -\frac{2}{5}$ (f) $x = -\frac{13}{14}, -\frac{25}{14}$

6. (a) \$37.92 (b) \$161.10

(c) $P = i/rt$, $r = i/Pt$, $t = i/Pr$

(d) \$5,120 (e) 6%

(f) 8%

8. (a) 18.86 sq. in. (b) $b = 2A/h$, $h = 2A/b$

(c) 11" (d) 1.43 cm

10. (a) 386.384 cm (b) $l = V/wh$, $w = V/lh$, $h = V/lw$

(c) 4.53 cm

12. $y = (\frac{3}{4}) - (\frac{x}{2})$, $y = -\frac{11}{4}$

Section 2

2. (a) $x > -3$ (b) $x \geq -21$

(c) $x > -\frac{49}{12}$ (d) $x \geq \frac{383}{342}$

(e) $x < -\frac{25}{437}$

4. (a) $-\frac{7}{3} \leq x \leq 3$ (b) $-\frac{5}{4} \leq x \leq -\frac{1}{4}$

(c) $-\frac{2}{7} < x < \frac{8}{7}$ (d) $x < -\frac{5}{2}$ or $x > \frac{7}{2}$

(e) $(2\sqrt{3}/3) - (\sqrt{14}/3) \leq x \leq (\sqrt{14}/3) + (2\sqrt{3}/3)$

(f) $-22 < x < \frac{58}{3}$ (g) $x \leq \frac{3}{8}$ or $x \geq \frac{11}{8}$

(h) $x < -\frac{134}{91}$ or $x > -\frac{52}{91}$

Section 3

2. $x = \frac{94}{29}$, $y = \frac{54}{29}$

4. $x = \frac{13}{71}$, $y = -\frac{1}{71}$

6. $x = -\frac{1197}{985} = -1.215$, $y = -\frac{18971}{19700} = -.963$

8. $x = \frac{280}{79}$, $y = -\frac{138}{79}$

10. no solution

12. $x = (12\sqrt{2} + 4\sqrt{3} - 2\sqrt{6} - 2)/5$, $y = (2\sqrt{2} - 4\sqrt{6} + 4\sqrt{3} - 4)/5$

14. $x = \frac{7237}{167}$, $y = -\frac{3649}{167}$

16. $y = -(\sqrt{18} - \sqrt{55})(\sqrt{42} + \sqrt{110})/68$

$x = (\sqrt{30} - \sqrt{35})(\sqrt{110} + \sqrt{42})/68$

Section 4

2. (a) $x = 2$, $x = -2$, $x = 5$, $x = -5$

(b) $y = 2$, $y = -1$, $y = -2$

(c) $x = -3$, $x = 3$, $x = -2$

(d) $x = 2$

(e) $x = -3$

(f) $x = 1$, $x = -1$, $x = 2$, $x = -2$

(g) $x = 3$, $x = -3$, $x = 2$, $x = -2$

(h) $x = 1$, $x = -1$, $x = 2$

Section 5

2. (a) $x = -2 + 2\sqrt{3}$, $x = -2 - 2\sqrt{3}$

(b) $x = (7 + \sqrt{61})/2$, $x = (7 - \sqrt{61})/2$

(c) $x = 1$, $x = -2$

(d) $x = -6 + \sqrt{41}$, $x = -6 - \sqrt{41}$

(e) $x = 1$, $x = -\frac{5}{2}$

(f) $x = 1$, $x = -\frac{5}{3}$

(g) $x = (-1 + \sqrt{17})/8$, $x = (-1 - \sqrt{17})/8$

(h) $x = -1$, $x = \frac{1}{3}$

6. (a) -2 (b) $\frac{7}{2}$

(c) $\frac{1}{2}$ (d) -6

(e) $-\frac{3}{4}$ (f) $-\frac{1}{3}$

(g) $-\frac{1}{8}$ (h) $-\frac{1}{3}$

8. (a) $-\frac{3}{4}$ (b) $-\frac{5}{2}$

(c) $-\frac{1}{14}$ (d) 2

(e) $-\frac{7}{10}$ (f) $-\frac{35}{14}$

(g) $-\frac{17}{28}$ (h) $\frac{15}{32}$

(i) $-\frac{6}{11}$ (j) $\frac{12}{13}$

10. C/A

Section 6

4. (a) $\sqrt{11} - 2 < x < \sqrt{11} - 2$

(b) $x < 3 - \sqrt{13}$ or $x > \sqrt{13} + 3$

(c) $-\sqrt{31} - 5 < x < \sqrt{31} - 5$

(d) $(5 - \sqrt{33})/2 < x < (\sqrt{31} + 5)/2$

(e) $-(\sqrt{31} + 3)/2 \times (\sqrt{31} - 3)/2$

(f) no solutions

(g) all real numbers are solutions

6. $Ap^2 + Bp + C \leq 0$

Section 7

2. $x = 6,\ x = \frac{3}{5}$

4. $x = \frac{1}{29}$

6. $x = (-5 + \sqrt{17})/2,\ x = -(5 + \sqrt{17})/2$

8. $x = -\frac{19}{5}$

10. $x = (-5 + \sqrt{113})/4,\ y = (9 - \sqrt{113})/8$ or

$x = -(5 + \sqrt{113})/4,\ y = (9 + \sqrt{113})/8$

12. $x = (9 + \sqrt{10})/6$, $y = (9 - 5\sqrt{10})/6$ or

$x = (9 - \sqrt{10})/6$, $y = (9 + 5\sqrt{10})/6$

14. $x = 1 + 3\sqrt{3}$, $y = 16 + 18\sqrt{3}$ or $x = 1 - 3\sqrt{3}$, $y = 16 - 18\sqrt{3}$

Section 9

2. $y = -\frac{9}{5}x$, $y = -45$
4. $11y - 15x = 35$, $y = \frac{185}{11}$
6. $18x + 3xy - 7y = 114$, $y = -\frac{66}{23}$
8. $y - 6x = 12$, $y = 72$
10. $5xy^2 + 2y^2 = 48$, $y = -\sqrt{\frac{13}{12}}$

CHAPTER 3

Section 1

2. (a) $x = 5$ (b) $y = 4$

4. (a) $(10, 15)$ (b) $(1, 8)$

   (c) $(\frac{9}{2}, 2)$ (d) $(-\frac{15}{2}, -16)$

   (e) $(2, \frac{9}{2})$

6. $\sqrt{(-1 - 2)^2 + (2 - 6)^2} = \sqrt{25} = 5 = \sqrt{(6 - 2)^2 + (9 - 6)^2}$

8. $\left(\frac{c + 2a}{3}, \frac{d + 2b}{3}\right)$

   $\left(\frac{2c + a}{3}, \frac{2d + b}{3}\right)$

10. $\frac{(x + 1)^2}{64} + \frac{(y - 3)^2}{55} = 1$

    $(\sqrt{(x - 2)^2 + (y - 3)^2} + \sqrt{(x + 4)^2 + (y - 3)^2} = 16)$

12. $12x^2 - 4y^2 = 63$

Section 3

4. (a) $y = -3x + 7$

   (b) $y = 8x - 3$

   (c) $y - 9 = 17(x - 7)$

   (d) $y - 6 = \frac{4}{5}(x - 2)$

   (e) $y - 7 = \frac{3}{2}(x - 4)$ or $y - 13 = \frac{3}{2}(x - 8)$

   (f) $y + 3 = \frac{8}{5}(x - 8)$ or $y - 5 = \frac{8}{5}(x - 13)$

   (g) $y = \frac{17}{4}x - 7$

   (h) $y = 8$

   (i) $y - 9 = 6(x - 4)$

   (j) $y + 4 = -\frac{1}{6}(x - 7)$

   (k) $y = 11$

   (1) $x = 9$

Section 4

2. $y = x - 5$

4. 57.9¢ per gallon

6. (a) $R = \frac{S}{40} + 125$

   (b) \$625

   (c) \$125

8. (a) $C = 15N + 20{,}000$ in dollars

   (b) \$245,000

   (c) \$20,000

10. (a) 24.5¢ (usually rounded to 25¢)

    (b) 11¢

Section 5

6. circle

Section 6

8. $\left(-\frac{2}{9}, \frac{211}{81}\right)$

10. Two points if $(B - \beta)^2 - 4(A - \alpha)(C - \gamma) > 0$ and $A - \alpha \neq 0$

12. 4p

Section 7

16. $-a < x < a$ implies $\frac{y^2}{b^2} = \left(\frac{y}{b}\right)^2 < 0$ which is impossible

Section 8

10. G(0) does not exist. The axes are asymptotes. If n is odd, the graph is much like that of $y = 1/x$. If n is even, the graph is much like that of $y = 1/x^2$.

Section 10

6. Axioms 0, 1, 2, 3, 4, 5, 6, 11 hold.

Section 12

2. (a) $\{x : x < -3, -3 < x < 2, \text{ or } x > 4\}$

(b) $\{x : x < -6.3, -6.3 < x < -4.7, \text{ or } x > 7.2\}$

(c) $\{x : -6 < x < -4 \text{ or } -1 < x < 3\}$

(d) $\{x : x < -\frac{1}{3}, \frac{1}{7} < x < \frac{4}{5}, \text{ or } x > \frac{7}{3}\}$

(e) $\{x : -\frac{1}{2} < x < \frac{1}{6} \text{ or } \frac{3}{4} < x < \frac{4}{5}\}$

(f) $\{x : \frac{2}{11} < x < \frac{3}{5}, \frac{3}{5} < x < \frac{7}{2}, \text{ or } x > \frac{7}{2}\}$

4. (a) $\{x : -2 < x < -1, 4 < x < 3, \text{ or } x > 5\}$

(b) $\{x : 1 < x < 2\}$

(c) $\{x : x < -4, 4 < x < 5, \text{ or } x > 7\}$

(d) $\{x : x < -4, 2 < x < 5, \text{ or } x > 7\}$

(e) $\{x : x < -4, -1 < x < 3, \text{ or } x > 3\}$

(f) $\{x : -5 < x < -2 \text{ or } -1 < x < 12\}$

CHAPTER 4

Section 1

2. 0.4266

4. 1.4771

6. 0.6644

8. 3.3010

10. 2.8451

12. 0.6990

14. 5.2888

16. 0.6320 - 1, $\overline{1}$.6320, 9.6320 - 10

18. 0.0880

20. 3.9542

22. 1.6990

24. $\overline{3}$.6990, 0.6990 - 3, 7.6990 - 10

28. If 0 < a < 1, log a < 0

30. If $10^n < a < 10^{n+1}$, then n < log a < n + 1

32. $\frac{\log 3}{\log 2} = 1.5850$ but log 3 - log 2 - 0.1761

Section 2

2. (a) 256.9 (b) 43.00
   (c) 0.5270 (d) 0.0101
   (e) 78500 (f) 0.0667
   (g) 0.08841 (h) 6821
   (i) 577.1 (j) 0.009740

4. (a) 416.6 (b) 3.014
   (c) 0.3779 (d) 21.02
   (e) 0.03112 (f) 0.005826
   (g) 8366 (h) 873.4
   (i) 0.004111 (j) 0.1635

Section 3

2. (a) 121 (b) 8.75
   (c) 8945 (d) 2.571
   (e) 5.524 (f) 3.206
   (g) 68.36 (h) 8.933
   (i) 1.059

4. (a) 0.9707 (b) 0.7136
   (c) 0.882 (d) 0.5433
   (e) 0.8945

Section 4

2. \$3825

4. \$3378, \$5987

6. difference is approximately \$4 [quarterly ≐ \$1343, annually ≐ \$1339]
   monthly ≐ \$1346

8. monthly ≐ \$1082.94

   24 times/year ≐ \$1083.14

   48 times/year ≐ \$1083.21

   360 times/year ≐ \$1083.24

Section 5

2. (a) 2.0960 (b) 1.1708
   (c) 0.7686 (d) 1.1073
   (e) 0.8503 (f) 0.8127
   (g) 1.2850 (h) 2.3026

Section 6

2. (a) $\mathbb{R}$ (b) $(0, \infty)$ (c) 1
   (d) $f(p) < f(q)$ (e) $f(p) > f(q)$

4. (a) $(0, \infty)$ (b) $\mathbb{R}$ (c) 0
   (d) $g(p) < g(q)$ (e) $g(p) > g(q)$

Section 7

2. $y = 20(0.5)^x$

4. 16180 bacteria, 59 hours

6. Using figures for total population from Statistical Abstract of the United States 1974, projected population for year 2000 ≐ 298.8 million.

8. 4.8 years

10. 166°, reaches 32° in approximately 135 minutes.

CHAPTER 5

Section 1

2.

θ	sin	cos	tan	cot	sec	csc
60°	$\sqrt{3}/2$	$1/2$	$\sqrt{3}$	$\sqrt{3}/3$	2	$2\sqrt{3}/3$
30°	$1/2$	$\sqrt{3}/2$	$\sqrt{3}/3$	$\sqrt{3}$	$2\sqrt{3}/3$	2

4. (a) -30, -160, 210, 330 (b) 60, -60, 300, -300

(c) -45, -225, 135, 315 (d) 30, 210, -330, -150

(e) 0 (f) -30, -160, 210, 330

6. $-1 \leq \sin t \leq 1$, $-1 \leq \cos\theta \leq 1$

8. $\mathbb{R}$, $\mathbb{R}$

Section 2

2. (a) 45° (b) 120° (c) -15°

(d) 225° (e) -630° (f) 270°

(g) 675° (h) -67.5° (i) 30°

(j) 150°

Section 3

12. $\cos x = \sqrt{1 - t^2}$, $\tan x = \dfrac{t}{\sqrt{1 - t^2}}$, $\cot x = \dfrac{\sqrt{1 - t^2}}{t}$

$\sec x = \dfrac{1}{\sqrt{1 - t^2}}$, $\csc x = \dfrac{1}{t}$

14. $\sin x = \sqrt{1 - t^2}$, $\tan x = \dfrac{\sqrt{1 - t^2}}{t}$, $\cot x = \dfrac{t}{\sqrt{1 - t^2}}$

$\sec x = \dfrac{1}{t}$, $\csc x = \dfrac{1}{\sqrt{1 - t^2}}$

16. $\cos x = -0.8$, $\tan x = -0.75$, $\cot x = -1.33$, $\sec x = -1.25$,

$\csc x = 1.66$

18. $\cot x \doteq -0.2$, $\csc x \doteq -1.02$, $\sin x \doteq -0.98$, $\sec x \doteq 5.1$,

$\cos x \doteq 0.2$

20. $\sin x = \sqrt{1 - t^2}$, $\tan x = \pm \dfrac{\sqrt{1 - t^2}}{t}$

Section 5

2. (a) 0.5446 (b) -0.05234 (c) 5.145

(d) 0.4245 (e) -0.7986 (f) -0.9272

(g) 1.367 (h) 1.167 (i) 2.747

(j) 0.5891 (k) 0.2254 (l) 0.6249

Section 8

2. (a) 60°, $\pi/3$ (b) 45°, $\pi/4$

(c) -60°, $-\pi/3$ (d) 135°, $3\pi/4$

(e) 103°, $\frac{77}{180}\pi$ (f) -41°, $-\frac{41}{180}\pi$

(g) 35°, $7\pi/36$ (h) 60°, $\pi/3$

4. (a) 0.8660 (b) 0.6614 (c) 0.4359

(d) $\frac{11}{7}$ (e) -0.6402 (f) 0.4364

6. (a) $\pi/6$ (b) $3\pi/8$ (c) does not exist

(d) does not exist (e) 0 (f) 54.6°

8. (a) $\sqrt{1 - x^2}$ (b) $\sqrt{1 - x^2}$

(c) $\frac{x}{\sqrt{1 - x^2}}$ (d) $\frac{\sqrt{1 - x^2}}{x}$

(e) $\frac{1}{\sqrt{1 - x^2}}$ (f) $\frac{1}{x}$

(g) $\frac{x}{\sqrt{1 + x^2}}$ (h) $\frac{1}{\sqrt{1 + x^2}}$

(i) $\frac{1}{\sqrt{1 + x^2}}$ (j) $\frac{\sqrt{1 - x^2}}{x}$

CHAPTER 6

Section 1

2. (a) 15°40' (b) 32°10' (c) 61°50'
   (d) 21°10' (e) 23°10' (f) 54°40'

4. (a) 16°7' (b) 60°51' (c) 26°42'
   (d) 22°25' (e) 57°41'36" (f) 70°17'38"

6. (a) 9.7270 - 10 (b) 9.8668 - 10 (c) 0.4982
   (d) 0.0950 (e) 0.0604 (f) 0.1047

*8. (a) 781.8 (b) 59.82 (c) 102.8
   (d) 80.61 (e) 30°2' (f) 60°41'

*10. (a) 5.077 (b) 0.008549 (c) 3.208 or -3.7405
   (d) 0.1486 or -0.1181 (e) 55°29' or -55°29'
   (f) 57°28' or -36°22'

Section 2

2. 67°23', 22°37'

4. 24°52', 65°8'

6. 17.01'

8. 14°56', 31.05 mph, 344°32'

10. 24.804 ft/min

Section 3

2. $c \doteq 48.99$, $B \doteq 5°41'$, $C \doteq 23°19'$

4. $B = 76$, $a \doteq 490.1$, $c \doteq 273.1$

6. $A = 66°34'$, $b \doteq 16.55$, $c \doteq 14.90$

8. $c = 30°2'$, $B = 114°49'$, $A = 35°9'$

10. $C \doteq 15°11'$, $A \doteq 44°49'$, $a \doteq 14.08$

12. $a = 17.32$, $B = 90°$, $A = 60°$

14. $c = 1$, $b = \frac{1}{2}$, $A = 60°$

16. $\cos A = \dfrac{b^2 + c^2 - a^2}{2bc}$

18. true speed ≑ 147.3 mph

true direction ≑ 20°10' east of north

20. heading ≑ 348°21', speed ≑ 27.21 mph

22. 495'

Section 4

2. 159.7

4. 5503

6. 2633000

10. 43970 sq. ft.

12. 90° each

14. 387.3 sq. in.

Section 5

2. $\frac{\pi}{4} + \frac{n\pi}{2}$ or $45° + n90°$

4. $19°28' + 360n$ or $160°32' + 360n$ or $.3398 + 2\pi n$ or

$2.8018 + 2\pi n$

6. $(2n + 1)\pi$ or $\pm (\pi/3) + 2n\pi$ or $(2n + 1)180°$ or $\pm 60° + n360°$

8. $24°28' + 360n$ or $155°32' + 360n$ or $.4270 + 2\pi n$ or $2.7146 + 2\pi n$

10. $135° + n(180°)$

$(3\pi/4) + n\pi$

12. $n\pi$ or $\pm(\pi/6) + n\pi$

14. no solution

16. no solution

CHAPTER 7

Section 1

2. (a) $\frac{38}{53} + \frac{27}{53} i$

(b) $\frac{52}{29} + \frac{43}{29} i$

(c) $\frac{100}{37} + \frac{66}{37} i$

(d) $\frac{\sqrt{15}}{12} + \frac{\sqrt{14}}{12} i$

(e) $\frac{\sqrt[3]{42} + \sqrt{55}}{\sqrt[3]{36} + 11} + \frac{(\sqrt{5}\sqrt[3]{6} - \sqrt{11}\sqrt[3]{7})i}{\sqrt[3]{36} + 11}$

(f) $\frac{\sqrt{17}\pi + e\sqrt{19}}{36} + \frac{(e\sqrt{17} - \pi\sqrt{19})i}{36}$

(g) $\frac{3}{5} - \frac{2}{5} i$

(h) $\frac{a^2 - b^2}{a^2 + b^2} + \frac{2abi}{a^2 + b^2}$

4. (a) $i^3 = -i$, $i^4 = 1$, $i^5 = i$, $i^6 = -1$, $i^7 = -i$, $i^8 = 1$, $i^9 = i$

(b) $i^n = 1$ if n divisible by 4

$= i$ if remainder is 1 on division by 4

$= -1$ if remainder is 2

$= -i$ if remainder is 3

12. (a) $\frac{\sqrt{2}}{2} + \frac{\sqrt{2}}{2} i$ or $\frac{-\sqrt{2}}{2} - \frac{\sqrt{2}}{2} i$

(b) $\pm\sqrt{5}\left(\frac{\sqrt{2}}{2} + \frac{\sqrt{2}}{2} i\right)$

(c) $\pm\sqrt{b}\ \frac{\sqrt{2}}{2} (1 + i)$

(d) $\pm \frac{\sqrt{2}}{2} \left(\sqrt{\sqrt{a^2 + b^2} + a} + i\sqrt{\sqrt{a^2 + b^2} - a}\right)$

14. They are complex conjugates.

Section 2

2. (a) $\sqrt{29}\left(\frac{2\sqrt{29}}{29} + \frac{5\sqrt{29}}{29} i\right)$

(b) $\sqrt{58}\left(\frac{7\sqrt{58}}{58} - \frac{3\sqrt{58}}{58} i\right)$

(c) $\sqrt{317}\left(\frac{11\sqrt{317}}{317} - \frac{14\sqrt{317}}{317}i\right)$

(d) $\frac{\sqrt{145}}{12}\left(\frac{8\sqrt{145}}{145} + \frac{9\sqrt{145}}{145}i\right)$

(e) $\sqrt{\frac{2545}{196}}\left(\frac{7}{2}\sqrt{\frac{196}{2545}} - \frac{6i}{7}\sqrt{\frac{196}{2545}}\right)$

(f) $\sqrt{5}\left(\frac{\sqrt{10}}{5} + \frac{\sqrt{15}}{5}i\right)$

(g) $\frac{\sqrt{65}}{10}\left(\frac{4\sqrt{65}}{65} - \frac{7\sqrt{65}}{65}i\right)$

(h) $\frac{\sqrt{653}}{10}\left(\frac{-13\sqrt{653}}{653} + \frac{22\sqrt{653}}{653}i\right)$

(i) $\sqrt{5}(0 + i) = \sqrt{5}(\cos\frac{\pi}{2} + i\sin\frac{\pi}{2})$

(j) $\sqrt{a}(0 + 1) = \sqrt{a}(\cos\frac{\pi}{2} + i\sin\frac{\pi}{2})$

4. (a) $\frac{5\sqrt{3}}{2} + \frac{5}{2}i$ (b) $3\sqrt{2} + 3\sqrt{2}i$

(c) $-\frac{11}{2} + \frac{11\sqrt{3}}{2}i$ (d) $-0.2\sqrt{3} - 0.2i$

(e) $0.3\sqrt{2} - 0.3\sqrt{2}i$ (f) $0 - 5i$

(g) $-\frac{3\sqrt{2}}{2} + \frac{3\sqrt{2}}{2}i$ (h) $-7\sqrt{2} + 7\sqrt{2}i$

(i) $0.1812 + 0.6761i$ (j) -1.3

8. (a) $10\sqrt{2} + 10\sqrt{2}i$ (b) $-0.0668 + 0.4753i$

(c) $-14.73 + 4.503i$ (d) $0.2500 + 0.2875i$

10. (a) $-1, \frac{1}{2} + \frac{i\sqrt{3}}{2}, \frac{1}{2} - \frac{i\sqrt{3}}{2}$

(b) $-i, \frac{\sqrt{3}}{2} + \frac{i}{2}, -\frac{\sqrt{3}}{2} + \frac{i}{2}$

(c) $i, -\frac{\sqrt{3}}{2} - \frac{i}{2}, \frac{1}{2} - \frac{\sqrt{3}}{2}i$

(d) $2, 2\left(-\frac{1}{2} + \frac{\sqrt{3}}{2}i\right), 2\left(-\frac{1}{2} - \frac{\sqrt{3}}{2}i\right)$

(e) $0.9848 + 0.1736i, -0.6428 + 0.7660i, -0.3420 - 0.9397i$

(f) $1.0136(0.9578 + 0.2874i), 1.0136(-0.7278 + 0.6858i),$
$1.0136(-0.2300 - 0.9732i)$

12. (a) $-\frac{3}{2} \pm \frac{\sqrt{11}}{2} i$

(b) $-\frac{3}{4} \pm \frac{\sqrt{65}}{4} i$

(c) $-\frac{1}{2} + \frac{\sqrt{2}}{4}\sqrt{\sqrt{17} + 1} + \frac{\sqrt{2}}{4}\sqrt{\sqrt{17} - 1}$ or

$-\frac{1}{2} - \frac{\sqrt{2}}{r}\sqrt{\sqrt{17} + 1} - \frac{\sqrt{2}}{4}\sqrt{\sqrt{17} - 1}$

(d) $\pm\sqrt{\frac{5}{3}}\left(-\frac{\sqrt{2}}{2} + \frac{\sqrt{2}}{2} i\right)$

(e) $-2 + \sqrt{2}\sqrt{\sqrt{2} + 1} + i\sqrt{2}\sqrt{\sqrt{2} - 1}$ or $-2 - \sqrt{2}\sqrt{\sqrt{2} + 1} - i\sqrt{2}\sqrt{\sqrt{2} - 1}$

(f) $1.6i$ or $-i$

(g) $\frac{(-3 \pm \sqrt{65})i}{4}$

(h) $\frac{\pi\sqrt{3}}{6} + \frac{\sqrt{6}}{12}\sqrt{\sqrt{\pi^4 + 1200} + \pi^2} + \frac{i\sqrt{6}}{12}\sqrt{\sqrt{\pi^4 + 1200} - \pi^2}$ or

$\frac{\pi\sqrt{3}}{6} - \frac{\sqrt{6}}{12}\sqrt{\sqrt{\pi^4 + 1200} + \pi^2} - \frac{i\sqrt{6}}{12}\sqrt{\sqrt{\pi^4 + 1200} - \pi^2}$

14. $|\bar{z}| = 1$

CHAPTER 8

Section 1

2. In the following, $n = 1, 2, 3, \ldots$

   (a) $\left\{\frac{1}{n + 1}\right\}$

   (b) $\left\{\frac{1}{2n + 1}\right\}$

   (c) $\{5 + n(n - 1)\}$

   (d) $\left\{\frac{1}{2(3)^{n-1}}\right\}$

   (e) $\left\{\frac{n + 2}{n + 1}\right\}$

   (f) $\{(-1)^{n+1}\}$

   (g) $\{2^{n-1}\}$

   (h) $\{5n\}$

   (i) $\left\{\frac{6n}{6n - 1}\right\}$

   (j) $\left\{\frac{n + (-1)^n}{n}\right\}$

4. (a) $a_1, a_1 + d, a_1 + 2d, a_1 + 3d, a_1 + 4d, a_1 + 5d$

   (b) $\{a_1 + (n - 1)d\}$

6. In #1, (a) and (c) are increasing; (b) is decreasing.

   In #2, (c), (g), and (h) are increasing; (a), (b), (d), (e), and (i) are decreasing.

8. If $r > 1$, the sequence is increasing. If $0 < r < 1$, it is decreasing.

Section 2

2. (a) $\frac{1}{3}$ (b) $\frac{2}{3}$ (c) 1 (d) $\frac{1}{6}$

   (e) $\frac{5}{6}$ (f) $\frac{5}{12}$ (g) 0 (h) 0

6. $b_n \to L$

Section 3

2. (a) $\frac{47}{99}$ (b) $\frac{32}{165}$ (c) $\frac{265}{99}$ (d) $\frac{34}{303}$

(e) $\frac{7}{495}$ (f) $\frac{5234}{9999}$ (g) $\frac{1333}{1980}$ (h) $\frac{4261}{4500}$

(i) $\frac{707}{999}$ (j) $\frac{1675}{198}$

Section 4

2. In #1(e) and (f), A_n converges to the following:

(e) $\to 12$ (f) $\to \frac{32}{33}$

4. $0 \le \frac{1}{2^{i^2}} \le \frac{1}{2^n}$ for every positive n. Therefore $\sum_1^n \frac{1}{2^{i^2}}$ converges.

76 77 78 79 80 9 8 7 6 5 4 3 2 1

ISBN 0-06-362972-0

Algebra and Trigonometry

HERBERT A. HOLLISTER
Bowling Green State University

Algebra and Trigonometry

Harper & Row, Publishers
New York, Hagerstown, San Francisco, London

This book is dedicated to Marian Simmons Schnell, J. Robert Billings, and the memory of Byron Bowman.

Sponsoring Editor: George J. Telecki
Project Editor: Brenda Goldberg
Designer: Emily Harste
Production Supervisor: Will C. Jomarrón
Compositor: Syntax International
Printer and Binder: The Maple Press Company
Art Studio: Danmark & Michaels Inc.

Cover Photograph: The Port of New York Authority

A preliminary edition of this title was published previously.

Algebra
and
Trigonometry

Library of Congress Cataloging in Publication Data

Hollister, Herbert A
Algebra and trigonometry.

Includes index.
1. Algebra. 2. Trigonometry. I. Title.
QA152.2.H65 512.13 76-20482
ISBN 0-06-042862-7

Contents

Preface

This book is intended for use in courses designed to prepare students for the study of calculus and other courses in which they need to know the fundamentals of algebra and trigonometry. The text provides a review of basic topics, but it is assumed that the student has studied elementary algebra and some geometry.

A one-quarter or one-semester course to prepare students for calculus can be taught choosing material from Chapters 1–6, with the emphasis on Chapters 2–5. There are several sections that are independent and can be skimmed or omitted if the instructor does not feel that they are of primary importance or if time is short. The topics to be covered will depend somewhat on the amount of review necessary.

I would like to express my appreciation to Janet Blair, Jane Francis, Jo Smith, Jim Franchello, Marc Amos, and Tom O'Conner for their help in testing the first draft and preliminary edition; to my typist, Connie Wetzel, for her outstanding work; to Imogene Krift and Marilyn Titkemeier for their work in preparing the preliminary edition, and to Mary Chambers, Linda Shellenbarger, and Kae Lea Main for their secretarial assistance.

Herbert A. Hollister

1
Preliminaries

It is assumed that the student using this text is acquainted with elementary algebra and geometry. This chapter provides a common ground of basic knowledge and terminology along with a review of the fundamental manipulative skills and the techniques needed throughout the rest of the text.

1. The Language of Sets

With the introduction of "modern math" into the elementary and secondary schools, some students were told that they were learning "set theory." This was not quite the case. Formal set theory is a rigorous, highly abstract part of mathematics, too sophisticated for study at the elementary or secondary level. Many current programs have, however, benefited from including some study of sets but some rather formal and rigorous programs have confused students with a great deal of unnecessary abstraction.

The *language* of sets is very useful in mathematics in that it enables us to condense and simplify many mathematical statements and hence make them easier to grasp. In this section we will develop enough of this language to help us simplify and clarify some of the concepts that will be developed later in the text. We will not attempt to develop set theory but will be content to discuss certain basic concepts and establish some notational conventions.

Intuitively, the word *set* means the same as the words "collection," "class," "aggregate," and "bunch." The "things" in a set are called its *elements*. Thus, for example, *Treasure Island* is an element of the set of all books written by Robert Louis Stevenson, and Bowling Green State University is an element of the set of all universities in Ohio. Just as we use letters to name numbers, we can use letters to name sets and elements of sets. Usually we use lower case letters to denote elements, and upper case letters to denote sets.

If A is a set and p is an element of A, we can abbreviate the sentence "p is an element of A" by "$p \in A$." The symbol "$\in$" is read "is an element of," "belongs to," or "is in."

Example 1

If E denotes the set of even numbers, then $6 \in E$. □

If A and B are sets such that every element of A is an element of B, we say that A is a *subset* of B. We say that A is a *proper* subset of B if A is a subset of B but A is not equal to B.

The sentence "A is a subset of B" can be abbreviated by "$A \subseteq B$." The symbol "$\subseteq$" is read "is a subset of" or "is contained in." Do not confuse the symbols "$\subseteq$" and "$\in$;" they have different meanings.

Example 2

If N denotes the set of all numbers and E denotes the set of even numbers, then $E \subseteq N$. In fact, E is a proper subset of N.

Note that E is *not* an element of N; the elements of N are numbers whereas E is a *set* of numbers. □

To negate "$\in$" and "$\subseteq$" we follow the customary practice of drawing a line through the symbol. That is, $p \notin A$ means p is not an element of A, and $A \nsubseteq B$ means that A is not a subset of B.

Once the elements of a set are determined, the set is completely determined. Thus one of the simplest methods of describing a set is just to list its elements, and we usually do this within a set of braces.

Example 3

The set of vowels is $\{a, e, i, o, u\}$. □

Sets can also be described by "set builder notation" using a set of braces, a letter, and a sentence containing that letter. The set so described is the set of objects that when substituted for that letter yield a true sentence.

Example 4

$\{x: x$ is a professional hockey player$\}$

is read

"The set of all x such that x is a professional hockey player." □

In the above example we would expect all possible replacements for x to be people. In some cases the possible replacements for x may not be clear from the context, and we specify them by requiring in our notation that x be an element of a specific set.

Example 5

Let B denote the set of all Bowling Green State University students. Then

$\{x \in B: x$ has a 3.20 average$\}$

is the set of all students at Bowling Green State University who have a 3.20 average. □

Using set builder notation or some other descriptive method, we might describe a set that has no elements.

Example 6

Let $\mathcal{N}$ denote the set of all numbers. Then $\{x \in \mathcal{N}: x \neq x\}$ is a set with no elements because every number is equal to itself. □

Example 7

Similarly, $\{y \in \mathcal{N}: y + 1 = y\}$ is a set with no elements because adding 1 to every number changes it. □

Since two sets are the same if they have the same elements, these two sets are the same, as are all sets with no elements. The set with no elements is called the *empty set* or the *null set* and is denoted by $\varnothing$.

If A and B are sets, then the *union of A and B* is denoted by $A \cup B$ and is defined by

$$A \cup B = \{x: x \in A \text{ or } x \in B\}$$

The union of more than two sets is defined analogously.

The *intersection* of two sets, A and B, is denoted by $A \cap B$ and is defined by

$$A \cap B = \{x: x \in A \text{ and } x \in B\}$$

The intersection of more than two sets is defined analogously.

Example 8

Suppose

$A = \{1, 2, 4, 5, 7\}$

and

$B = \{1, 4, 7, 9, 10\}$

Then

$A \cup B = \{1, 2, 4, 5, 7, 9, 10\}$

and

$A \cap B = \{1, 4, 7\}$ □

If $A \cap B = \varnothing$, then we say that A and B are *disjoint*.

If A and B are sets, then the *difference* of A and B is denoted by $A - B$ and is defined by

$$A - B = \{x: x \in A \text{ and } X \notin B\}$$

Example 9

Suppose M denotes the set of all men in the world, W denotes the set of all women in the world, and K denotes the set of all married people in the world. Then

$M \cap W = \varnothing$

that is, M and W are disjoint.

Note also that

$M \cap K$ is the set of married men
$M \cap W$ is the set of married women
$M - K$ is the set of single men

and

$W - K$ is the set of single women □

At this point a remark is in order regarding the use of parentheses. Parentheses are used in mathematics, as in grammar, for punctuation. When we enclose an expression in parentheses, we mean to consider the quantity or object inside the parentheses as one object rather than by its components, and other symbols are related to such quantities as they would be to single objects.

Example 10

By $A \cup (B - C)$ we mean the union of the set A with the set $B - C$. □

Example 11

By $5 \times (7 + 4)$ we mean the product of 5 with the sum of 7 and 4. □

Example 12

By $(8 + 10) \div 3$ we mean the quotient obtained by dividing the sum of 8 and 10 by 3; that is, $18 \div 3$. □

The concept of *equality is essentially a logical concept*. Two objects are equal if and only if *they are identical*, that is, they are the same object. Whenever we write $A = B$, we mean that A and B are two names for the same object. This narrow view of equality will prevail throughout the text.

If we know that $A = B$, then we can substitute B for A in any expression involving A and the result will have the same meaning because B *means the same* as A. It is this principle that enables us to "add the same number to both sides of an equation" or to substitute $3 + 2$ for 5 in any numerical expression. We can use this substitution principle with sets, numbers, or any objects we desire.

Example 13

If A, B, and C are sets such that $A = B$, then

$A \cup C = B \cup C$ and $A \cap C = B \cap C$ □

Example 14

If $A = B$ and $A \subseteq C$, then $B \subseteq C$. □

Example 15

If $A = B$ and A and C are disjoint, then B and C are disjoint. □

Example 16

If $3x - 4 = x$, then $(3x - 4) + 4 = x + 4$. □

Example 17

If $4x = 12$, then $\dfrac{4x}{4} = \dfrac{12}{4}$. □

There are a few basic principles that are very useful in dealing with sets. We list them below without proof but will illustrate some of them by examples and drawings.

(a) If A and B are sets such that $A \subseteq B$ and $B \subseteq A$, then $A = B$.

Statement (a) is one of the most commonly used tools from set theory. To show that two sets are the same, the easiest approach in many cases is to show that each is a subset of the other. In a formal development of set theory, this statement is one of the fundamental axioms, called the *axiom of extension.* The assertions below can be proved using this statement and the definitions.

(b) If A, B, and C are sets such that $A \subseteq B$ and $B \subseteq C$, then $A \subseteq C$.
(c) If A is a set, then $\varnothing \subseteq A$.

The only way that $\varnothing$ could fail to be a subset of A is if $\varnothing$ contains some element not in A. Since $\varnothing$ contains no elements, this is impossible. Thus $\varnothing \subseteq A$.

(d) If A and B are sets, then $A \cup B = B \cup A$, and $A \cap B = B \cap A$.

Example 18

Let $A = \{1, 2, 3, 4, 5\}$ and $B = \{3, 4, 5, 6, 7\}$. Then

$$A \cup B = \{1, 2, 3, 4, 5, 6, 7\} = B \cup A$$

and

$$A \cap B = \{3, 4, 5\} = B \cap A \quad \square$$

Statement (d) is true primarily because the order of listing elements in a set does not matter.

(e) If A, B, and C are sets, then

$$A \cup (B \cup C) = (A \cup B) \cup C, \qquad A \cap (B \cap C) = (A \cap B) \cap C,$$
$$A \cap (B \cup C) = (A \cap B) \cup (A \cap C), \text{ and}$$
$$A \cup (B \cap C) = (A \cup B) \cap (A \cup C)$$

In order to better illustrate statements (b) and (e), it is convenient to introduce the idea of a Venn diagram.

These diagrams frequently are used to describe the relationships among sets both intuitively and geometrically. The usual procedure is to first draw a rectangle to represent some set containing all elements under discussion and call this the *universal set* or *universe of discourse.* We then draw all sets under consideration as sets of points inside circles or other closed curves inside the rectangle.

Example 19

The union and intersection of sets A and B are represented by drawings in Figures 1–1 and 1–2. The union is shaded in Figure 1–1, and the intersection is shaded in Figure 1–2. $\square$

Figure 1–1 $A \cup B$

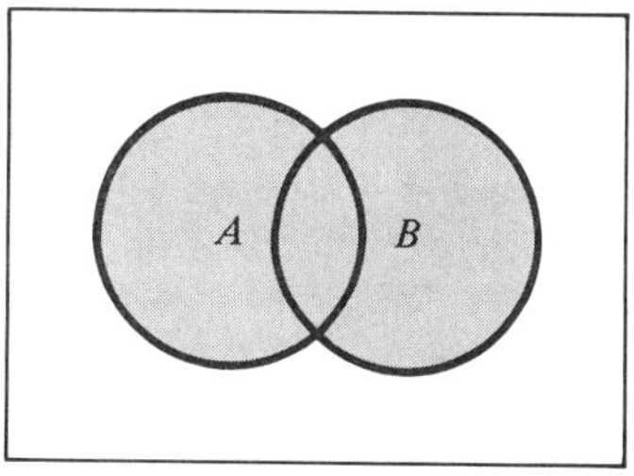

Figure 1–2 $A \cap B$

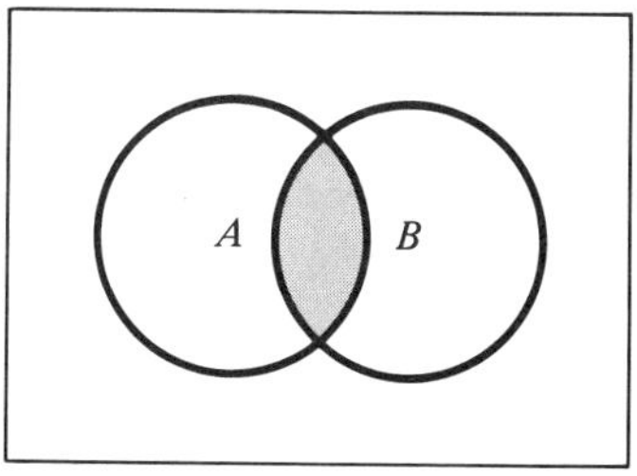

Example 20

The drawing in Figure 1–3 indicates that A and B are disjoint, and Figure 1–4 indicates that $A \subseteq B$. □

We can use Venn diagrams to help our intuition but we must realize that drawings do not constitute proofs.

Figure 1–3 $A \cap B = \phi$

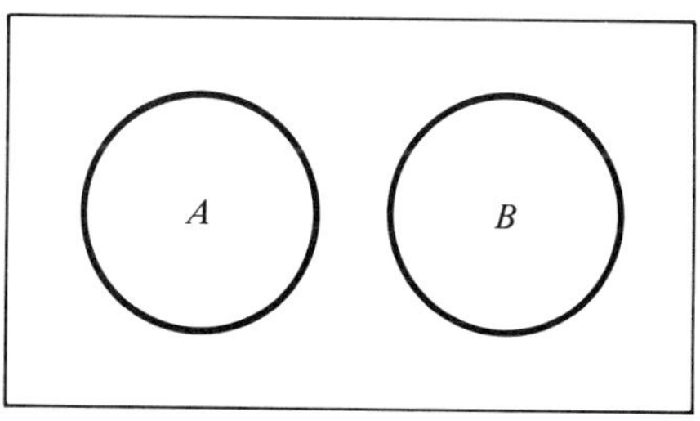

Figure 1-4 $A \subseteq B$

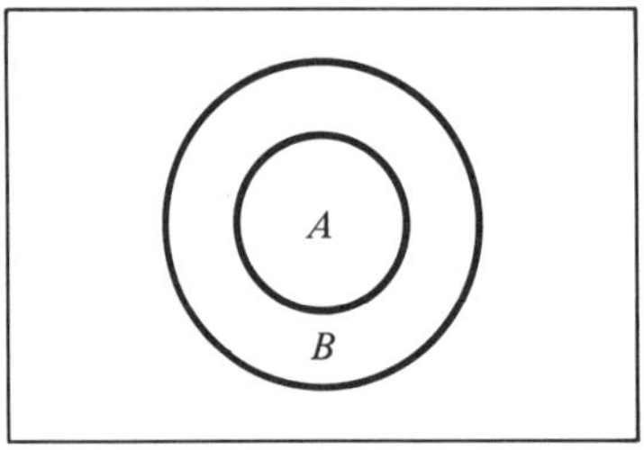

Example 21

By shading $A \cup (B \cap C)$ as in Figure 1–5 and shading $(A \cup B) \cap (A \cup C)$ as in Figure 1–6, we would expect that these sets were the same. We could not, however, claim to have a proof that such an equation would always hold. □

Figure 1-5 $A \cup (B \cap C)$

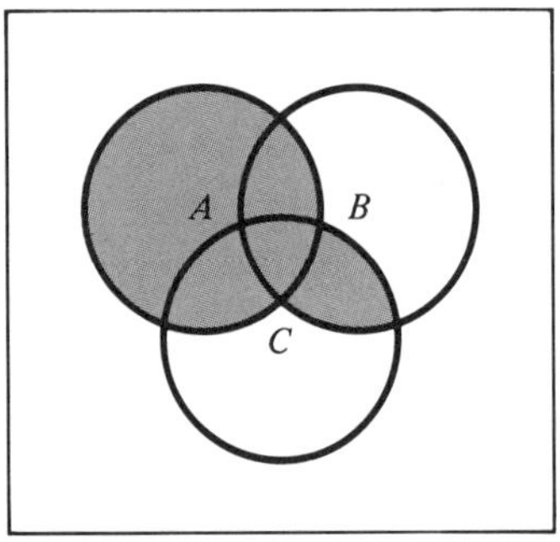

Figure 1-6 $(A \cup B) \cap (A \cup C)$

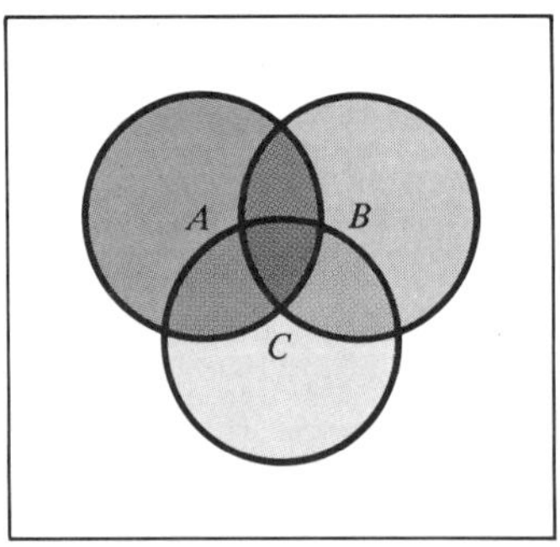

Since this is not a rigorous course in set theory, we will accept justifications made by means of Venn diagrams, providing the diagrams represent the general situation.

EXERCISES

1. Let $A = \{a, b, c, d, e, f\}$ and $B = \{a, c, e, g, h, k\}$. List the following sets.
   (a) $A \cup B$ (b) $A \cap B$ (c) $A - B$
   (d) $B - A$ (e) $A - (A \cap B)$ (f) $(A - B) \cup (B - A)$
2. Let A and B be arbitrary sets. Draw Venn diagrams to represent the following.
   (a) $A - B$ (b) $B - A$ (c) $A - (A \cap B)$ (d) $(A - B) \cup (B - A)$
3. Let A and B be arbitrary sets. Draw Venn diagrams to represent the following.
   (a) $(A \cup B) - C$ (b) $A \cup (B - C)$
   (c) $A \cap (B - C)$ (d) $A \cap (B \cap C)$
   (e) $(A \cap B) \cap C$ (f) $A \cap (B \cup C)$
   (g) $A \cup (B \cap C)$ (h) $(A - B) \cup (B - C) \cup (C - A) \cup (A \cap B \cap C)$
4. Let A, B, and C be arbitrary sets. Use Venn diagrams to illustrate and justify the following statements.
   (a) $A \cup (B \cup C) = (A \cup B) \cup C$ (b) $A \cap (B \cap C) = (A \cap B) \cap C$
   (c) $(A \cap B) \subseteq A$ (d) $A \subseteq (A \cup B)$
   (e) $A \cap (B \cup C) = (A \cap B) \cup (A \cap C)$
5. Let A, B, C, and D be sets. Decide whether the following are true or false. You may wish to use Venn diagrams to help make your decisions.
   (a) $A \cap (A \cup B) = A$
   (b) $A \cup (A \cap B) = A$
   (c) If $A \subseteq B$ and $B \cap C = \varnothing$, then $A \cap C = \varnothing$.
   (d) If $A \subseteq B$ and $A \subseteq C$, then $A \subseteq B \cap C$.
   (e) If $A \cup B \subseteq A \cap B$, then $A = B$.
   (f) $(A - B) \cap C = (A \cap C) - B$
   (g) If $A \cup B = A \cup C$, then $B = C$.
   (h) If $A \subseteq B$ and $C \subseteq D$, then $A \cap C \subseteq B \cap D$.
   (i) $(A - B) \cap B = \varnothing$
   (j) $A - (B \cap C) = (A - B) \cap (A - C)$
   (k) $A - (B \cap C) = (A - B) \cup (A - C)$
   (l) $A - (B \cup C) = (A - B) \cap (A - C)$

2. The Real Number System

In algebra, as in arithmetic, we are primarily concerned with numbers. In algebra, however, we are more concerned with general properties and behavior of numbers than we are with specific numbers. The set of numbers that we use in the first six chapters of this book is the real number system. In Chapter 6, we develop the complex number system which has many applications in higher mathematics, physics, and engineering. In the first six chapters of this book, the word "number" means "real number." We will not attempt to give a rigorous development of the real number system but will be content to provide the student with an intuitive, geometric, arithmetic grasp of the system and its properties. We will, however, organize the basic properties of the system into a set of axioms and a sequence of theorems. This will enable you to refer to the properties more easily and to see how one property may depend on others. A few proofs have been included to provide illustrations of how the theorems can be derived from the basic properties, the axioms. Proofs, after all, are just explanations of why a statement is true. Throughout this text we provide explanations of *why* certain statements are true and *why* certain methods can be used because we believe that you will remember best if you understand *why* the fact that you are studying is true. The explanations will have, as their foundation, the assertions listed in this section.

In writing numerical expressions we adopt the following standard notational conventions.

I. Multiplication will usually be indicated either by a dot · or by juxtaposition, that is, by placing the numbers side by side.

Example 1

By $5 \cdot 3$ we mean the product of 5 and 3. The expression (5)(3) would also mean their product. □

Example 2

If we represent two numbers by letters x and y, we usually write xy for the product of x and y. □

II. Multiplication and division take precedence over addition and subtraction.

Example 3

$5 \cdot \; + 4$ means to add the product of 5 and 3 to the number 4; that is, first multiply 5 times 3 and then add 4. If we wish to indicate the product of 5 with the sum of 3 and 4, we would use parentheses and write $5 \cdot (3 + 4)$. □

Example 4

$pq + mn$ means $(p \cdot q) + (m \cdot n)$. □

Example 5

$x + y/z$ means $x + (y/z)$. We do the division before the addition. □

III. Multiplication and division are to be performed in the order in which they occur as are addition and subtraction.

Example 6

$7 \cdot 8 \div 2$ means $(7 \cdot 8) \div 2 = 56 \div 2 = 28$. □

Example 7

$6 \div 3 \div 2$ means $(6 \div 3) \div 2 = 2 \div 2 = 1$. □

Example 8

$14 - 5 + 3$ means $(14 - 5) + 3 = 9 + 3 = 12$. □

Example 9

$17 - 6 - 4$ means $(17 - 6) - 4 = 11 - 4 = 7$. □

When you are writing expressions, make sure that you indicate clearly what you mean. If parentheses would clarify your meaning, use them. Notational conventions save us some writing but should not lead us into confusion.

Intuitively, the real numbers may be thought of as the points of a line as in Figure 1–7.

Figure 1–7

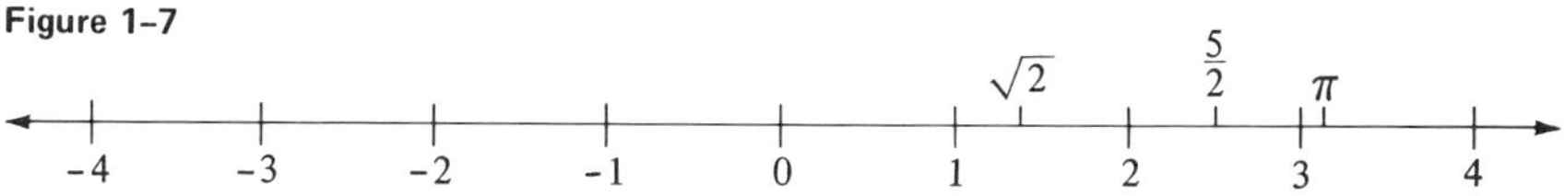

Numbers may also be considered as directed distances from the origin or "zero" point. Intuitively, a number p is less than a number q if p is "to the left" of q on the line. Besides comparing numbers, we expect to be able to do arithmetic, that is, to add, subtract, multiply, and divide numbers, and these processes along with the ordering of numbers are described and governed by certain fundamental principles.

These fundamental principles are presented in the form of axioms, lemmas, and theorems. The axioms are the fundamental assumptions that we make about the number system. Essentially we are saying "If the number system is to behave as our intuition and experience expect it to behave, then the following basic statements must hold." Once we choose the axioms, then the lemmas and theorems all can be proved by proceeding logically from the axioms. We will not provide proofs of most of these statements since we are primarily concerned with your ability to *use* these properties rather than your ability to derive or prove them. You should make sure that you understand the principles under discussion.

AXIOMS FOR THE REAL NUMBERS

The real numbers, denoted by R, is a set with two operations, addition and multiplication, denoted by $+$ and $\cdot$, respectively, which satisfies the following axioms.

AXIOM 0

If a and b are real numbers, then $a + b$ and $a \cdot b$ are real numbers. (Closure Axiom)

AXIOM 1

If a and b are real numbers, then

$$a + b = b + a \quad \text{and} \quad a \cdot b = b \cdot a \qquad \text{(Commutative Laws)}$$

Example 10

$$15 + 24 = 24 + 15 \quad \text{and} \quad (15)(24) = (24)(15). \ \square$$

AXIOM 2

If a, b, and c are real numbers, then

$$a + (b + c) = (a + b) + c$$

and

$$a \cdot (b \cdot c) = (a \cdot b) \cdot c \qquad \text{(Associative Laws)}$$

Example 11

$$16 + (13 + 20) = (16 + 13) + 20 \quad \text{because} \quad 16 + 33 = 29 + 20$$

$$5 \cdot (7 \cdot 3) = (5 \cdot 7) \cdot 3 \quad \text{because} \quad 5 \cdot 21 = 35 \cdot 3 \ \square$$

AXIOM 3

If a, b, and c are real numbers, then

$$a \cdot (b + c) = a \cdot b + a \cdot c$$

and

$$(a + b) \cdot c = a \cdot c + b \cdot c \qquad \text{(Distributive Laws)}$$

Example 12

$$5(20 + 7) = 5 \cdot 20 + 5 \cdot 7 \quad \text{since} \quad 5(27) = 100 + 35$$

In fact, it is this principle that lets us multiply $5 \cdot 27$ by multiplying $5 \cdot 20$ and $5 \cdot 7$, and then adding. □

AXIOM 4

There is a unique real number 0 such that if a is any real number, then $a + 0 = 0 + a = a$ and $a \cdot 0 = 0 \cdot a = 0$.

AXIOM 5

There is a unique real number 1, different from 0, with the property that if a is any real number, then $a \cdot 1 = 1 \cdot a = a$.

AXIOM 6

If a is a real number, then there is a unique real number, denoted by $-a$ and called the *negative of* a, with the property that if a is any real number, then $a + (-a) = (-a) + a = 0$.

Example 13

The negative of 15 is -15 because $15 + (-15) = 0$. The negative of -12 is 12 because $(-12) + (12) = 0$. □

AXIOM 7

If a and b are nonzero real numbers, then $a \cdot b \neq 0$.

AXIOM 8

There is a relation called "less than," denoted by "$<$," defined on R which satisfies the following conditions:

(i) For every pair of real numbers a and b, exactly one of the following holds: $a < b$, $a = b$, or $b < a$.
(ii) If a, b, and c are real numbers such that $a < b$ and $b < c$, then $a < c$.

The relation "less than" is used to define another relation "greater than" by saying that a is greater than b if and only if b is less than a. We denote "greater than" by "$>$."

If a and b are real numbers we say that $a \leq b$ if $a < b$ or $a = b$. The symbol "$\leq$" is read "less than or equal to." The meaning of the symbol "$\geq$" is defined in a similar manner.

A number is said to be *positive* if it is greater than 0 and *negative* if it is less than 0.

AXIOM 9

If a, b, and c are real numbers such that $a < b$, then $a + c < b + c$.

Example 14

Since $4 < 11$, $4 + 5 < 11 + 5$ and $4 + (-6) < 11 + (-6)$. □

Example 15

Since $-4 < 2$, $(-4) + 3 < 2 + 3$; that is, $-1 < 5$. □

AXIOM 10

If a, b, and c are real numbers such that $a < b$ and $0 < c$, then $ac < bc$.

Example 16

Since $5 < 7$ and $0 < 3$, $5 \cdot 3 < 7 \cdot 3$; that is, $15 < 21$. □

Example 17

Since $-6 < -2$ and $0 < 5$, $(-6)5 < (-2)5$; that is, $-30 < -10$. (The fact that $(-6)5 = -30$ and $(-2)5 = -10$ will be justified by Theorem 3.) □

If a and b are real numbers, then a/b is the unique real number whose product with b is a if such a unique real number exists. The number a/b is called *a divided by b*, or the quotient of a and b. We also use the usual division sign "÷." Remember that "$a \div b$" and "a/b" mean the same thing.

Example 18

$(72/9)9 = 72$ □

Example 19

$266/19 = 14$ because $14 \cdot 19 = 266$ □

Example 20

$(143/\pi)\pi = 143$ □

AXIOM 11

If a and b are real numbers and $b \neq 0$, then a/b exists in R; that is, there is a unique real number whose product with b is a.

AXIOM 12

If S is a nonempty subset of R and a is a real number such that every element of S is less than a, then there is a smallest real number b which is greater than or equal to every element of S.

Axiom 12 is called the "least upper bound property." An upper bound of a set is just a number greater than or equal to every number in the set and Axiom 12 asserts that any set having an upper bound in R must have a smallest or least upper bound in R. It is this property that assures us that every point on the line is associated with a real number and conversely, that every real number is associated with a point. Furthermore this property completely characterizes the real number system and distinguishes it from the rational number system and others. We will not use Axiom 12 in this text until Chapter 8.

Using the axioms, one can prove the following lemmas and theorems. We will not give proofs of most of them but will do so for the first three because many students have never seen a justification of the "Law of Signs." We suggest that the student attempt to provide proofs for some of the other assertions or at least parts of them. We will illustrate all of the assertions by examples.

LEMMA 1

If a is a real number, then $(-1)a = -a$.

Proof:

$a + (-1) \cdot a = 1 \cdot a + (-1) \cdot a$	because $a = 1 \cdot a$ by Axiom 5
$= [1 + (-1)] \cdot a$	by the Distributive Law, Axiom 3
$= 0 \cdot a$	because $1 + (-1) = 0$ by Axiom 6
$= 0$	by Axiom 4

Thus $a + (-1) \cdot a = 0$.

Since $-a$ is the only number whose sum with a is 0, $(-1) \cdot a$ must be $-a$, and, therefore, $(-1) \cdot a = -a$.

Example 21

$(-1)(7) = -7$ □

Example 22

$(-1)\pi = -\pi$ □

Do you recall the first time you were told that $(-3) \cdot (-5) = 15$? For many students, it does not seem possible that the product of two negative numbers could be positive. Some high school courses justify this by several intuitive and physical examples, but such arguments are quite artificial. That $(-3) \cdot (-5)$ must be 15 is a direct consequence of the first six axioms for R. Thus if we are to have a number system satisfying these basic properties, we must have $(-3)(-5) = 15$.

Before proving the general case of the Law of Signs in Theorem 3, we will prove a special case as a lemma which will then enable us to easily prove the more general case.

LEMMA 2

$(-1)\cdot(-1) = 1$

Proof:

$(-1)\cdot(-1) = -(-1)$	by Lemma 1 where (-1) is a
$= 1$	the negative of $(-a)$ is just a for any number a because of the uniqueness required in Axiom 4

THEOREM 3 (The Law of Signs)

If a and b are real numbers, then $-(a\cdot b) = (-a)\cdot b = a\cdot(-b)$ and $(-a)\cdot(-b) = a\cdot b$.

Proof:

$-(a\cdot b) = (-1)\cdot(a\cdot b)$	by Lemma 1
$= ((-1)\cdot a)\cdot b$	by associativity, Axiom 2
$= (-a)\cdot b$	by Lemma 1

Similarly,

$(-a)\cdot b = ((-1)\cdot a)\cdot b$	by Lemma 1
$= (a\cdot(-1))\cdot b$	by commutativity, Axiom 1
$= a\cdot((-1)\cdot b)$	by associativity, Axiom 2
$= a\cdot(-b)$	by Lemma 1

Therefore $-(a\cdot b) = (-a)\cdot b = a\cdot(-b)$ and the first part of the theorem is proved. The second part follows by applying Lemmas 1 and 2:

$(-a)\cdot(-b) = [(-1)\cdot a][(-1)\cdot b]$	by Lemma 1
$= [a\cdot(-1)][(-1)\cdot b]$	by commutativity, Axiom 1
$= a\cdot\{(-1)\cdot[(-1)\cdot b)]\}$	by associativity, Axiom 2
$= a\cdot\{[(-1)\cdot(-1)]\cdot b\}$	by associativity, Axiom 2
$= a\cdot[1\cdot b]$	by Lemma 2
$= a\cdot b$	by Axiom 5

Therefore $(-a)\cdot(-b) = a\cdot b$.

Example 23

$(3)(-4) = -(3\cdot 4) = -12$ □

Example 24

$(-5)(2) = -(5 \cdot 2) = -10$ □

Example 25

$(-4)(-6) = (4)(6) = 24$ □

In basic arithmetic, you define subtraction in terms of addition.

Example 26

$47 - 19 = 28$ because $28 + 19 = 47$ □

In general, $a - b$ is the number whose sum with b is a, that is, $(a - b) + b = a$.

Example 27

$21 - 15 = 6$ because $6 + 15 = 21$ □

Example 28

$19 - 26 = -7$ because $(-7) + (26) = 19$ □

THEOREM 4

If a and b are numbers, then $a - b = a + (-b)$.

Proof:

$$[a + (-b)] + b = a + [(-b) + b] = a + 0 = a$$

PROPOSITION 5

If a, b, and c are real numbers, then

$$a \cdot (b - c) = ab - ac \quad \text{and} \quad (a - b) \cdot c = ac - bc$$

Proof: Exercise

From Axiom 11, we know that division of a real number by a nonzero real number is always defined in R; that is, if x and y are real numbers and $y \neq 0$, then x/y will be a real number. Recall that, by definition, $(x/y)\,y = x$.

Is division by 0 possible? Consider, for example, the possibility of dividing 1 by 0. Is there a real number whose product with 0 is 1? Since the product of 0 with any number is 0, no such number exists. In general, if p is a nonzero real number, then there is *no* number whose product with 0 is p, and therefore, division of p by 0 is impossible. Is it possible to divide 0 by 0? Since $0 \cdot 0 = 0$ and $1 \cdot 0 = 0$, there is *more than one* real number whose product with 0 is 0. In order for 0/0 to have meaning, 0/0 would have to be the *unique* number whose product with 0 is 0, and there is no such unique number. In this discussion we have actually proved the following assertion.

PROPOSITION 6

If r is a real number, then $r/0$ has no meaning; that is, there is no unique real number whose product with 0 is r.

Our reason for choosing the fraction line for the division symbol becomes clear as we develop the properties of division. All the usual techniques for computing with fractions hold just because a fraction is, in fact, a quotient.

Frequently when a student first encounters fractions in arithmetic, he or she gets the idea that the arithmetic of fractions is different in some way from the arithmetic of whole numbers. That is, he or she expects to have different rules for operating with fractions than for operating with other numbers. Such an assumption or conclusion is a handicap in arithmetic. Fractions are, after all, just a way of representing numbers. Numbers represented by fractions have to behave according to the same laws as other numbers; that is, according to the principles governing the real numbers. Therefore, the basic properties of the arithmetic of numbers represented by fractions are the same as the basic properties of arithmetic for all numbers. The computational techniques developed are, in fact, just techniques for computing with numbers expressed in fractional form; that is, techniques for using numerators and denominators of fractions in the computation of sums, differences, products, and quotients of numbers expressed as fractions. Since the fraction line means division, all these techniques are derivable and provable using just the meaning of division and the basic properties of the arithmetic of the real number system. We have put these together in the following theorem so that the student may see the relationship between these techniques and division. We will discuss these principles in more detail when we study algebraic fractions.

THEOREM 7

Let a, b, c, and d be real numbers with $b \neq 0$ and $d \neq 0$.

The following all hold.

(i) $(a/b)b = a$
(ii) $a/1 = a$
(iii) $b/b = 1$
(iv) $a(1/b) = a/b$

(v) $(a/b)c = ac/b$
(vi) $(a/b)/d = a/bd$
(vii) $a/b = ad/bd$
(viii) $a/b = c/d$ if and only if $ad = bc$.
(ix) $(a/b)(c/d) = ac/bd$
(x) $(a/b) \div (c/d) = (a/b)(d/c)$ if $c \neq 0$.
(xi) $-(a/b) = (-a)/b = a/(-b)$
(xii) $a/b + c/b = (a + c)/b$
(xiii) $a/b - c/b = (a - c)/b$
(xiv) $a/b + c/d = (ad + bc)/bd$
(xv) $a/b - c/d = (ad - bc)/bd$

Proof:

We will not provide proofs of all of the above statements but will give justification for some of them and leave the rest for the student. All of the principles are illustrated by Example 29.

Principle (i) follows from the definition of division in that a/b is defined to be the number whose product with b is a.

Principle (iv) can be established just by showing that $a(1/b)$ is the number whose product with b is a. If we form the product, we see that this is, in fact, true.

$$[a(1/b)]b = a[(1/b)b] = a(1) = a$$

In order to establish principle (xii) we need to show that $a/b + c/b$ is the number whose product with b is $a + c$. Most of the other principles in this theorem can be established by similar arguments. Some of them require a bit more reasoning using the previously established principles. The student should attempt some of them.

Example 29

The following numerical statements illustrate the corresponding parts of Theorem 7.

(i) $(4/3)3 = 4$; $(35/5)5 = 35$
(ii) $-26/1 = -26$; $(4/5)/1 = 4/5$
(iii) $9/9 = 1$; $(2/3)/(2/3) = 1$
(iv) $6(1/5) = 6/5$; $(-7)(1/9) = -7/9$
(v) $(4/5)3 = 12/5$; $(-6/7)(4) = -24/7$
(vi) $(2/3)/5 = 2/15$; $(4/7)/3 = 4/21$
(vii) $2/3 = (2 \cdot 7)/(3 \cdot 7) = 14/21$; $9/4 = (9 \cdot 2)/(4 \cdot 2) = 18/8$
(viii) $2/3 = 8/12$ because $2 \cdot 12 = 3 \cdot 8$;
$4/7 \neq 5/8$ because $4 \cdot 8 \neq 7 \cdot 5$.

(ix) $(2/3)(4/5) = 8/15$; $(3/4)(2/3) = 6/12$

(x) $(2/3) \div (5/7) = (2/3)(7/5) = 14/15$

(xi) $-(4/5) = (-4)/5 = 4/(-5)$

(xii) $4/3 + 7/3 = (4 + 7)/3 = 11/3$

(xiii) $8/5 - 11/5 = (8 - 11)/5 = (-3)/5$

(xiv) $\dfrac{2}{3} + \dfrac{4}{5} = \dfrac{2 \cdot 5 + 3 \cdot 4}{3 \cdot 5} = \dfrac{10 + 12}{15} = \dfrac{22}{15}$

(xv) $\dfrac{5}{8} - \dfrac{2}{3} = \dfrac{5 \cdot 3 - 8 \cdot 2}{8 \cdot 3} = \dfrac{15 - 16}{24} = \dfrac{-1}{24}$ □

EXERCISES

1. Use the definition of division to decide whether the following are true or false.
   (a) $224/7 = 32$ (b) $323/17 = 19$ (c) $629/27 = 27$
   (d) $280/15 = 19$ (e) $115/5 = 23$ (f) $962/26 = 27$
2. Illustrate each part of Theorem 7 by letting $a = 14$, $b = 11$, $c = 7$, and $d = 5$.
3. Illustrate each part of Theorem 7 by letting $a = -18$, $b = 6$, $c = -2$, and $d = -3$.
4. Justify part (vii) of Theorem 7 by showing that $(a/b)(bd) = ad$. [*Hint*: $(a/b)(bd) = [(a/b)b]d$.]
5. Justify part (x) of Theorem 7 by showing that $[(a/b)(d/c)](c/d) = a/b$.
6. Justify part (xii) of Theorem 7 by showing that $(a/b + c/b)b = a + c$.
7. Justify part (xiv) of Theorem 7 by using parts (vii) and (xii). [*Hint*: $a/b = ad/bd$.]

*8. Complete the proof of Theorem 7.

3. The Order Properties of *R*

Along with being able to add, subtract, multiply, and divide real numbers, we also need to compare them; that is, to know when one number is less than another. This comparability is of primary importance in many applications. We may wish to compare bowling scores, test scores, prices of automobiles, costs of manufacturing swimsuits, or temperatures. The principles governing ordering and inequalities are given by Axioms 8, 9, 10, and 12, and by the consequences of these axioms.

Intuitively, Axiom 9 assures us that if we add the same number to both sides of an inequality, we obtain an inequality in the same direction or sense. Axiom 10 asserts that if we multiply both sides of an inequality by a *positive* number, we obtain an inequality in the same direction. The following theorem contains the other basic principles governing inequalities and these can be proved using the axioms. As before, we will not provide proofs but will illustrate each part of the theorem by examples and illustrations.

THEOREM 8

Let a, b, c, and d be real numbers. The following hold.

(i) If $a < b$, then $-b < -a$.
(ii) If $0 < a$ and $0 < b$, then $0 < ab$.
(iii) If $a < 0$ and $b < 0$, then $ab > 0$.
(iv) If $a < b$ and $0 < c$, then $a/c < b/c$.
(v) If $a < b$ and $c < 0$, then $ac > bc$ and $a/c > b/c$.
(vi) If $a \neq 0$, then $0 < a \cdot a$.
(vii) $0 < 1$
(viii) $c > 0$ if and only if $1/c > 0$.
(ix) Let $bd > 0$. Then $a/b < c/d$ if and only if $ad < bc$.

Example 1

The following numerical statements illustrate the corresponding parts of Theorem 8.

(i) Since $2 < 5$, $-5 < -2$.
Since $-3 < 7$, $-7 < 3$.
Since $-8 < -2$, $2 < 8$.
(ii) Since $0 < 7.2$ and $0 < 3$, $0 < 21.6$.
(iii) Since $2 < 5$ and $0 < 7$, $2/7 < 5/7$.
Since $-8 < -4$ and $0 < 2$, $-8/2 < -4/2$; that is, $-4 < -2$.
(iv) Since $3 < 5$ and $-2 < 0$, $3/(-2) > 5/(-2)$; that is, $-(3/2) > -(5/2)$.
Since $-10 < -4$ and $-8 < 0$, $-10/(-8) > -4/(-8)$; that is, $5/4 > 1/2$.
(v) Since $-6 < 0$ and $-2/3 < 0$, $(-6)(-2/3) > 0$; that is, $4 > 0$.
(vi) Since $5 \neq 0$, $0 < 25$.
Since $-3 \neq 0$, $0 < 9$.
(viii) Since $5 > 0$, $1/5 > 0$.
Since $1/19 > 0$, $19 > 0$.
Since $4/7 > 0$, $1/(4/7) > 0$; that is, $7/4 > 0$.
(ix) Since $2 \cdot 3 > 0$ and $7 \cdot 2 < 3 \cdot 9$, $7/3 < 9/2$.

To decide whether or not $42/13 < 51/15$, since $13 \cdot 15 > 0$, we compute $42 \cdot 15$ and $13 \cdot 51$ and compare. Since $42 \cdot 15 = 630 < 663 = 13 \cdot 51$, we know $42/13 < 51/15$. □

Proof of Theorem 8:

We will not prove every part but will justify some of them. The student should attempt to prove the others. Principle (i) is established by using Axiom 9 twice. Just add $-a$ to both sides and then add $-b$ to both sides.

To prove (ii), notice that if $a/c \not< b/c$, then either $a/c = b/c$ or $b/c < a/c$. Then since $0 < c$, either $(a/c)c = (b/c)c$ and $a = b$ or $(b/c)c < (a/c)c$ and $b < a$. Since both of these are impossible, it must be that $a/c < b/c$.

Number lines can also be used to illustrate parts of Theorem 8.

Example 2

(i) If a is to the left of b, then $-b$ is to the left of $-a$.

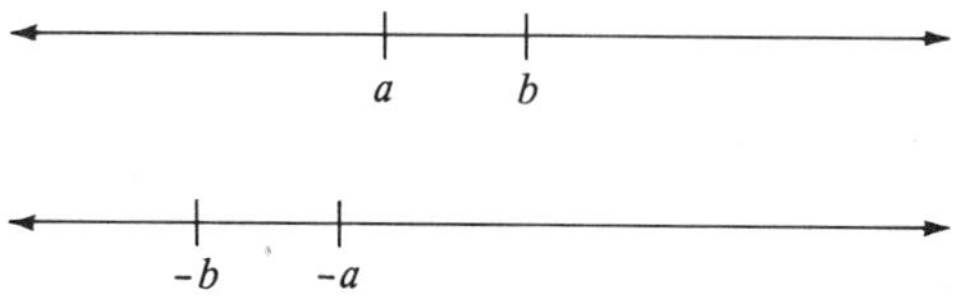

(ii) If a and b are both positive, so is ab.

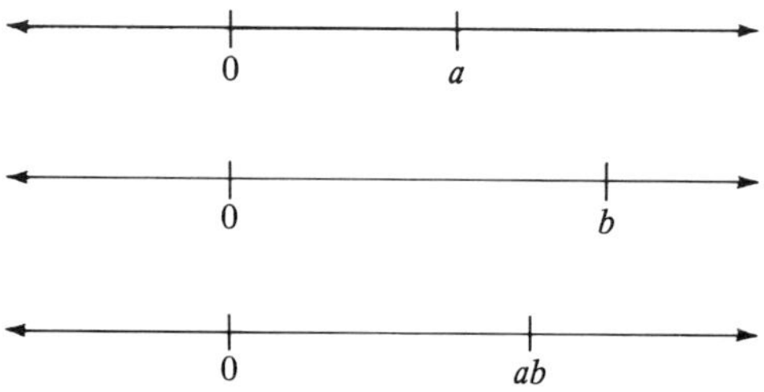

(iii) If both a and b are negative, then ab is positive.

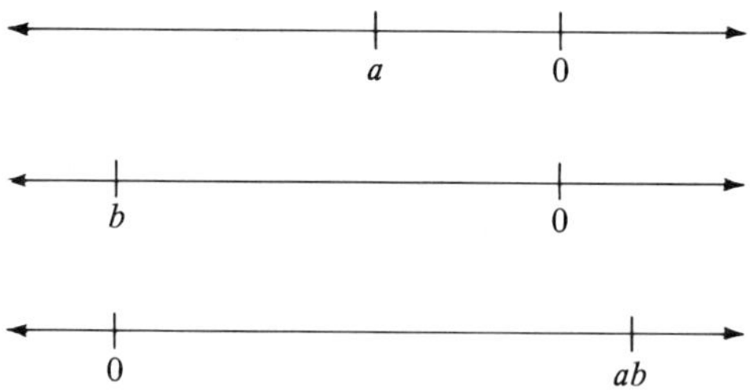

(iv) If a is to the left of b and c is positive, then a/c is to the left of b/c.

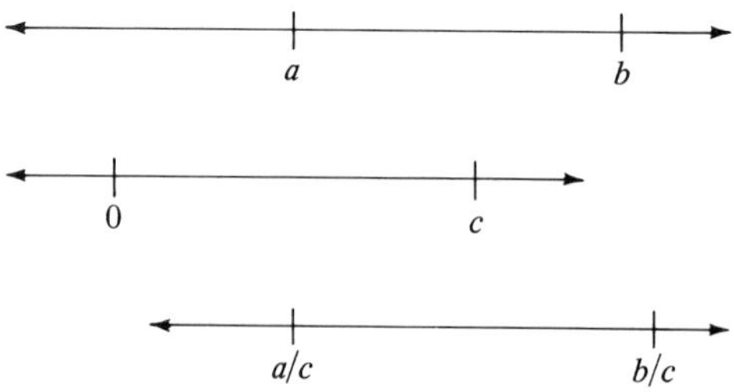

(v) If a is to the left of b and c is negative, then b/c is to the left of a/c.

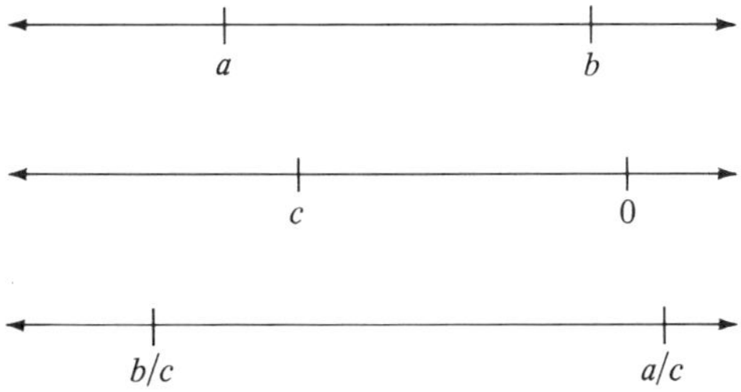

(vi) If $a \neq 0$, then $a \cdot a$ is positive. □

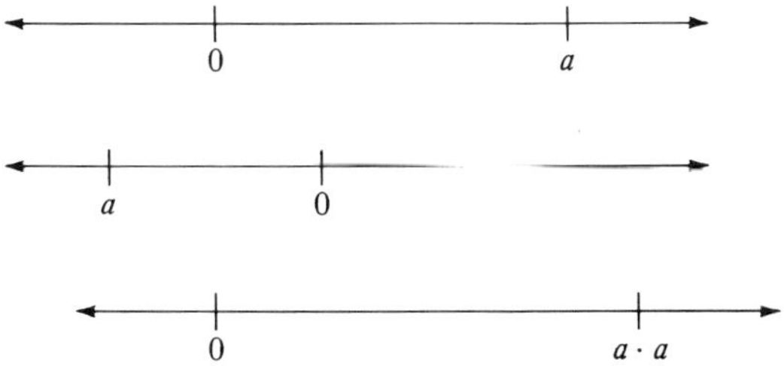

Axiom 10 and Principles (ii) and (iv) from Theorem 8 will be used a great deal in Chapter 2 to solve algebraic inequalities. They tell us that multiplying or dividing both sides of an inequality by a *positive* number *preserves* the direction of the inequality, whereas multiplying or dividing both sides of an inequality by a *negative* number *reverses* the direction of the inequality.

The concept of absolute value corresponds with the intuitive idea of the *size* of a number. The absolute value of a number can also be viewed as the distance of the number from 0 on the number line.

Example 3

(a) The absolute value of 7 is 7 because the distance from 0 to 7 is 7 units.

(b) The absolute value of -5 is 5 because the distance from 0 to -5 is 5 units. □

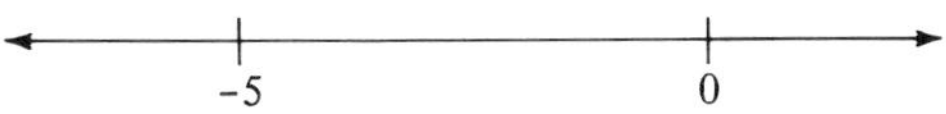

The concept is made precise by the following definition. If a is a real number, the *absolute value* of a is denoted by $|a|$ and is defined by

$$|a| = a, \quad \text{if } a \geq 0$$
$$|a| = -a, \quad \text{if } a < 0$$

Example 4

Since $\pi > 0$, $|\pi| = \pi$.
Since $-5 < 0$, $|-5| = -(-5) = 5$.
Since $4\pi - 13 < 0$, $|4\pi - 13| = -(4\pi - 13) = 13 - 4\pi$. □

Note that $|a| \geq 0$ for every number a. If a number is positive, then it is equal to its absolute value; if it is less than 0, then its absolute value is its negative. See Figure 1–8.

Figure 1-8 (a) $|a| = a$, (b) $|a| = -a$.

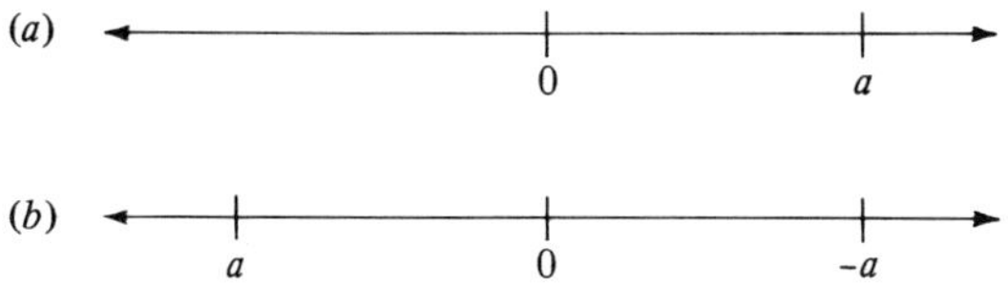

Example 5

Determine $|14/9 - 5/3|$.
We need to know whether this number is positive or negative.

$14/9 - 5/3 = 14/9 - 15/9 = -1/9$

Thus

$|14/9 - 5/3| = |-1/9| = 1/9$ □

The following theorem contains the basic principles governing the concept of absolute value.

THEOREM 9

Let p and q be real numbers. Then the following hold.

(i) $|p| = |-p|$
(ii) $-|p| \leq p \leq |p|$
(iii) $|pq| = |p| \cdot |q|$
(iv) If $p \geq 0$, then $|q| \leq p$ if and only if $-p \leq q \leq p$.
(v) $|p + q| \leq |p| + |q|$
(vi) $|p - q| \leq |p| + |q|$

We will not provide a proof of Theorem 9 but will, rather, illustrate portions by means of number lines.

Example 6

(i) $|4| = |-4|$ since both are 4 units from 0.

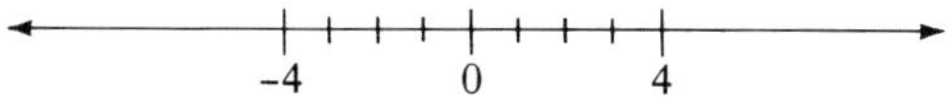

(iii)

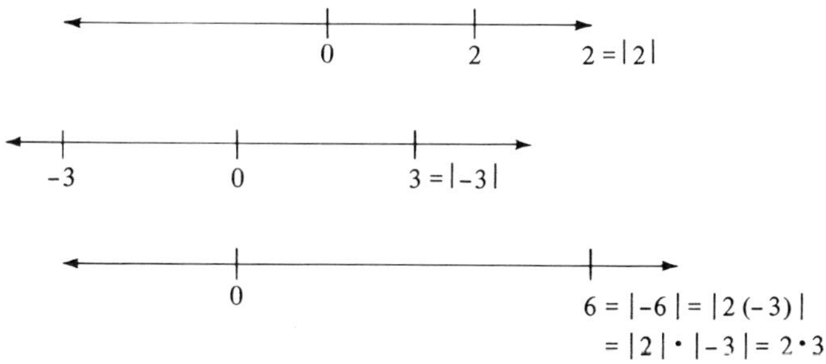

(iv) $-7 \leq q \leq 7$ if and only if $0 \leq |q| \leq 7$.

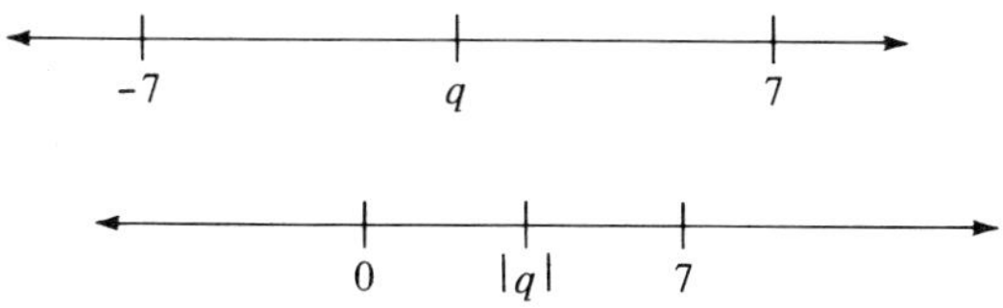

(v) $|2 + 3| = |5| = 5 = 2 + 3 = |2| + |3|$

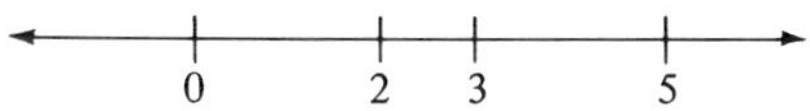

$|(-2) + 3| = |1| = 1 \leq |-2| + |3| = 2 + 3 = 5$

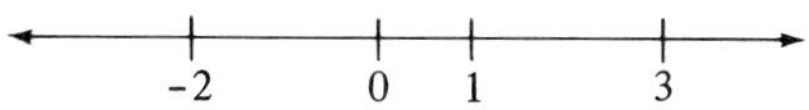

$|(-5) + 2| = |-3| = 3 \leq |-5| + |2| = 5 + 2 = 7$

$5 = |-5| = |(-3) + (-2)| = |-3| + |-2| = 3 + 2$

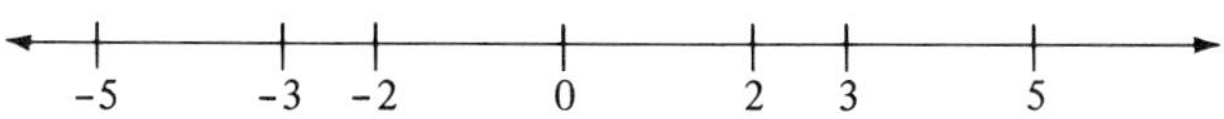

□

Figures 1–9 and 1–10 provide general illustrations of Principles (i) and (iv). We leave further general illustrations as exercises.

Figure 1-9

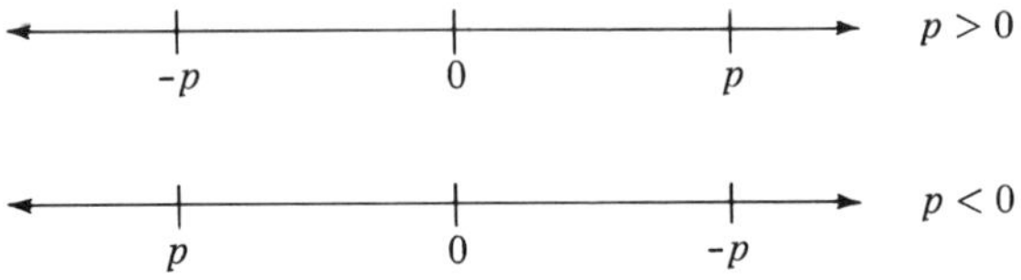

Figure 1-10

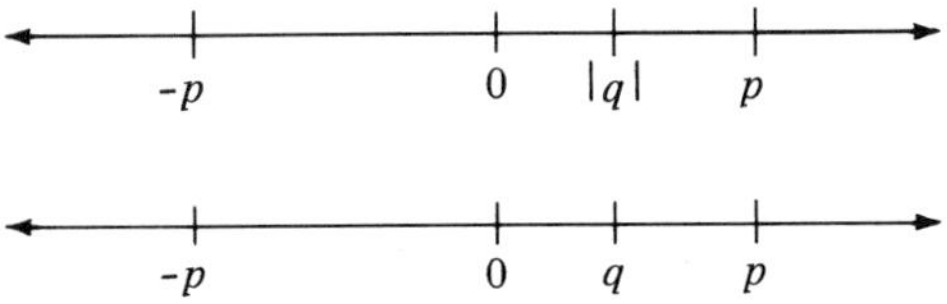

The theorem can be proved by using the definition and properties previously established; for example, (v) can be proved using (iv) and (ii).

EXERCISES

1. Use Theorem 8 and the axioms to decide if the following are true or false.
   (a) $16/32 < 94/188$ (b) $51/33 > 85/55$
   (c) $72/19 < 145/39$ (d) $4 \cdot 23/7 \cdot 12 > 531/1124$
   (e) $-312/54 < -727/124$ (f) $14/(-112) > 42/(-326)$
   (g) $96/-113 < 72/-101$ (h) $-47/31 > 59/47$
   (i) $16 + \pi < 32 + \pi$ (j) $-4 + \pi > -7 + \pi$
   (k) $\pi < 2\pi$ (l) $4739/18 < 4786/18$
   (m) $1/27 < 1/25$ (n) $5/37 > 5/39$
2. Use Axioms 8 and 9 to explain why the sum of two positive numbers is positive and why the sum of two negative numbers is negative.
3. Explain why the product of a positive number and a negative number is negative.
4. Decide whether the following are true or false.
   (a) $|3 - \pi| = 3 - \pi$ (b) $|7 - (183/476)| = 7 - (183/476)$
   (c) $|43 - 19| = |19 - 43|$ (d) $|\pi/.00001| = \pi/.00001$
   (e) $|7 - 9| = |7| - |9|$ (f) $|7\pi - 22| = 7\pi - 22$

*5. Use Theorem 8 and explain why $|a/b| = |a|/|b|$ for any numbers a and b (provided that $b \neq 0$).

*6. If $|a| < |b|$ and $a \neq 0$, what can be said about $|1/a|$ and $|1/b|$?

*7. Complete the proof of Theorem 8.

*8. Complete the proof of Theorem 9.

4. Classification of Numbers

It is convenient to classify numbers in the following traditional manner and to assign standard symbols as names for these sets of numbers.

The *natural numbers*, denoted by N, is another name for those numbers you may have called positive whole numbers; that is,

$$N = \{1, 2, 3, 4, \ldots\}$$

The set N can be axiomatized by the following assertions:

(i) N is identical to R
(ii) 1 is an element of N
(iii) If $a \in N$ and $b \in N$, then $(a + b) \in N$ and $(a \cdot b) \in N$.
(iv) If K is any subset of R satisfying (i), (ii), and (iii), then $N \subseteq K$.

This is another way of saying that N is the smallest set of numbers containing 1 that is closed under addition and multiplication (closed means that whenever two numbers are in the set, their sum and product are also in the set).

The *integers*, denoted by Z, is another name for what you may have called whole numbers; that is,

$$Z = \{\ldots, -2, -1, 0, 1, 2, 3, \ldots\}$$

The *rational numbers*, denoted by Q, are those real numbers which can be expressed as quotients (ratios) of integers:

$$Q = \{a/b\colon a \in Z, b \in Z, \text{and } b \neq 0\}$$

Notice that $N \subseteq Z \subseteq Q \subseteq R$.

Real numbers that are not rational are said to be *irrational*; the number π is an example of such a number. In Section 5 we will discuss many other irrational numbers.

An integer p is said to be *prime* if $p > 1$ and p is not evenly divisible by any positive integers other than itself and 1; that is, if a and b are positive integers such that $ab = p$, then $a = p$ and $b = 1$ or $b = p$ and $a = 1$.

Example 1

The number 5 is prime because the only positive integers dividing 5 evenly are 1 and 5. □

Using our definitions of N and Z, one can, with *quite a lot of work*, prove the following *Fundamental Theorem of Arithmetic*.

THEOREM 10

Every integer greater than 1 is either prime or can be expressed uniquely as the product of prime integers.

Example 2

If we write $70 = 2 \cdot 5 \cdot 7$, then we have expressed 70 as a product of primes. Theorem 10 assures us that this can be done in only one way (except, of course, for the order of the primes). □

Example 3

$504 = 2 \cdot 2 \cdot 2 \cdot 3 \cdot 3 \cdot 7$ □

Example 4

$462 = 2 \cdot 3 \cdot 7 \cdot 11$ □

A rational number p/q is said to be expressed in *lowest terms* if

(i) q is positive
(ii) p and q are not both evenly divisible by a common prime.

By repeated use of Theorem 7, part (vii) $[a/b = ad/bd]$ and the Fundamental Theorem of Arithmetic, it follows that any rational number may be expressed in lowest terms and the expression will be unique.

Example 5

The fraction $\frac{4}{6}$ is not in lowest terms because 4 and 6 are both divisible by 2. By part (vii) of Theorem 7 $(a/b = ad/bd)$ we know $\frac{4}{6} = \frac{2 \cdot 2}{3 \cdot 2} = \frac{2}{3}$. □

Example 6

The fraction $\frac{14}{27}$ is in lowest terms because the only primes dividing 14 are 2 and 7 and neither of these divides 27. □

Notice that in reducing fractions to lowest terms, we need only look for primes that divide the numerator and denominator.

Example 7

$$\frac{105}{231} = \frac{35 \cdot 3}{77 \cdot 3} = \frac{35}{77} = \frac{5 \cdot 7}{11 \cdot 7} = \frac{5}{11} \quad \square$$

EXERCISES

1. Is Q closed under addition and multiplication? Justify your answer.
2. List the primes between 1 and 100.
3. Decide whether or not 1021 is prime.
4. Use Theorem 10 to decide whether or not $7 \cdot 17 \cdot 19 \cdot 13 = 23 \cdot 31 \cdot 71$.
5. Reduce the following to lowest terms.
   (a) 15/35 (b) 24/90 (c) 121/77 (d) 16/100
   (e) 49/27 (f) 76/57 (g) 85/119 (h) 130/169

*6. What is the largest divisor you would have to try to decide whether or not 211 is prime? Justify your answer.

*7. Can you generalize your result from Exercise 6?

5. Exponents and Radicals

Before we are able to solve problems with algebra, we need to develop the ability to manipulate and compute with algebraic expressions. The key to developing such skills is in thoroughly understanding exactly what the symbols mean. In Section 2 we saw that the techniques for computing with fractions depend only on the principles governing division because the fraction line is just the division symbol. The techniques for computing with exponents and radicals are, likewise, dependent upon the meanings of the symbols.

If a is a number and n is a positive integer, then by a^n we mean the number which can be expressed by $a \cdot a \cdot a \cdots a$, where a appears n times.

By a^1 we mean just a and by a^2 we mean $a \cdot a$. The expression a^n is read "a to the nth power." The expression a^2 is usually read "a squared" or "the square of a," and a^3 is frequently read "a cubed" or the "cube of a." To compute a^n or just to write a^n is called "raising a to the nth power."

Example 1

$4^3 = 4 \cdot 4 \cdot 4 = 64$ □

Example 2

$(1/3)^2 = (1/3)(1/3) = 1/9$ □

For all numbers a and all positive integers m and n,

I. $a^m \cdot a^n = a^{m+n}$

since

$$\underset{m \text{ times}}{(a \cdot a \cdots a)}\underset{n \text{ times}}{(a \cdot a \cdots a)} = \underset{m+n \text{ times}}{(a \cdot a \cdots a)}$$

Furthermore,

$$(a^m)^n = \underbrace{(a^m)(a^m)\cdots(a^m)}_{n \text{ times}} = \underbrace{\underbrace{(a\cdot a\cdots a)}_{m\text{ times}}\underbrace{(a\cdot a\cdots a)}_{m\text{ times}}\underbrace{(a\cdot a\cdots a)}_{m\text{ times}}}_{n\text{ times}}$$

Since a appears m times in each of the grouped expressions and there are n of these, a appears mn times. Thus for any number a and any positive integers m and n,

II. $(a^m)^n = a^{mn}$

The next law of exponents is illustrated by the following two special cases:

$$(xy)^2 = (xy)(xy) = xxyy = x^2y^2$$
$$(xy)^3 = (xy)(xy)(xy) = xxxyyy = x^3y^3$$

Both of these equations depend on the associative and commutative laws. In general, if x and y are any numbers and n is a positive integer, then

$$(xy)^n = (xy)(xy)\cdots(xy)$$

Since xy appears n times on the right side, x and y each must appear n times and from commutativity and associativity it follows that:

$$\begin{aligned}(xy)^n &= (xy)(xy)\cdots(xy)\\ &= (xx\cdots x)(yy\cdots y)\\ &= x^n y^n\end{aligned}$$

Thus for any numbers x and y and any positive integer n,

III. $(xy)^n = x^n y^n$

A similar argument shows that if x and y are numbers with $y \neq 0$ and n is a positive integer, then

IV. $\left(\dfrac{x}{y}\right)^n = \dfrac{x^n}{y^n}$

Thus far we have defined a^x if $a \in R$ and $x \in N$ and we now extend the definition to the case $a \neq 0$ and $x \in Z$.

If $a \in R$ and $a \neq 0$, then $a^0 = 1$.

This definition was formulated so that our previous laws of exponents would extend to zero exponents. We would expect

$$a^2 \cdot a^0 = a^{2+0} = a^2$$

and the only way to have this hold is to require a^0 to be 1. In order to extend the law to negative exponents, we would expect to have

$$a^m a^{-m} = a^{m+(-m)} = a^0 = 1$$

In order for this equation to hold, a^{-m} must be $1/a^m$. We make the definition accordingly.

If $a \neq 0$ and m is a positive integer, then $a^{-m} = \dfrac{1}{a^m}$.

It is not too difficult now to show that all of the laws previously established extend to negative exponents. The proofs are done in a case by case manner and we leave the verifications to the student. Thus if a and b are nonzero numbers and x and y are integers, then:

I. $a^x a^y = a^{x+y}$
II. $(a^x)^y = a^{xy}$
III. $(ab)^x = a^x b^x$
IV. $\left(\dfrac{a}{b}\right)^x = \dfrac{a^x}{b^x}$

The process of taking roots is not quite the opposite of raising to powers but is close to a reversal of the process.

If $r \geq 0$, $\sqrt{r}$ denotes the nonnegative number whose square is r. If $r \geq 0$, $\sqrt{r}$ is called the *square root* of r or the *principal square root* of r. Axiom 12 for R can be used to show that for each $r \geq 0$, exactly one such number does, in fact, exist.*

Notice that if $y \in R$, then $\sqrt{y^2} = |y|$ because $\sqrt{y^2}$ must be nonnegative. This is why the root taking process is not quite the reverse of taking powers. If we square a negative number and then take the square root, we obtain a positive number.

Example 3

$\sqrt{(-2)^2} = 2$ □

Example 4

$\sqrt{(-.4)^2} = .4$ □

* In this section, we frequently say that some number does or does not exist. All numbers under discussion in Chapters 1–6 are real numbers and we really mean that some real number does or does not exist. In Chapter 7, we shall see that in the complex number system, numbers such as $\sqrt{-1}$ do, in fact, exist.

An expression such as $\sqrt{a}$ is called a *radical* and a here is called the *radicand.* The symbol $\sqrt{}$ is called a radical sign.

As with exponents, these symbols are just bits of notation and computational techniques involving them are based on their meaning.

If a and b are nonnegative numbers, can we find a simpler notation for the product of their square roots; that is, can we simplify $\sqrt{a}\sqrt{b}$? Notice first that

$$\begin{aligned}[(\sqrt{a})(\sqrt{b})]^2 &= (\sqrt{a})(\sqrt{b}) \cdot (\sqrt{a})(\sqrt{b}) \\ &= (\sqrt{a})(\sqrt{a}) \cdot (\sqrt{b})(\sqrt{b}) \\ &= ab\end{aligned}$$

Thus $(\sqrt{a})(\sqrt{b})$ is a nonnegative number (since $\sqrt{a}$ and $\sqrt{b}$ are nonnegative) whose square is ab. Since there is only one nonnegative number whose square is ab, it follows that

$$(\sqrt{a})(\sqrt{b}) = \sqrt{ab}$$

Example 5

$$\sqrt{2}\sqrt{3} = \sqrt{6} \ \square$$

Example 6

$$\sqrt{6}\sqrt{2} = \sqrt{12} \ \square$$

Example 7

$$\sqrt{5}\sqrt{7} = \sqrt{35} \ \square$$

In Section 3 we defined a rational number to be a real number that can be expressed as a fraction with an integral numerator and denominator. Not all real numbers are rational, in fact most are not. The following argument shows that $\sqrt{2}$ is irrational.

The argument is indirect in that we demonstrate that $\sqrt{2}$ cannot be rational by showing that the assumption $\sqrt{2}$ is rational leads to a contradiction.

If $\sqrt{2}$ is rational, then there are integers p and q such that $p/q = \sqrt{2}$ and furthermore we can assume that p/q is in lowest terms. (If it is not, we can reduce it to an expression that is in lowest terms and go on from there.)

Since

$$\frac{p}{q} = \sqrt{2}$$

$$\left(\frac{p}{q}\right)^2 = (\sqrt{2})^2$$

Thus

$$\frac{p^2}{q^2} = 2$$

and furthermore

$$p^2 = 2q^2$$

Since

$$p^2 = 2q^2$$

p^2 must be even.

If p^2 is even, p must be even. (You may have to think about that for a minute.)

Then $p = 2k$ for some integer k. Substituting, we have

$$(2k)^2 = 2q^2$$

Thus $2 \cdot 2 \cdot k^2 = 2q^2$ and therefore $2k^2 = q^2$.

Then q^2 also is even and, furthermore, q must be even. Since p is even and q is even, both are divisible by 2 and p/q is not in lowest terms. This contradicts our assumption and therefore $\sqrt{2}$ cannot be rational.

The following general result can be established but we will not give a proof in this text.

If a is a nonnegative integer, then $\sqrt{a}$ *is rational if and only if* $\sqrt{a}$ *is an integer.*

Example 8

Since $2 < \sqrt{6} < 3$, $\sqrt{6}$ is not an integer and therefore is irrational. □

If a is an integer such that $\sqrt{a}$ is an integer, then a is said to be a *perfect square*. The numbers 4, 9, 16, 25, 49, and 100 are examples of perfect squares.

Axiom 12 can be used to show that if r is any real number and n is an *odd* positive integer, then the equation $x^n = r$ has exactly one solution. If r is a positive real number and n is an *even* positive integer, it can be shown that the equation $x^n = r$ has two solutions, one positive and one negative. With these facts in mind, we make the following definitions.

(i) If r is a real number and n is an *odd* positive integer, then $\sqrt[n]{r}$ is the unique solution of the equation $x^n = r$.

(ii) If r is a positive real number and n is an *even* positive integer, then $\sqrt[n]{r}$ is the *positive* solution of the equation $x^n = r$.

The number $\sqrt[n]{r}$ is called the *nth root of r or the principal nth root of r.*

Example 9

$\sqrt[3]{-1} = -1$ since $(-1)^3 = -1$ and no other solutions exist. □

Example 10

$\sqrt[4]{81} = 3$ since $3^4 = 81$ and $(-3)^4 = 81$ and no other solutions exist. □

Example 11

$\sqrt[6]{64} = 2$ since $2^6 = 64$ and $(-2)^6 = 64$ and no other solutions exist. □

Example 12

$\sqrt[5]{32} = 2$ since $2^5 = 32$ and no other solutions exist in R. □

An expression such as $\sqrt[n]{a}$ is again called a *radical*, n is the *index* of the radical, and a is the *radicand*. We have already treated $\sqrt[2]{a}$ since $\sqrt[2]{a} = \sqrt{a}$, the square root of a. The expression $\sqrt[3]{a}$ is frequently called the *cube root* of a instead of the third root.

As before, radicals are notational devices and the computational techniques involving them stem from the meaning of the symbols. The techniques are, in fact, mostly generalizations of those we developed for square roots.

Notice, by definition, that $(\sqrt[n]{a})^n = a$ providing $\sqrt[n]{a}$ exists. To write $(\sqrt[4]{-3})^4 = -3$ is nonsense because $\sqrt[4]{-3}$ does not exist. If n is even, then $\sqrt[n]{a^n} = |a|$ as in the case of square roots but if n is odd, then $\sqrt[n]{a^n} = a$ because odd powers of negative numbers are negative.

Example 13

$\sqrt[4]{(-3)^4} = 3$ but $\sqrt[3]{(-3)^3} = -3$. □

If a and b are real numbers such that $\sqrt[n]{a}$ and $\sqrt[n]{b}$ exist, what can we say about the product $(\sqrt[n]{a})(\sqrt[n]{b})$?

From our laws for exponents, we have

$$(\sqrt[n]{a}\sqrt[n]{b})^n = (\sqrt[n]{a})^n(\sqrt[n]{b})^n = ab$$

Thus $\sqrt[n]{a}\sqrt[n]{b}$ is a solution to the equation $x^n = ab$. If n is odd, $\sqrt[n]{a}\sqrt[n]{b}$ is the only solution. If n is even, $\sqrt[n]{a}$ and $\sqrt[n]{b}$ are both nonnegative and therefore

$\sqrt[n]{a}\sqrt[n]{b}$ is the nonnegative solution to the equation $x^n = ab$. Thus in either case

$$\sqrt[n]{a}\sqrt[n]{b} = \sqrt[n]{ab}$$

We can now develop simplification techniques for radicals.

Example 14

$$\sqrt[3]{40} = \sqrt[3]{8 \cdot 5} = \sqrt[3]{8}\sqrt[3]{5} = 2\sqrt[3]{5} \ \square$$

Example 15

$$\sqrt[4]{8.1} = \sqrt[4]{81(.1)} = \sqrt[4]{81}\sqrt[4]{.1} = 3\sqrt[4]{.1}$$

or

$$\sqrt[4]{8.1} = \sqrt[4]{(.0081)(1000)} = \sqrt[4]{.0081}\sqrt[4]{1000} = .3\sqrt[4]{1000} \ \square$$

Example 16

$\sqrt[5]{a^{10}} = \sqrt[5]{(a^2)^5} = a^2$ for any number a. $\square$

Example 17

$\sqrt[7]{a^9} = \sqrt[7]{a^7 \cdot a^2} = \sqrt[7]{a^7}\sqrt[7]{a^2} = a\sqrt[7]{a^2}$ for any number a. $\square$

If $\sqrt[n]{a}$ and $\sqrt[n]{b}$ exist and $\sqrt[n]{b} \neq 0$, then

$$\left(\frac{\sqrt[n]{a}}{\sqrt[n]{b}}\right)^n = \frac{(\sqrt[n]{a})^n}{(\sqrt[n]{b})^n} = \frac{a}{b}$$

If n is odd, only one solution to the equation $x^n = a/b$ exists and if n is even, all numbers in the above equation are nonnegative. In either case,

$$\sqrt[n]{a/b} = \sqrt[n]{a}/\sqrt[n]{b}$$

Since the process of taking roots is close to being a reversal of taking powers, how are radical expressions related to expressions involving exponents? We saw some examples in our simplifications and these techniques can be expanded.

Example 18

$$\sqrt[5]{2^6 5^{12}} = \sqrt[5]{(2^5 5^{10})(2 \cdot 5^2)} = \sqrt[5]{2^5 5^{10}}\sqrt[5]{2 \cdot 5^2} = 2 \cdot 5^2\sqrt[5]{2 \cdot 5^2} = 50\sqrt[5]{50} \ \square$$

Example 19

$$\sqrt[4]{a^7b^5} = \sqrt[4]{(ab)^4a^3b} = \sqrt[4]{(ab)^4}\sqrt[4]{a^3b} = ab\sqrt[4]{a^3b} \;\square$$

Example 20

$$\sqrt[3]{x^5y^8} = \sqrt[3]{(x^3y^6)(x^2y^2)} = \sqrt[3]{(xy^3)^3}\sqrt[3]{x^2y^2} = xy^2\sqrt{x^2y^2} \;\square$$

Example 21

$$\sqrt[5]{p^7q^{31}} = \sqrt[5]{p^5q^{30}}\sqrt[5]{p^2q} = pq^6\sqrt[5]{p^2q} \;\square$$

The general method used in simplifying a radical is as follows:

(i) Factor the radicand so that one factor is a perfect nth power.
(ii) Split the radical into two radicals using these factors.
(iii) Rewrite as the product of the nth root of the first factor and the nth root of the second factor.

The usual practice is to attempt to choose the factorization so that the second factor has no perfect nth powers as factors. This is not necessary, it is just more efficient in most cases.

Example 22

$$\begin{aligned}\sqrt[4]{a^7b^9c^6} &= \sqrt[4]{a^4b^8c^4}\sqrt[4]{a^3bc^2}\\ &= |ab^2c|\sqrt[4]{a^3bc^2} \;\square\end{aligned}$$

Example 23

$$\sqrt[3]{x^8y^4z^9} = \sqrt[3]{x^6y^3z^9}\sqrt[3]{x^2y} = x^2yz^3\sqrt[3]{x^2y} \;\square$$

When we extended the definition of exponents from positive integers to integers we formulated the definitions so as to extend the laws of exponents. With this in mind, how should we define $a^{1/2}$? We would expect

$$(a^{1/2})^2 = a^{(1/2)\cdot 2} = a^1 = a$$

and so $a^{1/2}$ should be a number whose square is a. We cannot expect such a number to exist if $a < 0$ but if $a \geq 0$, this number is $\sqrt{a}$. Similarly if $a \in R$ and $q \in N$, we would expect

$$(a^{1/q})^q = a^{(1/q)\cdot q} = a^1 = a$$

Thus if $\sqrt[q]{a}$ exists, the only way to define $a^{1/q}$ and to extend the laws of exponents is to require $a^{1/q}$ to be $\sqrt[q]{a}$. If $\sqrt[q]{a}$ does not exist, there is no adequate way to define $a^{1/q}$.

If $a \in R$ and $q \in N$, then $a^{1/q} = \sqrt[q]{a}$ if this real number exists. Otherwise, $a^{1/q}$ is undefined.

If $p \in N$, $q \in N$, and $a^{1/q}$ exists, then if the laws of exponents are to be extended, we must have

$$(a^{1/q})^p = a^{(1/q)p} = a^{p/q}$$

If we define $a^{p/q} = (a^{1/q})^p$, then we are letting our definition of $a^{p/q}$ depend on the form of the rational number p/q. In some forms this might have meaning and in other forms it might not. For example,

$$(-8)^{1/3} = \sqrt[3]{-8} = -2$$

but

$$(-8)^{2/6} = [(-8)^{1/6}]^2 = (\sqrt[6]{-8})^2$$

which does not exist because $\sqrt[6]{-8}$ does not exist. In order to avoid such difficulties we define $a^{p/q}$ as follows:

If r is a positive rational number and,
expressed in lowest terms, $r = p/q$,
then $a^r = (a^{1/q})^p$ providing $a^{1/q}$ exists.

Notice that if $a \geq 0$, no difficulties arise since $a^{1/q}$ always exists; in fact, if $a \geq 0$ with p and q positive integers, $a^{p/q} = (a^{1/q})^p$ even if p/q is not in lowest terms. (We leave this verification as an exercise.) *If* $a < 0$ *and* p/q *is in lowest terms,* $a^{p/q}$ *exists only if* q *is odd.* It can also be shown that if p and q are any positive integers with q odd, then $a^{p/q} = (a^{1/q})^p$. The only real difficulty arises when we have $a < 0$ and q even with p/q not in lowest terms. To avoid difficulties, remember:

1. If $a \geq 0$, $a^{p/q}$ is defined for all positive rational numbers p/q.
2. If $a < 0$, $a^{p/q}$ is undefined when q is even and p/q is in lowest terms.
3. If $a < 0$, $a^{p/q}$ is defined if q is odd.

We define a meaning for negative rational exponents by

$$a^{-r} - 1/a^r$$

if $a \neq 0$, $r > 0$, and a^r has meaning.

Example 24

$$(-32)^{-2/5} = 1/(-32)^{2/5} = 1/(\sqrt[5]{-32})^2 = 1/(-2)^2 = 1/4 \ \square$$

The laws of exponents now extend as expected:

If $a \in R$, $b \in R$, $x \in Q$, and $y \in Q$, then

I. $a^x a^y = a^{x+y}$
II. $(a^x)^y = a^{xy}$
III. $(ab)^x = a^x b^x$
IV. $(a/b)^x = a^x/b^x$

providing all expressions given have meaning.

We will not provide proofs of these laws but leave the verification to the interested student.

If $a > 0$ and $x \in R$, Axiom 12 may be used to define a meaning for a^x. If we let

$$M = \{a^{p/q}: p/q \in Q \text{ and } p/q \le x\}$$

then a^x is defined to be the least upper bound of M. A great deal of work is involved proving that such numbers exist and that the laws of exponents are extended by this definition. Such a development is beyond the scope and intent of this text.

EXERCISES

1. Compute.
   (a) 9^3 (b) 9^{-3} (c) $9^{1/2}$ (d) $9^{-1/2}$
   (e) $9^{3/2}$ (f) $9^{-3/2}$ (g) $(9^{2/3})^{3/2}$ (h) $(9^{-2/3})^{3/2}$
2. Compute.
   (a) $(-27)^{2/3}$ (b) $(-27)^{-2/3}$ (c) $(-27)^{1/2}$ (d) $(-27)^{4/2}$
   (e) $(-27)^{4/6}$ (f) $(-27)^{6/4}$ (g) $[(-27)^{1/2}]^4$ (h) $[(-27)^{1/3}]^6$
3. Simplify.
   (a) $5^4 \cdot 5^{-2}$ (b) $5^{2/3} \cdot 5^{-1/6}$ (c) $5^2 \cdot 5^3 \cdot 5^{-7}$
   (d) $5^{1/2} \cdot 5^{2/3} \cdot 5^{5/6}$ (e) $5^{-.02} \cdot 5^{4.8}$ (f) $(5^{.03})^7$
   (g) $5^{-4/3} \cdot 5^{2/7}$ (h) $(5^{3/4})^{7/6}$
4. Simplify.
   (a) $6^3 \cdot 6^{-4}$ (b) $6^{2/5} \cdot 6^{-3/10}$ (c) $6^2 \cdot 6^4 \cdot 6^{-5}$
   (d) $6^{2/3} \cdot 6^{3/4} \cdot 6^{5/12}$ (e) $6^{-.07} \cdot 6^{5.9}$ (f) $(6^{.14})^3$
   (g) $6^{-5/3} \cdot 6^{4/7}$ (h) $(6^{3/8})^{11/6}$
5. Simplify.
   (a) $\sqrt{50}$ (b) $\sqrt{96}$ (c) $\sqrt{60x^9y^4}$
   (d) $\sqrt{72x^5y^9}$ (e) $\sqrt{32x^7y^8z^{13}}$ (f) $\sqrt{68a^9b^6c^8}$
   (g) $\sqrt{(-5)^2x^{10}y^2}$ (h) $\sqrt{242a^{11}b^{13}}$
6. Simplify.
   (a) $\sqrt{56}$ (b) $\sqrt{150}$ (c) $\sqrt{12x^5y^4}$
   (d) $\sqrt{40x^7y^6}$ (e) $\sqrt{45x^{17}y^{21}}$ (f) $\sqrt{48a^{13}b^4c^{10}}$
   (g) $\sqrt{(-6)^2x^4y^6}$ (h) $\sqrt{238x^{13}y^{15}}$
7. Simplify.
   (a) $\sqrt[3]{54}$ (b) $\sqrt[4]{48x^{14}y^6}$ (c) $\sqrt[5]{.000001}$ (d) $\sqrt[3]{-72}$
   (e) $\sqrt[3]{-17}$ (f) $\sqrt[3]{5/9}$ (g) $\sqrt[4]{3/8}$ (h) $\sqrt[3]{5/y}$
   (i) $\sqrt[3]{16x^8y^{10}z}$ (j) $\sqrt[7]{1024a^{25}b^{16}}$
8. Simplify.
   (a) $\sqrt[3]{-40}$ (b) $\sqrt[4]{96x^6y^9}$ (c) $\sqrt[5]{.0000001}$ (d) $\sqrt[3]{512}$
   (e) $\sqrt[3]{-19}$ (f) $\sqrt[3]{7/4}$ (g) $\sqrt[4]{7/9}$ (h) $\sqrt[3]{3/x}$
   (i) $\sqrt[3]{54a^{11}b^7c}$ (j) $\sqrt[7]{256x^{32}y^{54}z^{10}}$
9. Simplify.
   (a) $\sqrt{2}\sqrt{6}$ (b) $\sqrt[3]{10}\sqrt[3]{12}$ (c) $\sqrt[3]{40}\sqrt[3]{25}$ (d) $\sqrt{12}\sqrt{10}\sqrt{5}$
   (e) $\sqrt[3]{.09}\sqrt[3]{.3}$ (f) $\sqrt{6}/\sqrt{3}$ (g) $\sqrt[3]{10}/\sqrt[3]{.01}$ (h) $\sqrt[5]{9}\sqrt[5]{54}$

10. Simplify.
(a) $\sqrt{3}\sqrt{6}$ (b) $\sqrt[3]{12}\sqrt[3]{6}$ (c) $\sqrt[3]{20}\sqrt[3]{50}$ (d) $\sqrt{14}\sqrt{10}\sqrt{35}$
(e) $\sqrt[3]{.04}\sqrt[3]{.2}$ (f) $\sqrt{6}/\sqrt{2}$ (g) $\sqrt[3]{100}/\sqrt[3]{.1}$ (h) $\sqrt[5]{8}\sqrt[5]{12}$
11. By multiplying both the numerator and denominator of the following expressions by a suitable number, express the following as fractions having no radical in the denominator.

Example

$$\frac{\sqrt{5}}{\sqrt{2}} = \frac{(\sqrt{5}\sqrt{2})}{(\sqrt{2}\sqrt{2})} = \frac{\sqrt{10}}{2} \quad \square$$

(a) $\dfrac{\sqrt{7}}{\sqrt{3}}$ (b) $\dfrac{\sqrt{2}}{\sqrt{7}}$ (c) $\dfrac{\sqrt{2}}{\sqrt{6}}$ (d) $\dfrac{\sqrt[3]{5}}{\sqrt[3]{2}}$ (e) $\dfrac{\sqrt{a}}{\sqrt{b}}$

(f) $\dfrac{\sqrt[3]{a}}{\sqrt[3]{b}}$ (g) $\dfrac{\sqrt[4]{10}}{\sqrt[4]{8}}$ (h) $\dfrac{\sqrt[3]{.2}}{\sqrt[3]{.01}}$ (i) $\dfrac{\sqrt{2}}{\sqrt[3]{3}}$ (j) $\dfrac{\sqrt[3]{6}}{\sqrt{2}}$

12. Follow the instructions given for Exercise 11.

(a) $\dfrac{\sqrt{5}}{\sqrt{3}}$ (b) $\dfrac{\sqrt{7}}{\sqrt{2}}$ (c) $\dfrac{\sqrt{3}}{\sqrt{6}}$ (d) $\dfrac{\sqrt[3]{7}}{\sqrt[3]{3}}$ (e) $\dfrac{\sqrt{x}}{\sqrt{y}}$

(f) $\dfrac{\sqrt[3]{x}}{\sqrt[3]{y}}$ (g) $\dfrac{\sqrt[4]{6}}{\sqrt[4]{8}}$ (h) $\dfrac{\sqrt[3]{.6}}{\sqrt[3]{.01}}$ (i) $\dfrac{\sqrt{3}}{\sqrt[3]{2}}$ (j) $\dfrac{\sqrt[3]{6}}{\sqrt{2}}$

13. Simplify.
(a) $(9x^4y^2)^{1/2}$ (b) $(30x^5y^7)^{1/2}$
(c) $(27a^{13}b^8)^{1/3}$ (d) $(4a^2b^4)^{3/2}$
(e) $(2a^2b^5)^{1/3}(3a^4b^3)^{1/2}$ (f) $(5x^4y^5)^{1/3} \div (3x^2y^4)^{1/3}$
(g) $(8a^{1/3}b^{2/5})^2 \div (6a^{1/5}b^{2/3})^3$ (h) $(x^{3/4}y^{2/3})^6 \div (x^{2/3}y^{3/4})^5$
14. Simplify.
(a) $(25a^6b^3)^{1/2}$ (b) $(18x^9y^6)^{1/2}$
(c) $(8x^{11}y^9)^{1/3}$ (d) $(9a^4b^2)^{5/2}$
(e) $(5a^3b^4)^{1/3} \cdot (3a^5b^3)^{1/2}$ (f) $(6x^5y^7)^{1/3} \div (5x^4y^2)^{1/3}$
(g) $(5a^{2/3}b^{4/5})^3 \div (6a^{3/5}b^{5/6})^2$ (h) $(x^{2/3}y^{4/5})^6 \div (x^{2/5}y^{5/6})^3$

6. Polynomials and Factoring

An algebraic expression which consists of only one numeral or variable or is an indicated product of several numerals or variables is called a *monomial.* Notice that monomials do not involve addition, subtraction, or division. An expression which is the sum of two monomials is called a *binomial*; an expression which is the sum of three monomials is called a *trinomial*; and any expression consisting of one monomial or the sum of several monomials is called a *polynomial.* The monomials which are added together are called the *terms* of the polynomial and the constants appearing are called the *coefficients* of the polynomial. Expressions such as $4x^2 - 3xy - 7y^2$ will also be called polynomials because subtraction can always be rewritten using addition.

Example 1

$$4x^2 - 3xy - 7y^2 = 4x^2 + (-3)xy + (-7)y^2 \ \square$$

A polynomial in which the variable parts of the monomials are the same can be simplified by using the distributive law and completing the indicated computation with the numerical factors.

Example 2

$$5xy^2 + 7xy^2 = (5 + 7)xy^2 = 12xy^2 \ \square$$

Example 3

$$3xyz - 7xyz = (3 - 7)xyz = -4xyz \ \square$$

Example 4

$$8x^2 - 3x^2 + 4x^2 = (8 - 3 + 4)x^2 = 9x^2 \ \square$$

In practice we do not need to write down the intermediate steps but can do the computation mentally. The process of simplifying such polynomials is sometimes called "collecting like terms," "like terms" being monomials with the same variable parts.

As we saw in previous examples, the product of a polynomial and a monomial can be rewritten using the distributive laws and our methods for simplifying monomials. We can rewrite the product of more complicated polynomials by applying these procedures several times.

Example 5

$$\begin{aligned}(x + 3)(y + 2) &= (x + 3)y + (x + 3)2 \\ &= xy + 3y + 2x + 6 \ \square\end{aligned}$$

If possible, we should collect like terms in the simplification.

Example 6

$$\begin{aligned}(x^2 + 2x)(x^3 + 3x^2) &= (x^2 + 2x)(x^3) + (x^2 + 2x)(3x^2) \\ &= (x^5 + 2x^4) + (3x^4 + 6x^3) \\ &= x^5 + 2x^4 + 3x^4 + 6x^3 \\ &= x^5 + 5x^4 + 6x^3 \ \square\end{aligned}$$

Example 7

$$\begin{aligned}(a + 2b)(2c - 3d) &= a(2c - 3d) + 2b(2c - 3d) \\ &= 2ac - 3ad + 4bc - 6bd \ \square\end{aligned}$$

Example 8

$$\begin{aligned}(2x + 3)(x - 5) &= 2x(x - 5) + 3(x - 5)\\ &= 2x^2 - 10x + 3x - 15\\ &= 2x^2 - 7x - 15 \ \square\end{aligned}$$

Example 9

$$\begin{aligned}(a + 2b)(3a + b) &= a(3a + b) + 2b(3a + b)\\ &= 3a^2 + ab + 6ab + 2b^2\\ &= 3a^2 + 7ab + 2b^2 \ \square\end{aligned}$$

In the last two examples, the variables involved were the same in both binomials and the result was a trinomial. There are four multiplications involved in such computations and the pattern is illustrated by the diagrams in the following examples.

Example 10

$$\begin{aligned}(2x + 5y)(3x + 7y) &= 6x^2 + 14xy + 15xy + 35y^2\\ &= 6x^2 + 29xy + 35y^2 \ \square\end{aligned}$$

Example 11

$$\begin{aligned}(5a - 7b)(2a + 11b) &= 10a^2 + 55ab - 14ab - 77b^2\\ &= 10a^2 + 41ab - 77b^2 \ \square\end{aligned}$$

Example 12

$$\begin{aligned}(x - 5)(3x - 7) &= 3x^2 - 7x - 15x + 35\\ &= 3x^2 - 22x + 35 \ \square\end{aligned}$$

The general procedure is:

(i) Compute the product of the first terms in the binomials.
(ii) Compute the product of the first term in the first binomial with the second term in the second binomial, and the product of the second term in the first binomial with the first term in the second binomial. If they are like terms, combine them.
(iii) Compute the product of the last terms in the binomials.
(iv) Write the sum of all the products.

After some practice, you should be able to do step (ii) mentally.

Multiplication of algebraic expressions is governed by the associative, commutative, and distributive laws. We have seen that indicated products can be simplified by using these laws. Many times, however, an expression is more useful if it is written in a factored form, that is, as a product of simpler factors rather than as a polynomial with several terms. We will see in Section 7 how the factored form of an expression can be used to simplify computations with algebraic fractions, and in Chapter 2 we will see how to solve equations by using a factored form of a polynomial.

To *factor* an expression is to rewrite it as a product of simpler expressions. The first step in factoring a polynomial with integral coefficients is to look for monomial factors and to rewrite the original polynomial as a product of a monomial and a polynomial with integral coefficients in which the terms do not have any common variable or integral factors (other than 1 or -1, of course).

Example 13

$$4x^3y^2 + 2x^2y^3 + 6x^4y^4 = 2x^2y^2(2x + y + 3x^2y^2)$$

The terms $2x$, y, and $3x^2y^2$ have no common variable or integral factor. □

Example 14

$$3ab^3 - 12a^2b^2 + 9a^2b = 3ab(b^2 - 4ab + 3a) \ \square$$

Example 15

$$4p^4q^3 - 12p^2q^5 + 28p^3q^2 = 4p^2q^2(p^2q - 3q^3 + 7p) \ \square$$

Example 16

$$5x^2 + 10x - 15 = 5(x^2 + 2x - 3) \ \square$$

Example 17

$$4y^3 - 20y^2 - 24y = 4y(y^2 - 5y - 6) \ \square$$

A polynomial with nonintegral coefficients can frequently be simplified by rewriting it as a product of a monomial and a polynomial with integral coefficients.

Example 18

$$\tfrac{2}{3}x^2y^3 - \tfrac{5}{6}x^3y + \tfrac{1}{6}xy^2 = \tfrac{1}{6}xy(4xy^2 - 5x + y) \ \square$$

Example 19

$$\sqrt{2}a^4b^5 + 3\sqrt{2}a^7b^3 = \sqrt{2}a^4b^3(b^2 + 3a^3) \ \square$$

Certain special products appear frequently enough to make studying them worthwhile for later work.

$$(a + b)^2 = (a + b)(a + b) = a^2 + 2ab + b^2$$
$$(a - b)^2 = (a - b)(a - b) = a^2 - 2ab + b^2$$
$$(a + b)(a - b) = a^2 - b^2$$

Thus any expression which takes one of these three forms can be easily factored.

Example 20

$$x^2 + 2x + 1 = (x + 1)^2 \ \square$$

Example 21

$$y^2 + 6y + 9 = (y + 3)^2 \ \square$$

Example 22

$$z^2 - 4z + 4 = (z - 2)^2 \ \square$$

Example 23

$$x^2 - 9 = (x + 3)(x - 3) \ \square$$

Example 24

$$y^2 - 1 = (y - 1)(y + 1) \ \square$$

Example 25

$$4x^2 - 9y^2 = (2x + 3)(2x - 3) \ \square$$

Example 26

$$9y^2 + 6y + 1 = (3y + 1)^2 \ \square$$

These special forms enable us to factor many trinomials in which two of the terms are perfect squares.

(i) The difference of squares $a^2 - b^2$ can be expressed as $(a + b)(a - b)$.
(ii) $a^2 +$ ______ $+ b^2$ is a perfect square if the missing term is either $2ab$ or $-2ab$.

$$a^2 + 2ab + b^2 = (a + b)^2$$
$$a^2 - 2ab + b^2 = (a - b)^2$$

Certainly we cannot expect all trinomials to take one of these forms. What is the general method for factoring trinomials? First, the reader should realize that most trinomials are not factorable. Our problem is to factor those that are. The first step, remember, is to look for monomial factors.

If we are attempting to factor a polynomial into the form $(a + b)(c + d)$, then we do so by first factoring terms. If our expression is written in decreasing order of powers of some variable, the possibilities become more apparent.

Example 27

Factor $x^2 - 5x - 6$.

If we consider the possible factors of the first and last terms we have x and x; or x^2 and 1; or their negatives for factors of x^2, and we have 2, -3; -2, 3; 1, -6; and -1, 6 as factors of -6.

We can try all the possibilities to see if any work.

$$(x^2 + 2)(1 - 3) = (x^2 + 2)(-2) = -2x^2 - 4$$

From this we see that choosing x^2 and 1 will always yield an expression with a term involving x^2 but no term involving just x. Thus to factor this, we must obtain an expression of the form

$$(x \qquad)(x \qquad)$$

Now what are the possibilities? If we use 2 and -3 we obtain

$$(x + 2)(x - 3) = x^2 - x - 6$$

Using -2 and 3 yields

$$(x - 2)(x + 3) = x^2 + x - 6$$

Trying -1 and 6 yields

$$(x - 1)(x + 6) = x^2 + 5x - 6$$

But 1 and -6 yields

$$(x + 1)(x - 6) = x^2 - 5x - 6$$

which is what we wanted. □

Example 28

Factor $4x^2 + 15x + 9$.

Among the factors of $4x^2$ are $4x$, x and $2x$, $2x$ and their negatives. The integral factorizations of 9 are 9, 1; -9, -1; 3, 3; and -3, 3. Although both $4x^2$ and 9 are perfect squares, the "middle term" is not such that this expression is a perfect square since $(2x \pm 3)^2$ has $\pm 12x$ for a middle term. Thus we have ruled out $(2x + 3)^2$ and $(2x - 3)^2$ as possibilities. Notice that the middle term has a positive sign. Thus any expression of the form $(\quad - \quad)(\quad - \quad)$ is ruled out. We now try the rest of the possibilities.

$$(2x + 1)(2x + 9) = 4x^2 + 20x + 9$$
$$(4x + 1)(x + 9) = 4x^2 + 37x + 9$$
$$(4x + 9)(x + 1) = 4x^2 + 13x + 9$$
$$(4x + 3)(x + 3) = 4x^2 + 15x + 9$$

Thus $(4x + 3)(x + 3)$ is the desired form. □

Factoring expressions is not a step-by-step procedure which you can memorize as you did the multiplication process in arithmetic. You must consider all the possibilities and try those that might work.

There are, however, some guidelines to follow in factoring:

(i) Look for a "largest" monomial factor and rewrite the original expression as a product of this monomial and a polynomial in order of increasing or decreasing powers of some variable such that the sign of the first term of the polynomial is positive.

(ii) Examine the polynomial to see if it fits any special forms you know, such as $a^2 - b^2$, $a^2 + 2ab + b^2$ or $a^2 - 2ab + b^2$.

(iii) If the polynomial is a trinomial, consider all possibilities of the form $(a \quad b)(c \quad d)$ where ac is the first term and bd is the last term. We know ac has a positive sign and we only need consider a and c with positive signs. In considering all possibilities, some can be ruled out by the signs of the terms. If the last term is positive, b and d must agree in sign and both signs will be the same as the sign of the middle term. If the last term has a negative sign, b and d will disagree in sign and the choice of signs will depend on the middle term.

Example 29

$$\begin{aligned} 3z^3y^2 + 6z^2y^3 + 3zy^4 &= 3zy^2(z^2 + 2zy + y^2) \\ &= 3zy^2(z + y)^2 \ \square \end{aligned}$$

Example 30

$$\begin{aligned} 5a^2b^4 - 20a^4b^2 &= 5a^2b^2(b^2 - 4a^2) \\ &= 5a^2b^2(b + 2a)(b - 2a) \ \square \end{aligned}$$

In both of these examples, after we found the monomial factor, the remaining polynomial was in one of the special forms mentioned earlier.

Example 31

$$\begin{aligned} 4y^2 - 10y - 24 &= 2(2y^2 - 5y - 12) \\ &= 2(2y \qquad)(y \qquad) \end{aligned}$$

The integral possibilities for second terms are, pairwise, ± 1, ± 12; ± 2, ± 6; and ± 3, ± 4. We know the signs must disagree because the last term has a negative sign. Trying the possibilities we see that

$$2y^2 - 5y - 12 = (2y + 3)(y - 4)$$

Thus

$$4y^2 - 10y - 24 = 2(2y + 3)(y - 4) \ \square$$

Example 32

$$\begin{aligned} 2x^3 - 9x^2 + 10x &= x(2x^2 - 9x + 10) \\ &= x(2x \qquad)(x \qquad) \end{aligned}$$

The only integral possibilities for second terms are, pairwise, ± 1, ± 10 and ± 2, ± 5. Since the last term has a positive sign, the second terms agree in sign. Both must have a negative sign because the middle term has a negative sign. Thus $2x^3 - 9x^2 + 10x = x(2x - \qquad)(x - \qquad)$ and we have yet to try the possibilities. After trying them we see that

$$2x^3 - 9x^2 + 10x = x(2x - 5)(x - 2) \ \square$$

In the previous examples, all the coefficients considered were integers. We can use numbers other than integers in factoring polynomials and, in fact, will find such factorizations important. Polynomials involving rational numbers can be rewritten as the product of a rational number and a polynomial with integral coefficients or they can be factored using the given form.

Example 33

$$\tfrac{1}{2}x^2 - \tfrac{3}{2}x - 2 = \tfrac{1}{2}(x^2 - 3x - 4) = \tfrac{1}{2}(x - 4)(x + 1)$$

or

$$(\tfrac{1}{2}x^2 - \tfrac{3}{2}x - 2) = (\tfrac{1}{2}x - 2)(x + 1)$$

The first method is more common because possible integral factors are easier to determine than possible rational factors. $\square$

Example 34

$$x^2 - 5 = (x + \sqrt{5})(x - \sqrt{5})$$

In this example, the factorization followed one of the familiar forms although the factors involved irrational numbers whereas the original polynomial did not. $\square$

Certain special products can be used to simplify fractions involving sums or differences of radicals in the denominator. The process is called "rationalizing the denominator."

Example 35

$$\frac{5}{\sqrt{3}-\sqrt{2}} = \frac{5(\sqrt{3}+\sqrt{2})}{(\sqrt{3}-\sqrt{2})(\sqrt{3}+\sqrt{2})} = \frac{5(\sqrt{3}+\sqrt{2})}{3-2} = 5(\sqrt{3}+\sqrt{2}) \quad \square$$

Example 36

$$\frac{\sqrt{5}-\sqrt{3}}{\sqrt{5}+\sqrt{3}} = \frac{(\sqrt{5}-\sqrt{3})(\sqrt{5}-\sqrt{3})}{(\sqrt{5}+\sqrt{3})(\sqrt{5}-\sqrt{3})}$$
$$= \frac{5-2\sqrt{3}\sqrt{5}+3}{5-3} = \frac{8-2\sqrt{15}}{2} = 4-\sqrt{15} \quad \square$$

Example 37

$$\frac{4}{\sqrt{11}-\sqrt{5}} = \frac{4(\sqrt{11}+\sqrt{5})}{(\sqrt{11}-\sqrt{5})(\sqrt{11}+\sqrt{5})} = \frac{4(\sqrt{11}+\sqrt{5})}{11-5}$$
$$= \frac{4(\sqrt{11}+\sqrt{5})}{6} = (2/3)(\sqrt{11}+\sqrt{5})$$

The basic idea is to multiply both the numerator and denominator by a suitable number or expression so that the resulting denominator contains no radicals. □

Certain special products enable us to factor some expressions involving higher powers. The reader should verify the following five special products by direct computation.

$$a^3 - b^3 = (a-b)(a^2+ab+b^2)$$
$$a^3 + b^3 = (a+b)(a^2-ab+b^2)$$
$$a^4 - b^4 = (a-b)(a^3+a^2b+ab^2+b^3) = (a-b)(a+b)(a^2+b^2)$$
$$a^5 - b^5 = (a-b)(a^4+a^3b+a^2b^2+ab^3+b^4)$$
$$a^5 + b^5 = (a+b)(a^4-a^3b+a^2b^2-ab^3+b^4)$$

In general, if n is any positive integer,

$$a^n - b^n = (a-b)(a^{n-1}+a^{n-2}b+a^{n-3}b^2+\cdots+ab^{n-2}+b^{n-1})$$

If n is an *odd* positive integer,

$$a^n + b^n = (a+b)(a^{n-1}-a^{n-2}b+a^{n-3}b^2-a^{n-4}b^3+\cdots$$
$$-ab^{n-2}+b^{n-1})$$

These general formulas can be proved formally by using a technique known as mathematical induction but the patterns that appear in the verifications of the first five special products indicate what really happens.

In general, these forms do not provide complete factorizations because the second expression can frequently be factored further. They do, however, give us a starting point.

Thus far we have discussed addition, subtraction, and multiplication of polynomials but have not touched on division. Just as we cannot expect one integer to divide evenly into another, we cannot expect one polynomial to evenly divide another. If we try to divide 34 by 5 in the integers, we see that we get a quotient of 6 and a remainder of 4:

$$34 = 6 \cdot 5 + 4$$

We expect the remainder to be smaller than the divisor. If we limit ourselves to polynomials containing only one variable, a similar result holds but, of course, it makes no sense to talk about the remainder being smaller than the divisor. The usual method for long division can be extended for use with polynomials.

Example 38

$$\begin{array}{r|l}
 & 4x^3 + 5x^2 - 4x + 20 \\ \hline
x^2 + 2x - 1 & 4x^5 + 13x^4 + 2x^3 + 7x^2 + 5x - 4 \\
 & 4x^5 + 8x^4 - 4x^3 \\ \hline
 & 5x^4 + 6x^3 + 7x^2 \\
 & 5x^4 + 10x^3 - 5x^2 \\ \hline
 & -4x^3 + 12x^2 + 5x \\
 & -4x^3 - 8x^2 + 4x \\ \hline
 & 20x^2 + x - 4 \\
 & 20x^2 + 40x - 20 \\ \hline
 & -39x + 16
\end{array}$$

Thus the remainder is $-39x + 16$. We cannot say that it is smaller than $x^2 + 2x - 1$ but the highest power of x appearing in the remainder is the first whereas the second appears in the divisor. □

The polynomial $a_n x^n + a_{n-1}x^{n-1} + \cdots + a_1 x + a_0$ is said to have *degree n* providing $a_n \neq 0$. Intuitively, the degree of a polynomial in one variable, say x, is the exponent of the highest power of x appearing with a nonzero coefficient. We will frequently find it convenient to denote a polynomial in one variable x by symbols such as $f(x)$, $g(x)$, and $p(x)$. If c is a number and $f(x)$ is a polynomial, $f(c)$ is the number obtained by substituting c for x in the polynomial. The expression "$f(x)$" is read "f of x" and the expression "$f(c)$" is read "f of c."

Example 39

If

$$f(x) = x^2 - 3x + 5$$

then

$$f(2) = 4 - 6 + 5 = 3$$

and

$$f(7) = 49 - 21 + 5 = 33$$

The sentence "$f(2) = 3$" is read "f of 2 equals 3." □

Example 40

If

$$g(x) = 3x^3 - 2x^2 + 3x - 7$$

then

$$g(2) = 3(8) - 2(4) + 3(2) - 7 = 15$$

and

$$g(1) = -3 \ \square$$

If we have two polynomials

$$b_m x^m + b_{m-1} x^{m-1} + \cdots + b_1 x + b_0 \quad \text{and} \quad a_n x^n + \cdots + a_1 x + a_0$$

and we try to divide the second by the first, either it divides evenly or the long division technique will produce a remainder of degree less than m. This can be proved rigorously and the proof is really an extension of the long division method. We state this result more formally below.

DIVISION ALGORITHM FOR POLYNOMIALS

If $f(x)$ and $g(x)$ are polynomials in the one variable x with $g(x) \neq 0$, then there exist unique polynomials $q(x)$ and $r(x)$ such that

$$f(x) = g(x)q(x) + r(x)$$

with the degree of $r(x)$ less than the degree of $g(x)$ or else $r(x) = 0$.

Notice that if $g(x)$ is of degree 1, then $r(x)$ is either 0 or a polynomial of degree 0, that is, a constant. Thus if a is a number and $f(x)$ a polynomial, there exist a polynomial $q(x)$ and a number r such that

$$f(x) = (x - a)q(x) + r$$

If we evaluate $f(x)$ at a we have

$$\begin{aligned} f(a) &= (a - a)q(a) + r \\ &= 0 \cdot q(a) + r \\ &= 0 + r \\ &= r \end{aligned}$$

Thus the remainder after dividing $f(x)$ by $(x - a)$ is just $f(a)$. Furthermore, since $(x - a)$ is a factor of $f(x)$ if and only if $r = 0$, $(x - a)$ is a factor of $f(x)$ if and only if $f(a) = 0$.
These two results are usually stated more formally as follows:

THE REMAINDER THEOREM

If $f(x)$ is a polynomial and $a \in R$, the remainder after dividing $f(x)$ by $(x - a)$ is $f(a)$.

THE FACTOR THEOREM

If $f(x)$ is a polynomial and $a \in R$, $(x - a)$ is a factor of $f(x)$ if and only if $f(a) = 0$.

Example 41

$x - 2$ is not a factor of $g(x) = 2x^2 + 3x - 4$ because $g(2) = 10 \neq 0$. □

Example 42

$x - 2$ is a factor of $f(x) = x^3 + 5x^2 - 3x - 22$ because $f(2) = 0$. If we divide $x^3 + 5x^2 - 3x - 22$ by $x - 2$ we obtain a quotient of $x^2 + 7x + 11$ and, therefore,

$$x^3 + 5x^2 - 3x - 22 = (x - 2)(x^2 + 7x + 11)$$

Further attempts to factor $x^2 + 7x + 11$ with rational coefficients fail. □

These results can be used to determine the factors of a polynomial which are of the form $(x - a)$. If $f(a) = 0$, $x - a$ is a factor and we can factor simply by dividing. If $f(x)$ is a polynomial with integral coefficients and we are looking only for factors with integral coefficients, then in order for $x - a$ to be a factor, a must divide the constant term of $f(x)$. In other words, if $f(x) = b_n x^n + b_{n-1}x^{n-1} + \cdots + b_1 x + b_0$ and $x - a$ is a factor of $f(x)$, then a must be an integral factor of b_0.

Example 43

Factor $f(x) = x^3 - 19x + 30$ into a product of polynomials with integral coefficients. The factors of 30 are ± 1, ± 30, ± 2, ± 15, ± 3, ± 10,

± 5, ± 6. Since $f(2) = 0$, $f(3) = 0$, and $f(-5) = 0$, $(x - 2)$, $(x - 3)$, and $(x + 5)$ are all factors. If we compute we see that

$$(x - 2)(x - 3)(x + 5) = x^3 - 19x + 30 \quad \square$$

We know that

$$(a + b)^2 = a^2 + 2ab + b^2$$
$$(a + b)^3 = a^2 + 3a^2b + 3ab^2 + b^3$$
$$(a + b)^4 = a^4 + 4a^3b + 6a^2b^2 + 4ab^3 + b^4$$

Furthermore,

$$(a + b)^5 = a^5 + 5a^4b + 10a^3b^2 + 10a^2b^3 + 5ab^4 + b^5$$

and

$$(a + b)^6 = a^6 + 6a^5b + 15a^4b^2 + 20a^3b^3 + 15a^2b^4 + 6ab^5 + b^6$$

The *binomial theorem* is a generalization of these results which provides a formula for writing the general expansion of $(a + b)^n$. The formula uses a special notation, called *factorial* notation, for certain integers. If n is a nonnegative integer, n *factorial* is denoted by $n!$ and defined by

$$0! = 1$$
$$1! = 1$$
$$n! = n(n - 1) \cdots 2 \cdot 1, \qquad \text{for } n > 1$$

Thus

$$2! = 2 \cdot 1 = 2$$
$$3! = 3 \cdot 2 \cdot 1 = 6$$
$$4! = 4 \cdot 3 \cdot 2 \cdot 1 = 24$$
$$5! = 5(4!) = 5 \cdot 24 = 120$$

Notice that for $n \geq 1$,

$$n! = n[(n - 1)!]$$

The binomial theorem gives the expansion of $(a + b)^n$ with the coefficients expressed using factorial notation.

THE BINOMIAL THEOREM

If a and b are numbers and n is a nonnegative integer, then

$$(a + b)^n = a^n + na^{n-1}b + \frac{n(n - 1)}{2}a^{n-2}b^2 + \cdots + \frac{n!}{(n - k)!k!}a^{n-k}b^k + \cdots + b^n$$

Thus the $(k+1)$st term of the expansion is

$$\frac{n!}{(n-k)!k!}a^{n-k}b^k$$

The theorem can be proved by mathematical induction after the following lemma has been established.

LEMMA

If n and k are positive integers with $k \leq n$, then

$$\frac{n!}{[n-(k-1)]!(k-1)!}+\frac{n!}{(n-k)!k!}=\frac{(n+1)!}{[(n+1)-k]!k!}$$

Proof:

The result is established by a straightforward computation; we need only to notice that

$$[n-(k-1)]! = [(n-k)+1]! = [(n-k)+1](n-k)!$$
$$k! = k(k-1)!$$

and

$$(n+1)! = (n+1)n!$$

A common denominator for the fractions on the left is

$$[n-(k-1)]!k!$$

Thus

$$\begin{aligned}\frac{n!}{[n-(k-1)]!(k-1)!}+\frac{n!}{(n-k)!k!} &= \frac{n!(k)}{[n-(k-1)]!(k-1)!(k)} \\ &\quad + \frac{n![n-(k-1)]}{[n-(k-1)](n-k)!k!} \\ &= \frac{n![k+n-(k-1)]}{[n-(k-1)]!k!} \\ &= \frac{n!(n+1)}{[(n+1)-k]!k!} \\ &= \frac{(n+1)!}{[(n+1)-k]!k!}\end{aligned}$$

**Proof of the Binomial Theorem*:

This proof uses mathematical induction.

Let

$$S = \left\{t \in N: (a+b)^t = a^t + \cdots + \frac{t!}{(t-k)!k!}a^{t-k}b^k + \cdots + b^t\right\}$$

Since

$$(a + b)^1 = a + b$$

1 is in S.

Now suppose $w \in S$. Then

$$\begin{aligned}(a + b)^{w+1} &= (a + b)(a + b)^w \\ &= (a + b)\left[a^w + \cdots + \frac{w!}{[w - (k - 1)]!(k - 1)!} a^{w-(k-1)}b^{k-1} \right. \\ &\quad \left. + \frac{w!}{(w - k)!k!} a^{w-k}b^k + \cdots + b^w\right]\end{aligned}$$

because w is in S.

Then multiplying out yields

$$\begin{aligned}(a + b)^{w+1} &= a^{w+1} + wa^w b + \cdots \\ &\quad + \frac{w!}{(w - k)!k!} a^{w-k+1}b^k + \cdots + ab^w + a^w b + \cdots \\ &\quad + \frac{w!}{[w - (k - 1)]!(k - 1)!} a^{w-(k-1)}b^{(k-1)+1} + \cdots + b^{w+1} \\ &= a^{w+1} + wa^w b + \cdots \\ &\quad + \left(\frac{w!}{(w - k)!k!} + \frac{w!}{[w - (k - 1)]!(k - 1)!}\right) a^{(w+1)-k}b^k \\ &\quad + \cdots + b^{w+1}\end{aligned}$$

Using the lemma we now have

$$(a + b)^{w+1} = a^{w+1} + \cdots + \frac{(w + 1)!}{[(w + 1) - k]!k!} a^{(w+1)-k}b^k + \cdots + b^{w+1}$$

Thus $w + 1 \subset S$.

By the principle of mathematical induction, $S = N$ and the theorem is proved.

The coefficients $\dfrac{n!}{(n - k)!k!}$ are sometimes denoted by ${}_nC_k$ or by $\dbinom{n}{k}$ and are called the *binomial coefficients*. In studying permutations and combinations, it can be shown that ${}_nC_k$ is the number of ways of choosing k objects from a set of n objects.

EXERCISES

1. Multiply and simplify.
   (a) $(2x - 7)(4x + 3)$ (b) $(8x + 5)(2x + 7)$
   (c) $(6x - 5)(x - 11)$ (d) $(2y + 5)(3y + 10)$
   (e) $(4a - 7b)(5a + 9b)$ (f) $(6x - 7y)(2x + 11y)$

(g) $(8x - 9z)(2x + 5z)$ (h) $(6x - 7y)(5x - 13y)$

2. Multiply and simplify.
   (a) $(2x + 9)(3x - 5)$ (b) $(4x + 7)(5x - 3)$
   (c) $(2y + 11)(3y - 5)$ (d) $(7z + 10)(2z - 3)$
   (e) $(6b - 11a)(4b + 5a)$ (f) $(9x + 11y)(2x + 3y)$
   (g) $(3t + 4z)(5t - 19z)$ (h) $(6x + 17y)(2x + 3y)$
3. Multiply and simplify.
   (a) $(x^2 + 2x + 1)(x^2 - 3)$
   (b) $(x^2 - xy + y^2)(x + y)$
   (c) $(4x^2 - 3x + 5)(x^2 + 2x - 1)$
   (d) $(a^2 + ab + b^2)(a^2 - 2ab + 3b^2)$
   (e) $(x^3 + 2x^2 - 3x + 5)(x^2 + 2x - 7)$
   (f) $(a^3 - 4a^2b + 5ab^2 - 7b^3)(a^3 + 6a^2b - 4ab^2 + 5b^3)$
4. Multiply and simplify.
   (a) $(x^2 + 4x + 2)(x^2 - 5)$
   (b) $(x^2 + xy + y^2)(x - y)$
   (c) $(5x^2 - 4x + 3)(x^2 + 3x - 1)$
   (d) $(x^2 - xy + y^2)(x^2 + 2xy + 5y^2)$
   (e) $(x^3 - 3x^2 + 5x + 2)(x^2 - 7x + 4)$
   (f) $(a^3 + 5a^2b - 3ab^2 + 9b^3)(a^3 - 7a^2b + 3ab^2 + 2b^3)$
5. Factor.
   (a) $x^2 - 4xy + 4y^2$ (b) $b^{12} - 64a^6$ (c) $a^3 + 5a^2b + 6ab^2$
   (d) $2x^3 - 14x^2 - 16x$ (e) $a^4 - 16$ (f) $3x^3 - 6x^2 - 9x$
   (g) $x^3 - 1$ (h) $y^3 - 8$ (i) $x^3 + 1$
   (j) $x^4 - 1$ (k) $x^3 - 2$ (l) $y^5 - 1$
   (m) $a^5 + 1$ (n) $x^5 + 32$ (o) $x^7 - 128$
6. Factor.
   (a) $a^2 + 4ab + 4b^2$ (b) $x^6 - 64y^6$ (c) $x^4 + 5x^2 - 6x^2y^2$
   (d) $3a^4 + 3a^3b - 36a^2b^2$ (e) $x^4 - 81$ (f) $5x^3 + 10x^2 - 15x$
   (g) $a^3 - 1$ (h) $y^3 - 27$ (i) $8a^3 + 1$
   (j) $16x^4 - 1$ (k) $y^3 - 3$ (l) $x^5 - 1$
   (m) $b^5 - 11$ (n) $y^5 - 32$ (o) $x^7 + 128$
7. Factor.
   (a) $20x^2 - 11xy - 3y^2$ (b) $x^2 + 2xy - 8y^2$
   (c) $6a^2 - 5a - 4$ (d) $6b^2 - 5b - 6$
   (e) $5x^2 + xy - 6y^2$ (f) $10x^2 - 9xy - 9y^2$
   (g) $4x^5 - 8x^4 + 4x^3$ (h) $12x^3y + 2x^2y^2 - 70xy^3$
   (i) $3x^6y^2 - 3x^2y^6$ (j) $x^6 - 28x^3 + 27$
   (k) $x^2 - 7$ (remember: $7 = \sqrt{7} \cdot \sqrt{7}$) (l) $y^2 - 3$
   (m) $x^2 - \frac{1}{6}x - \frac{1}{6}$ (n) $x^2 + \frac{22}{15}x + \frac{8}{15}$
8. Factor.
   (a) $10x^2 + xy - 21y^2$ (b) $16x^2 - 6xy + 7y^2$
   (c) $20x^2 + 2x - 6$ (d) $20x^2 + 3x - 9$
   (e) $16x^2 + 28x - 55$ (f) $6x^2 + 19x - 20$
   (g) $18x^4 - 54x^3 + 40x$ (h) $6x^2y^2 + 45xy^2 - 72y^2$
   (i) $5x^6y^3 - 5x^3y^6$ (j) $x^6 - x^3 - 12$
   (k) $x^2 - 10$ (l) $y^2 - 11$
   (m) $x^2 - \frac{9}{5}x - \frac{2}{5}$ (n) $x^2 + x + \frac{1}{4}$

9. Simplify.

(a) $\dfrac{3}{\sqrt{7}-\sqrt{5}}$ (b) $\dfrac{1}{\sqrt{11}+\sqrt{7}}$ (c) $\dfrac{\sqrt{10}+\sqrt{3}}{\sqrt{10}-\sqrt{3}}$

(d) $\dfrac{\sqrt{19}-\sqrt{15}}{\sqrt{19}+\sqrt{15}}$ (e) $\dfrac{\sqrt{a}-\sqrt{b}}{\sqrt{a}+\sqrt{b}}$ (f) $\dfrac{\sqrt{6}}{\sqrt{7}-\sqrt{3}}$

(g) $\dfrac{1}{\sqrt[3]{5}-\sqrt[3]{2}}$ (h) $\dfrac{1}{\sqrt[3]{x}-\sqrt[3]{y}}$

10. Simplify.

(a) $\dfrac{5}{\sqrt{7}-\sqrt{3}}$ (b) $\dfrac{1}{\sqrt{13}+\sqrt{5}}$ (c) $\dfrac{\sqrt{5}+\sqrt{7}}{\sqrt{5}-\sqrt{7}}$ (d) $\dfrac{\sqrt{13}-\sqrt{2}}{\sqrt{13}+\sqrt{2}}$

(e) $\dfrac{\sqrt{x}-\sqrt{y}}{\sqrt{x}+\sqrt{y}}$ (f) $\dfrac{\sqrt{7}}{\sqrt{5}-\sqrt{3}}$ (g) $\dfrac{1}{\sqrt[3]{7}-\sqrt[3]{5}}$ (h) $\dfrac{1}{\sqrt[3]{a}+\sqrt[3]{b}}$

11. Factor the following polynomials using the factor theorem.
(a) $x^3 - 12x^2 + 41x - 42$ (b) $x^3 + 6x^2 + 21x + 36$
(c) $x^3 - 3x^2 - 3x - 4$ (d) $x^5 - 9x^3 - 8x^2 + 72$
(e) $x^5 + 5x^4 - 6x^3 - 27x^2 - 135x + 162$

12. Use the factor theorem to factor the following polynomials.
(a) $x^3 + 2x^2 - 23x - 60$ (b) $x^3 + 3x^2 - 3x - 14$
(c) $x^4 - 3x^3 - 7x^2 + 7x + 10$ (d) $x^4 - x^3 - 21x^2 + x + 20$
(e) $x^4 + 5x^3 - 10x^2 - 20x + 24$

13. Show that for any positive integer n and any positive integer k, $k \le n$, ${}_nC_k = {}_nC_{(n-k)}$.

14. Compute.
(a) ${}_4C_0 + {}_4C_1 + {}_4C_2 + {}_4C_3 + {}_4C_4$
(b) ${}_5C_0 + {}_5C_1 + {}_5C_2 + {}_5C_3 + {}_5C_4 + {}_5C_5$
(c) ${}_6C_0 + {}_6C_1 + {}_6C_2 + {}_6C_3 + {}_6C_4 + {}_6C_5 + {}_6C_6$

15. What do you think is ${}_nC_0 + {}_nC_1 + {}_nC_2 + \cdots + {}_nC_n$? Can you prove your statement? [*Hint*: $2 = 1 + 1$.]

16. Show that for any positive integer n

$${}_nC_0 - {}_nC_1 + {}_nC_2 - {}_nC_3 + {}_nC_4 - \cdots + (-1)^n {}_nC_n = 0$$

[*Hint*: $0 = 1 + (-1)$.]

17. Use the binomial theorem to derive a general formula for $(x - y)^n$.

7. The Arithmetic of Algebraic Fractions

In Section 2 we pointed out that the fraction line is just a division symbol. By $\frac{2}{3}$ we mean the quotient of 2 by 3, that is, the unique number whose product with 3 is 2. Since algebraic expressions are just representations of numbers, the laws for adding, subtracting, multiplying, and dividing are the same for fractions involving algebraic expressions as they are for fractions that involve only numerals. The challenge to meet when dealing with algebraic fractions is to express the result in the simplest form.

All simplification techniques are based on the following fundamental idea:

If a, b, and c are numbers with $b \neq 0$ and $c \neq 0$, then

$$\frac{ac}{bc} = \frac{a}{b}$$

An equivalent phrasing of this principle is:

We can multiply or divide both the numerator and denominator of a fraction by the same nonzero number without changing the value of the fraction.

Example 1

$$\frac{2}{3} = \frac{2 \cdot 5}{3 \cdot 5} = \frac{10}{15} \quad \square$$

Example 2

$$\frac{9}{12} = \frac{3 \cdot 3}{4 \cdot 3} = \frac{3}{4} \quad \square$$

Example 3

$$\frac{4\sqrt{2}}{3\sqrt{2}} = \frac{4}{3} \quad \square$$

Example 4

$$\frac{3x^2}{5x^2} = \frac{3}{5} \quad \square$$

Example 5

$$\frac{(x + 2)(x + 3)}{(x - 7)(x + 3)} = \frac{x + 2}{x - 7} \quad \square$$

In dealing with fractions in arithmetic, this principle is used in two ways:

1. To reduce fractions to lowest terms.
2. To rewrite summands so that addition or subtraction can be completed using fractions with a common denominator.

Example 6

$$\frac{4}{6} = \frac{2 \cdot 2}{2 \cdot 3} = \frac{2}{3} \quad \square$$

Example 7

$$\frac{2}{3} + \frac{4}{5} = \frac{2 \cdot 5}{3 \cdot 5} + \frac{4 \cdot 3}{5 \cdot 3} = \frac{10}{15} + \frac{12}{15} = \frac{22}{15} \quad \square$$

We use this principle in the same ways when dealing with algebraic fractions. In applying the principle to algebraic fractions, factoring is very helpful. The only expressions that will evenly divide both the numerator and denominator of a fraction are expressions that are factors of both of them.

Example 8

$$\frac{x^2 - y^2}{xy + y^2} = \frac{(x + y)(x - y)}{(x + y)y} = \frac{x + y}{y}$$

dividing numerator and denominator by $x + y$ □

Example 9

$$\frac{x^2 - 5x + 6}{x^2 - 4} = \frac{(x - 2)(x - 3)}{(x - 2)(x + 2)}$$

$$= \frac{x - 3}{x + 2}$$

dividing numerator and denominator by $x - 2$ □

Example 10

$$\frac{a^2 - ab}{ab - b^2} = \frac{a(a - b)}{(a - b)b} = \frac{a}{b}$$

dividing by $a - b$ □

Example 11

$$\frac{(a + b)^4(x - y)^2}{(a + b)^2(x - y)^3} = \frac{(a + b)^2}{x - y}$$

dividing numerator and denominator both by $(a + b)^2(x - y)^2$ □

Example 12

$$\frac{4y^2 - 4y - 8}{2y^2 - 12y + 16} = \frac{4(y^2 - y - 2)}{2(y^2 - 6y + 8)}$$

$$= \frac{4(y + 1)(y - 2)}{2(y - 4)(y - 2)}$$

$$= \frac{2(y + 1)}{y - 4}$$

In this case we divided by $2(y - 2)$. □

The general procedure for simplifying algebraic fractions is just the same as the procedure for reducing numerical fractions to lowest terms: Divide the numerator and denominator by common factors and continue until no common factors remain. In practice it is usually simplest to factor both the numerator and denominator as much as possible before trying any division.

After gaining some experience, you will not write down all intermediate steps but will just "keep track" of the division by "crossing out" or "cancelling" common factors as you divide the numerator and denominator by them.

Example 13

$$\frac{x^2 - 5x - 6}{2x^2 - 2} = \frac{(x - 6)(x + 1)}{2(x^2 - 1)} = \frac{(x - 6)\cancel{(x + 1)}}{2(x - 1)\cancel{(x + 1)}} = \frac{x - 6}{2(x - 1)} \quad \square$$

Example 14

$$\frac{(a + 1)\cancel{(a - 2)}\cancel{(a + 3)}}{\cancel{(a - 2)}(a - 3)\cancel{(a + 3)}} = \frac{a + 1}{a - 3} \quad \square$$

Example 15

$$\frac{3y^2 - 3}{6y^2 + 12y + 6} = \frac{3(y^2 - 1)}{6(y^2 + 2y + 1)} = \frac{\cancel{3}\cancel{(y + 1)}(y - 1)}{2\cancel{6}(y + 1)^{\cancel{2}\,1}} = \frac{y - 1}{2(y + 1)} \quad \square$$

In the last example we divided the numerator and denominator by 3 and by $(y + 1)$. This method of "keeping track" was to change the exponent on $(y + 1)$ in the denominator rather than crossing out because $(y + 1)$ remains as a factor of the denominator.

Example 16

$$\frac{\cancel{(a + b)^2}(a - b)^{\cancel{3}\,2}}{(a + b)^{\cancel{4}\,2}\cancel{(a - b)}} = \frac{(a - b)^2}{(a + b)^2} \quad \square$$

Remember that this "crossing out" or "cancelling" is not an algebraic process! It is merely a notational convenience that enables us to eliminate quite a bit of writing. When you are simplifying a fraction the only principle you are using is the fact that you can multiply or divide the numerator and denominator of a fraction by a nonzero number without changing the value of the fraction.

Since algebraic expressions are just representations of numbers, the arithmetic of algebraic fractions is the same as the arithmetic of numerical fractions. The following basic governing principles were given in Section 2.

If a, b, c, and d are numbers such that $b \neq 0$ and $d \neq 0$, then

(i) $(a/b) = ad/bd$
(ii) $(a/b)(c/d) = ac/bd$
(iii) $(a/b)/(c/d) = ad/bc$ if $c \neq 0$
(iv) $a/b + c/b = (a + c)/b$

(v) $a/b - c/b = (a - c)/b$
(vi) $a/b + c/d = (ad + bc)/bd$
(vii) $a/b - c/d = (ad - bc)/bd$

In dealing with algebraic fractions we are presented with two challenges:

(a) When adding or subtracting fractions, how can we find a relatively simple common denominator?
(b) What procedures yield the simplest form of an answer?

We call these *challenges* because we could add or subtract using principles (vi) and (vii) and not worry about the simplicity of the form, although doing so might require quite a bit more algebraic computation.

We need to develop methods for dealing with the arithmetic of algebraic fractions which will require a minimum of algebraic computation and yield expressions in simplified forms. As you might expect, factoring will play an important role.

If we add two numerical fractions, we usually first find a least common denominator.

Example 17

$$\frac{5}{8} + \frac{3}{20} = \frac{5 \cdot 5}{8 \cdot 5} + \frac{3 \cdot 2}{20 \cdot 2} = \frac{25}{40} + \frac{6}{40} = \frac{31}{40} \quad \square$$

Other common denominators would also work. In fact, any integer evenly divisible by both 8 and 20 would suffice but computation is usually held to a minimum if we choose the smallest common denominator. Since algebraic expressions are often more complicated than numerical expressions, you might expect finding "least common denominators" to be even more useful in algebra than in arithmetic and you would be right.

Example 18

$$\begin{aligned}\frac{x + 2y}{x^2 - y^2} + \frac{x - 3y}{x^2 + 2xy + y^2} &= \frac{x + 2y}{(x + y)(x - y)} + \frac{x - 3y}{(x + y)^2} \\ &= \frac{(x + 2y)(x + y)}{(x + y)^2(x - y)} + \frac{(x - 3y)(x - y)}{(x + y)^2(x - y)} \\ &= \frac{(x^2 + 3xy + 2y^2) + (x^2 - 4xy + 3y^2)}{(x + y)^2(x - y)} \\ &= \frac{2x^2 - xy + 5y^2}{(x + y)^2(x - y)} \quad \square\end{aligned}$$

In this example we found a common denominator by first factoring the original denominators and then choosing as our common denominator the polynomial (in factored form) that contained as factors all the polynomials which appeared as factors in either denominator. This is, in fact, the general procedure used in adding or subtracting algebraic fractions:

(a) Factor all denominators.

(b) Choose as a common denominator the product of all factors appearing in any of the denominators.

If, say, $(x - 3)^4$ appears in one of the denominators, then $(x - 3)^4$ is a factor and $(x - 3)$ must appear to the fourth or higher powers in our common denominator. Thus $(x + y)^2$ appears in the common denominator in Example 18.

To talk about a "least common denominator" has no meaning since we have no way of saying that one algebraic expression is "less than" another. What we are finding is a "least complicated common denominator."

Multiplication and division of fractions are governed by principles (ii) and (iii). The primary challenge in such computations is to simplify the answers and the basic technique is to factor the numerators and denominators of the original fractions, indicate the factored form of the product or quotient, and then simplify.

Example 19

$$\begin{aligned}\frac{x + y}{x^2 + 5xy + 6y^2} \cdot \frac{x^2 + 4xy + 3y^2}{x + 4y} &= \frac{x + y}{(x + 2y)(x + 3y)} \cdot \frac{(x + 3y)(x + y)}{(x + 4y)} \\ &= \frac{(x + y)\cancel{(x + 3y)}(x + y)}{(x + 2y)\cancel{(x + 3y)}(x + 4y)} \\ &= \frac{(x + y)^2}{(x + 2y)(x + 4y)} \quad \square\end{aligned}$$

Example 20

$$\begin{aligned}\frac{x - 2y}{x^2 + 6xy + 9y^2} \div \frac{x^2 - 4y^2}{x^2 + 4xy + 3y^2} &= \frac{x - 2y}{(x + 3y)^2} \div \frac{(x + 2y)(x - 2y)}{(x + y)(x + 3y)} \\ &= \frac{x - 2y}{(x + 3y)^2} \cdot \frac{(x + y)(x + 3y)}{(x + 2y)(x - 2y)} \\ &= \frac{\cancel{(x - 2y)}(x + y)\cancel{(x + 3y)}}{(x + 3y)^{\cancel{2}\,1}(x + 2y)\cancel{(x - 2y)}} \\ &= \frac{x + y}{(x + 3y)(x + 2y)} \quad \square\end{aligned}$$

Complex fractions are algebraic fractions in which the numerator and/or denominator contain algebraic fractions.

Example 21

$$\frac{\dfrac{4x + 3}{2x + 1}}{\dfrac{x^2 + 1}{x^2 - 3}} \quad \square$$

Example 22

$$\frac{\dfrac{6x^2 - 1}{x^2 - 2x + 7}}{\dfrac{x^3 + 3x - 7}{x^2 + 2}} \quad \square$$

We can simplify such expressions by treating the fraction linc as the division symbol it is, that is, by rewriting the fraction as an indicated division and completing the computation.

Example 23

$$\frac{\dfrac{x^2 - x - 12}{x^2 - 1}}{\dfrac{x^2 - 9}{x^2 + 7x + 6}} = \frac{x^2 - x - 12}{x^2 - 1} \div \frac{x^2 - 9}{x^2 + 7x + 6}$$

$$= \frac{(x - 4)(x + 3)}{(x + 1)(x - 1)} \div \frac{(x + 3)(x - 3)}{(x + 1)(x + 6)}$$

$$= \frac{(x - 4)\cancel{(x + 3)}}{\cancel{(x + 1)}(x - 1)} \cdot \frac{\cancel{(x + 1)}(x + 6)}{\cancel{(x + 3)}(x - 3)}$$

$$= \frac{(x - 4)(x + 6)}{(x - 1)(x - 3)} = \frac{x^2 + 2x - 24}{x^2 - 4x + 3} \quad \square$$

Example 24

$$\frac{\dfrac{2x^2 - 11x - 21}{6x^2 - x - 1}}{\dfrac{2x^2 - x - 6}{2x^2 + 3x - 2}} = \frac{2x^2 - 11x - 21}{6x^2 - x - 1} \div \frac{2x^2 - x - 6}{2x^2 + 3x - 2}$$

$$= \frac{\cancel{(2x + 3)}(x - 7)}{(3x + 1)\cancel{(2x - 1)}} \cdot \frac{\cancel{(2x - 1)}(x + 2)}{\cancel{(2x + 3)}(x - 2)}$$

$$= \frac{(x - 7)(x + 2)}{(3x + 1)(x - 2)} = \frac{x^2 - 5x - 14}{3x^2 - 5x - 2} \quad \square$$

We can also simplify complex fractions by multiplying or dividing the numerator and denominator by the same expression. We can minimize computation if we use a common denominator as the multiplier.

Example 25

$$\frac{\dfrac{x^2+1}{2x+3}}{\dfrac{x^3-7}{2x+3}} = \frac{\dfrac{x^2+1}{\cancel{2x+3}}\cancel{(2x+3)}}{\dfrac{x^3-7}{2x+3}\cancel{(2x+3)}} = \frac{x^2+1}{x^3-7} \quad \square$$

Example 26

$$\frac{\dfrac{2x^2-3}{x+4}}{\dfrac{x^2+7}{2x+1}} = \frac{\dfrac{2x^2-3}{\cancel{x+4}}\cancel{(x+4)}(2x+1)}{\dfrac{x^2+7}{\cancel{2x+1}}(x+4)\cancel{(2x+1)}}$$

$$= \frac{(2x^2-3)(2x+1)}{(x^2+7)(x+4)} = \frac{4x^3+2x^2-6x-3}{x^3+4x^2+7x+28} \quad \square$$

Example 27

$$\frac{\dfrac{2x+5}{x^2+2x+1}}{\dfrac{x-3}{x^2-1}} = \frac{\dfrac{2x+5}{(x+1)^2}}{\dfrac{x-3}{(x+1)(x-1)}}$$

$$= \frac{\dfrac{2x+5}{\cancel{(x+1)^2}}\cancel{(x+1)^2}(x-1)}{\dfrac{(x-3)}{\cancel{(x+1)}\cancel{(x-1)}}(x+1)^{\cancel{2}}\cancel{(x-1)}}$$

$$= \frac{(2x+5)(x-1)}{(x-3)(x+1)} = \frac{2x^2+3x-5}{x^2-2x-3} \quad \square$$

EXERCISES

1. Simplify.

(a) $\dfrac{x^2+xy-6y^2}{x^2-xy-2y^2}$ (b) $\dfrac{x^2-y^2}{x^3-y^3}$

(c) $\dfrac{x^2-7xy+10y^2}{x^2-5xy+6y^2}$ (d) $\dfrac{3x^2-12y^2}{15x^2-30xy+60y^2}$

(e) $\dfrac{x^4-1}{x^3-1}$ (f) $\dfrac{x^5+x^4+x^3+x^2+x+1}{x^3+x^2+x+1}$

(g) $\dfrac{2x^2 + x - 15}{x^2 - x - 12}$ (h) $\dfrac{30x^3 + 81x^2 - 84x}{30x^5 + 6x^3 - 24x^2}$

2. Simplify.

(a) $\dfrac{x^2 + 2xy + y^2}{x^2 - y^2}$ (b) $\dfrac{x^2 + 5x + 6}{x^2 + 4x + 4}$ (c) $\dfrac{x^2 + 7x + 12}{x^2 + 6x + 8}$

(d) $\dfrac{3x^2 - 12y^2}{15x^2 - 60xy - 60y^2}$ (e) $\dfrac{y^5 - 1}{y^4 - 1}$ (f) $\dfrac{x^2 - 2x - 35}{x^2 - 9x + 14}$

(g) $\dfrac{2x^2 + 13xy + 20y^2}{3x^2 + 11xy - 4y^2}$ (h) $\dfrac{a^3 - b^3}{a^2 - b^2}$

3. Complete the indicated computations and simplify.

(a) $\dfrac{5x}{x + 1} + \dfrac{2x}{x - 1}$ (b) $\dfrac{2x}{x^2 - 1} - \dfrac{1}{x - 1}$

(c) $\dfrac{x^2 + xy - 6y^2}{6x^2 - xy - y^2} \cdot \dfrac{2x - y}{x - 2y}$ (d) $\dfrac{x^2 + 3x + 9}{3x + 9} \cdot \dfrac{x^2 - 9}{x^3 - 27}$

(e) $\dfrac{x + 1}{3x} + \dfrac{x - 1}{2x}$ (f) $(x^2 - 4y^2) + \dfrac{x + 2y}{x - 2y}$

(g) $\dfrac{2x}{x^2 - 4x - 5} \div \dfrac{3x}{x^2 - 1}$ (h) $\dfrac{5}{x^2 - 1} \div \dfrac{3}{x^3 - 1}$

4. Complete the indicated computations and simplify.

(a) $\dfrac{7y}{2y + 1} + \dfrac{3y}{y - 1}$ (b) $\dfrac{6x}{x^2 - 4} + \dfrac{3}{x + 2}$

(c) $\dfrac{x^2 + 3xy - 4y^2}{x^2 - 2xy - 8y^2} \cdot \dfrac{x + 2y}{x - y}$ (d) $\dfrac{x^2 - x + 1}{2x + 2} \cdot \dfrac{x^2 - 1}{x^3 + 1}$

(e) $\dfrac{2x + 3}{4y} - \dfrac{2x - 1}{3y}$ (f) $x^2 - 9 - \dfrac{x + 3}{x - 3}$

(g) $\dfrac{4y}{y^2 - 5y + 6} \div \dfrac{3y^2}{y^2 - 9}$ (h) $\dfrac{7}{y^2 - 4} \div \dfrac{5}{y^3 - 8}$

5. Complete the indicated computations and simplify.

(a) $\dfrac{2x + 3}{x^2 - 7x + 10} + \dfrac{4x - 1}{x^2 + 5x + 6}$

(b) $\dfrac{4}{x - 1} + \dfrac{5}{x} + \dfrac{6}{x + 1}$

(c) $\dfrac{3}{x^2 + 2x + 1} + \dfrac{4}{x^2 - 2x + 1} + \dfrac{5}{x^2 - 1}$

(d) $\dfrac{2x - 3}{x^2 - 3x - 4} + \dfrac{x^2 + 1}{x^2 - 16}$

(e) $\dfrac{2x - 1}{x + 2} + \dfrac{x - 1}{x - 2} + \dfrac{x^2 - 3x}{x^2 - 4} + \dfrac{x^2 - 2x - 3}{(x - 2)^2}$

(f) $\dfrac{2x + 1}{x^2 - 4} - \dfrac{x^2 - 1}{x - 2}$

(g) $\dfrac{x^2 + 3x - 4}{x^2 - 3x - 4} + \dfrac{x^2 + 5x + 6}{x^2 + 4x + 3}$

(h) $\dfrac{2x}{x + 5} + \dfrac{x + 1}{x^2 - 25} + \dfrac{3x - 1}{x^3 - 125}$

6. Complete the indicated computations and simplify.

(a) $\dfrac{4x + 1}{x^2 - 3x - 10} - \dfrac{3x + 1}{x^2 + 6x + 8}$

(b) $\dfrac{7}{x + 2} + \dfrac{11}{x} - \dfrac{5}{x + 3}$

(c) $\dfrac{7}{x^2 + 10x + 25} - \dfrac{4}{x^2 - 10x + 25} + \dfrac{3}{x^2 - 25}$

(d) $\dfrac{4x + 1}{x^2 - 3x - 18} - \dfrac{x^2 + 3}{x^2 - 9}$

(e) $\dfrac{6y - 1}{y - 1} - \dfrac{3y}{y + 1} + \dfrac{y^2 + 4y}{y^2 - 1} + \dfrac{y^2 - 3y + 4}{(y - 1)^2}$

(f) $\dfrac{5x - 3}{x^2 - 9} - \dfrac{4x - 5}{x + 3}$

(g) $\dfrac{x^2 + x + 1}{x^2 - x - 12} - \dfrac{x^2 + 3x + 1}{x^2 - 9x + 20}$

(h) $\dfrac{2y + 3}{y^3 + 8} - \dfrac{y - 2}{y + 2} + \dfrac{3y}{y^2 - 4}$

7. Simplify.

(a) $\dfrac{\dfrac{2x + 5}{x^2 - 1}}{\dfrac{3x - 1}{x^2 + 2x + 1}}$ (b) $\dfrac{\dfrac{x - 3}{x^2 + 5x - 6}}{\dfrac{x - 2}{x^2 + 3x - 4}}$ (c) $\dfrac{\dfrac{x^2 - 1}{x^2 + 4x + 4}}{\dfrac{x^2 - 2x - 1}{x^2 - 4}}$

(d) $\dfrac{\dfrac{x - 7}{x^2 + 3x + 2}}{\dfrac{2x + 1}{x^2 + 4x + 5}}$ (e) $\dfrac{x - \dfrac{1}{x}}{x + \dfrac{1}{x}}$ (f) $\dfrac{\dfrac{x}{y} + 1}{\dfrac{x^2}{y^2} - 1}$

(g) $\dfrac{\dfrac{x^3 - 8}{x^2 - 1}}{\dfrac{x^2 - 4}{x^3 - 1}}$ (h) $\dfrac{\dfrac{y^3 - 3y^2 + 3y - 1}{y^2 - 4y + 4}}{\dfrac{y^3 - 8}{y^2 - 2y + 1}}$

8. Simplify.

(a) $\dfrac{\dfrac{3x + 1}{x^2 - 9}}{\dfrac{5x + 2}{x^2 - 7x + 12}}$ (b) $\dfrac{\dfrac{4x - 1}{x^2 + 9x - 10}}{\dfrac{x + 4}{x^2 - 4x + 3}}$ (c) $\dfrac{\dfrac{x - 2}{x^2 + 7x + 6}}{\dfrac{x - 4}{x^2 - 9}}$

(d) $\dfrac{\dfrac{x^2 - 4}{x^2 - 2x + 1}}{\dfrac{x^2 + 4x + 4}{x^2 - 1}}$

(e) $\dfrac{a + \dfrac{2}{a}}{a - \dfrac{3}{a}}$

(f) $\dfrac{\dfrac{a}{b} - 1}{\dfrac{a^2}{b^2} - 1}$

(g) $\dfrac{\dfrac{y^3 + 1}{y^2 - 4}}{\dfrac{y^3 + 8}{y^2 - 1}}$

(h) $\dfrac{\dfrac{x^3 + 3x^2 + 3x + 1}{x^2 + 4x + 4}}{\dfrac{x^3 + 8}{x^2 + 2x + 1}}$

2

Equations and Inequalities

To many people, algebra means "solving" equations. Such an impression is not quite accurate but has an historical foundation. Teaching techniques for solving equations and inequalities has always been one of the most important objectives of a course in algebra and in this chapter we present the most common and important of these methods.

1. Linear Equations in One Variable

An equation is just a simple sentence using the linking verb "equals." The left side of the equation is the subject of the sentence and the right side is the predicate noun. If we write

"The sum of three and two equals five."

the subject of the sentence is "The sum of three and two," the verb is "equals," and the predicate noun is "five." The use of mathematical symbols enables us to write this sentence in a shorter form:

"$3 + 2 = 5$"

As you might expect, we will write most equations symbolically. When we are dealing with applications, many equations will arise from verbal statements about problem situations. In such situations, our task will be to formulate verbal equations and then to write them using algebraic symbols.

Recall that there are two basic types of symbols that represent numbers. *Numerals* such as 2, 5, 1342, 47/93, π, and $\sqrt{7}$ name particular numbers and can be viewed as proper nouns. A second way of representing numbers is by *variables*, such as a, b, x, y, z, p, and q, which can be viewed as pronouns. Thus, for example, when we write

$$3 + x = 5$$

the symbol "x" is a pronoun representing a number. Our knowledge of arithmetic assures us that 2 is the only number that x can represent and still make the sentence true.

If we write

$$a + b = b + a$$

we have a sentence that holds for all possible numbers that we could substitute for a and b. An equation that holds for *all meaningful substitutions* is called an *identity*. An equation that does not hold for all meaningful substitutions is called a *conditional equation.*

In our study of the properties of R, we made several statements which were identities. When we write

$$(x + y)(x - y) = x^2 - y^2$$

we are writing an identity. The equation

$$\frac{x^2 - y^2}{x - y} = x + y$$

is also an identity. In this case, if we substitute 2 for x and 2 for y, the left side becomes meaningless and therefore substituting 2 for x and 2 for y is not a meaningful substitution. The equation holds for all substitutions that do not yield a zero denominator on the left side. The equations we wrote in our studies of factoring and algebraic fractions are all examples of identities.

In this chapter we will be primarily concerned with conditional equations.

Example 1

The equations $(-4) + x = 19$ and $7y + 3 = 34$ both are conditional equations because, although substituting 5 for x and 2 for y would yield meaningful statements, both statements are false. □

To *solve* a conditional equation containing one variable means to determine all numbers which when substituted for the variable will yield a true statement. These numbers are called *solutions* or *roots* of the equation. If we substitute 23 for x in the first equation of Example 1, we get a true statement and 23 is a solution of the equation. Similarly if we substitute 31/7 for y in the second equation, we again get a true statement.

One fundamental principle underlies most of the techniques for solving equations. Intuitively it is:

If you do the same thing to both sides of a true equation,
the resulting statement is also true.

Example 2

$3 + 2 = 5$	is known to be true. Thus
$(3 + 2) \cdot 7 = 5 \cdot 7$	is also true, as are
$(3 + 2) - 8 = 5 - 8$	and

$$\frac{3 + 2}{9} = \frac{5}{9} \quad \square$$

Equality means "sameness" and when we write

"$3 + 2 = 5$"

we mean that "$3 + 2$" names the same number that "5" does. If we have two names for the same number, we can interchange them in any algebraic expression and obtain two expressions that represent the same number. It is this substitution principle that underlies the "do the same thing to both sides" idea. In high school mathematics, students are frequently given a list of properties of equality:

If $a = b$, then $a + c = b + c$.
If $a = b$, then $a - c = b - c$.
If $a = b$, then $ac = bc$.
If $a = b$, then $a/c = b/c$ providing c is not 0.

These are not separate properties or axioms dealing with equality but are just consequences of the substitution principle which follows from the meaning of equality. For example, if a means the same as b, then $a + c$ must mean the same as $b + c$. If it is easier for you to think in terms of these four properties, then do so, but you may find that it is easier to think in terms of the substitution principle.

Most of the time, you will need to apply the principle several times in order to solve the equation.

Example 3

$5x = 2x + 8$	
$(-2x) + (5x) = (-2x) + [(2x + 8)]$	Adding $(-2x)$ to both sides
$[(-2) + 5]x = [(-2x) + 2x] + 8$	Associative and distributive laws
$3x = 0 + 8$	Arithmetic
$3x = 8$	
$\frac{3x}{3} = \frac{8}{3}$	Dividing both sides by 3
$x = \frac{8}{3}$	Arithmetic

Checking:

$$5\left(\frac{8}{3}\right) = 2\left(\frac{8}{3}\right) + 8 \qquad \text{is equivalent to}$$

$$\frac{40}{3} = \frac{16}{3} + 8 \qquad \text{which is true } \square$$

Example 4

$13x+2=4x+14$	
$(-4x)+(13x+2)=(-4x)+(4x+14)$	Adding $-4x$ to both sides
$[(-4x)+13x]+2=[(-4x)+4x]+14$	Associativity
$9x+2=14$	Arithmetic and the distributive law
$(9x+2)-2=14-2$	Subtracting 2 from both sides
$9x=12$	Arithmetic
$\frac{9x}{9}=\frac{12}{9}$	Dividing both sides by 9
$x=\frac{4}{3}$	

Checking: The left side is

$$13\left(\frac{4}{3}\right) + 2 = \frac{52}{3} + \frac{6}{3} = \frac{58}{3}$$

The right side is

$$4\left(\frac{4}{3}\right) + 14 = \frac{16}{3} + 14 = \frac{16 + 42}{3} = \frac{58}{3} \ \square$$

The reader should provide the justification for each step in the following examples. At times two steps are combined into one.

Example 5

$$\sqrt{3}x + 2 = 8$$
$$\sqrt{3}x = 8 - 2$$
$$\sqrt{3}x = 6$$
$$x = \frac{6}{\sqrt{3}}$$
$$x = \frac{6\sqrt{3}}{3}$$
$$x = 2\sqrt{3}$$

Checking:

$$(\sqrt{3})(2\sqrt{3}) + 2 = 2 \cdot 3 + 2 = 8 \square$$

Example 6

$$\begin{aligned} .24x + 1.7 &= .13x - .5 \\ .24x - .13x &= -.5 - 1.7 \\ .11x &= -2.2 \\ 11x &= -220 \\ x &= \frac{-220}{11} \\ x &= -20 \end{aligned}$$

Checking:

$$\begin{aligned} .24(-20) + 1.7 &= -4.8 + 1.7 = -3.1 \\ .13(-20) - \ \ .5 &= -2.6 - \ \ .5 = -3.1 \square \end{aligned}$$

The methods followed in Examples 3 to 6 are typical of the methods you should be able to use. In several cases, two or more steps were combined into one. If you cannot follow this procedure, then fill in the single steps. After you have become more accustomed to working with equations, such techniques will come naturally to you. Do not strive for them. *There is no great virtue in brevity.* Write down each step that you cannot do *easily* in your head.

Notice that in each example, we substituted the solution in the original equation to see if it was, in fact, a solution. *We cannot be sure that we have a solution until it has been tested.*

Frequently we encounter equations, such as formulas, which include several quantities, all or some of which are represented by letters. Such equations are sometimes called "literal equations."

$A = lw$	the formula for the area of a rectangle
$p = 2l + 2w$	the formula for the perimeter of a rectangle
$d = rt$	the relationship between distance, rate, and time
$i = Prt$	the formula for computing interest when P denotes principle, r denotes rate, and t denotes time
$C = 2\pi r$	the formula for the circumference of a circle where r denotes the radius and π is a special constant (that is sometimes approximated by 3.1416 or by 22/7).

It is sometimes quite useful to solve such equations for one quantity in terms of the others.

Example 7

The formula $A = lw$ expresses the area in terms of the length and width. We may, however, wish to obtain a formula that expresses length in terms of area and width. To do so we can just solve the equation $lw = A$ for l by treating l as a variable and A and w as constants. In this case, if we divide both sides by w we have

$$\frac{lw}{w} = \frac{A}{w} \quad \text{and} \quad l = \frac{A}{w} \; \square$$

Example 8

Suppose we wish to express the length of a rectangle in terms of the width and perimeter. We know that

$$2l + 2w = p$$

Subtracting $2w$ from both sides yields

$$2l = p - 2w$$

Dividing both sides by 2 yields

$$l = \frac{p - 2w}{2} \quad \text{or} \quad l = \frac{p}{2} - w \; \square$$

Example 9

To express the radius of a circle in terms of the circumference, we solve $C = 2\pi r$ for r:

$$2\pi r = C$$

$$\frac{2\pi r}{2\pi} = \frac{C}{2\pi}$$

$$r = \frac{C}{2\pi} \; \square$$

In solving literal equations we use exactly the same techniques as in solving numerical equations. We just treat one of the letters as a variable and the others as we would any number.

EXERCISES

1. Solve the following equations. Write down each step. Check your results.
   (a) $3x + 4 = 11$ (b) $\sqrt{2x} - 5 = 3$ (c) $3y + 9 = 11$
   (d) $4z - 6 = \sqrt{2z} + 9$ (e) $\sqrt[3]{4x} - 7 = \sqrt[3]{2x} + 3$
2. Solve the following equations. Be sure to check your solutions.

(a) $.2x + 3 = .5x - 7$
(b) $7y - .5 = .2y + 4$
(c) $8x - 7 = \frac{1}{3}x + \frac{2}{3}$
(d) $\frac{4}{5}x + \frac{7}{11} = \frac{2}{3}x - \frac{5}{9}$
(e) $\sqrt[3]{4}x + .31 = \sqrt[3]{9}x - .72$

3. Solve the following equations. Check your solutions.
   (a) $1.6x + 9.2 = .42x - 7.93$
   (b) $\frac{3}{7}x + \frac{7}{3} = \frac{7}{3}x + \frac{3}{7}$
   (c) $y/3 + 2y/5 = \frac{7}{10}$
   (d) $\frac{4}{5}x - \sqrt{2/3} = \frac{9}{7}x + \sqrt{5/9}$
   (e) $31.46x + 92.3 = 7.41x - 15$
4. Solve the following equations. [*Hint*: $|y| = c$ if and only if $y = c$ or $y = -c$.]
   (a) $|x + 1| = 4$
   (b) $|x - 2| = 7$
   (c) $|x + 7| = 3$
   (d) $|2x + 1| = 9$
   (e) $|5x - 7| = 9$
   (f) $|14x + 19| = 6$
5. (a) Solve the formula $d = rt$ for r in terms of d and t.
   (b) If a runner runs 880 yd in 1 min 56 sec, what is his average rate per second?
   (c) If an airplane travels 1150 mi in 2.3 hr, what is the average rate to the nearest mile per hour?
   (d) Solve the formula for t in terms of d and r.
   (e) How long does it take an automobile to go .01 mi at 55 mph? Express your answer to the nearest second. (It may be of interest that .01 mile is 52.8 ft.)
   (f) How long does it take an airplane to go 5 mi at 500 mph? Again, express your answer to the nearest second.
   (g) Spacecraft carrying astronauts to and from the moon re-enter our atmosphere at 25,000 mph. How long does it take them to travel 10 mi? Express your answer to the nearest second.
6. (a) Determine the interest on \$1300 invested at 7 percent per year for 5 months.
   (b) What is the interest on \$3800 at 9.25 percent per year for $5\frac{1}{2}$ months?
   (c) Solve the formula $i = Prt$ for P, for r, and for t in terms of the other variables.
   (d) If you are receiving \$64 every 3 months on savings receiving 5 percent per year, how much have you invested in your savings account?
   (e) If you are receiving \$126 every 6 months on \$4200, what is your annual rate of interest?
   (f) If you are paying \$45 interest every month on a loan of \$7000, what is your annual rate of interest?
7. (a) Determine the perimeter of a rectangle of length 5.8 in. and width 3.2 in.
   (b) What is the width of a rectangle with perimeter 85 cm and length 31 cm?
   (c) What is the width of a rectangle with length 8.2 in. and area 20.5 sq in.?
8. The area of a triangle is given by the formula $A = \frac{1}{2}bh$.
   (a) What is the area of a triangle with base 9.2 cm and height 4.1 cm?
   (b) Solve the formula for b in terms of A and h. Solve it for h in terms of b and A.
   (c) What is the base of a triangle with height 8 in. and area 44 sq in.?
   (d) What is the height of a triangle with area 8.2 sq cm and base 11.5 cm?
9. The area of a trapezoid is given by the formula $A = \frac{1}{2}(b_1 + b_2)h$ where h denotes the height and b_1 and b_2 are the bases.

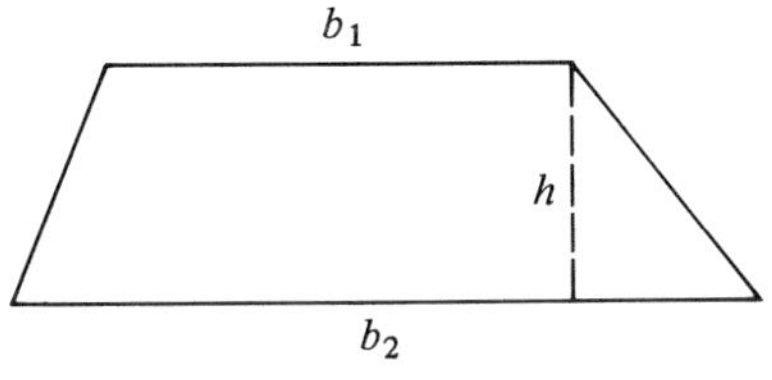

(a) Determine the area of a trapezoid with height 9.2 cm having bases of length 7.4 cm and 6.4 cm.
(b) Solve the formula for h, b_1, and b_2 in terms of the other variables.
(c) If the area of a trapezoid is 95.8 sq in. and the bases are 4.4 and 3.2 in., respectively, what is its height?
(d) If the area of a trapezoid is 230 sq cm, its height is 12 cm, and one base is 27 cm, what is the other base?

10. The volume of a rectangular solid is given by $V = lwh$, where l denotes length, w denotes width, and h denotes height.

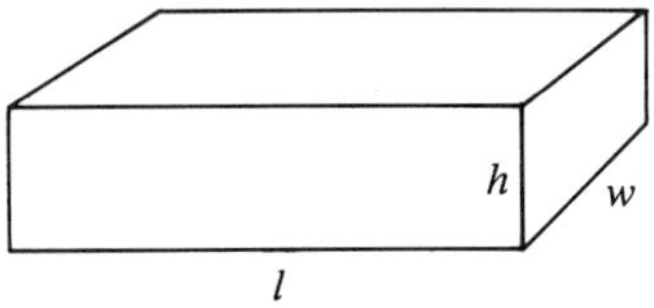

(a) What is the volume of a box 12.4 cm long, 8.2 cm wide, and 3.8 cm high?
(b) Solve the formula for l, for w, and for h.
(c) If the volume of the box is 68 cu cm, its length is 5 cm, and its width is 3 cm, what is its height?

11. Solve the equation $6x + 12y = 8$ for x in terms of y. If $y = 3$, what is x?
12. Solve the equation $2x + 4y = 3$ for y in terms of x. If $x = 7$, what is y?

2. Solving Linear Inequalities

In Chapter 1 we pointed out that comparability was one of the important properties of numbers. Sometimes we do not care exactly what the numbers are but how they are related. For example, in a football game the scores themselves are not the important thing. We just want our team's score to be greater than the other team's score. If we are taking an automobile trip, we do not want our tank to contain just enough gasoline but rather we make sure that it contains at least enough.

Since an equation or equality is a sentence of the form $A = B$, we might expect an inequality to be a statement of the form $A \neq B$. Since, however, all numbers are comparable, if A and B are numbers and $A \neq B$, then either $A < B$ or $B < A$. For our purposes, the study of inequalities will be concerned only with statements of the form

$$A < B, \quad A \leq B, \quad A \geq B, \quad \text{or} \quad A > B$$

If an inequality contains one variable, say x, *to solve the inequality means to determine all numbers which when substituted for x yield a true statement.* In general, we can expect an inequality to have more solutions than is expected of an equation. For example, consider the following two sentences:

$2x < 6 \qquad 2x = 6$

Any number less than 3 is a solution of the inequality, whereas 3 is the only solution of the equation.

The techniques for solving inequalities are based on the three principles discussed in Chapter 1.

(i) If a, b, and c are numbers such that $a < b$, then $a + c < b + c$ and $a - c < b - c$.
(ii) If a, b, and c are numbers such that $a < b$ and c is positive, then $ac < bc$ and $a/c < b/c$.
(iii) If a, b, and c are numbers such that $a < b$ and c is negative, then $bc < ac$ and $b/c < a/c$.

Similar statements hold if we replace $<$ in the above statements by $>$, $\leq$, or $\geq$.

These statements can be condensed as follows:

A: If we add or subtract the same number to or from both sides of an inequality, we obtain an inequality in the *same* direction.
B: If we multiply or divide both sides of an inequality by a *positive* number, we obtain an inequality in the *same* direction.
C: If we multiply or divide both sides of an inequality by a *negative* number, we obtain an inequality in the *opposite* direction.

The basic techniques for solving inequalities are founded on these principles.

Example 1

$x + 3 < 7$

$(x + 3) - 3 < 7 - 3$ Subtracting 3 from both sides yields an inequality in the same direction

$x < 4$ □

The solution to an inequality is a set of numbers. We could say now that the solution to the inequality $x + 3 < 7$ is the set $\{x: x < 4\}$. This is a lot of extraneous writing and really does not convey much meaning. The set of solutions to the original inequality is the set of numbers satisfying $x < 4$ and we will consider the statement $x < 4$ as sufficient in describing the solution.

The solution set of an inequality may be easier to grasp if we draw a graph or picture of the set on a number line. This is done by drawing a number line in the usual fashion and then indicating the solution set by a heavier line. (See Figure 2–1.)

Figure 2–1

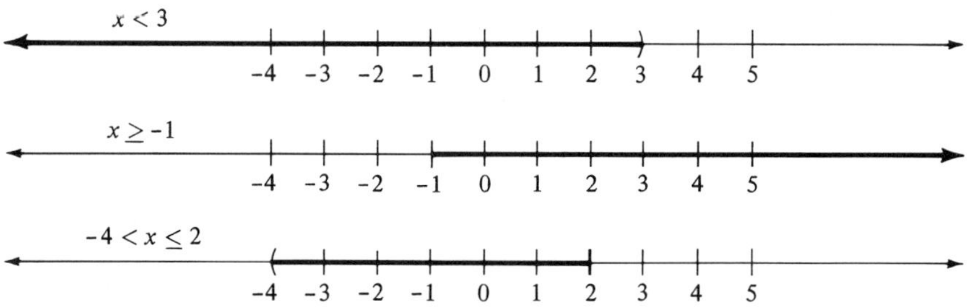

Notice that we indicate that the end point is not included by using a form such as "—)" or "(—" and that it is included is indicated by a form such as "—|" or "|—."

Example 2

$$4x + 7 \geq 9$$

$$(4x + 7) - 7 \geq 9 - 7$$

Subtracting 7 from both sides preserves the inequality.

$$4x \geq 2$$

$$\frac{4x}{4} \geq \frac{2}{4}$$

Dividing both sides by 4 preserves the inequality.

$$x \geq \tfrac{1}{2} \quad \square$$

Example 3

$$5x - 3 < 7x + 5$$

$$(5x - 3) + 3 < (7x + 5) + 3$$

Adding 3 to both sides preserves the inequality.

$$5x < 7x + 8$$

$$5x - 7x < 7x + 8 - 7x$$

Subtracting $7x$ from both sides preserves the inequality.

$$-2x < 8$$

$$\frac{-2x}{-2} > \frac{8}{-2}$$

Dividing both sides by -2 reverses the direction of the inequality.

$$x > -4 \quad \square$$

Example 4

$$
\begin{aligned}
.5y + 6 &< .4y - 2 \\
(.5y + 6) - 6 &< .4y - 2 - 6 \\
.5y &< .4y - 8 \\
.5y - .4y &< .4y - 8 - .4y \\
.1y &< -8 \\
y &< -80 \;\square
\end{aligned}
$$

The reader should provide reasons.

The graphs of the solutions for Examples 2–4 are given in Figure 2–2.

Figure 2–2

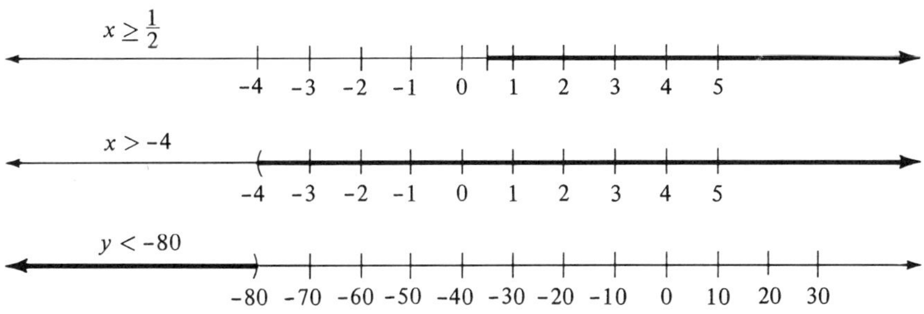

In solving equations we stressed the idea that it is necessary to check solutions. However, if all the steps and procedures followed are reversible and if our computation is correct, the solutions obtained will be solutions of the original equation. Most of the time it is more efficient to check our solutions rather than make sure that all steps taken are reversible. In solving inequalities, the fact that there are usually infinitely many solutions makes checking impossible. The steps taken in solving inequalities, however, are all reversible if we use only the three principles discussed above. In particular, adding the same number to both sides of an inequality can be reversed by adding the negative of the same number to both sides. Multiplying or dividing both sides by a nonzero number is also reversible. Thus the final inequality is logically equivalent to the original inequality and both have the same solutions.

All of the inequalities considered thus far are of the form

$$ax + b < cx + d$$

$$ax + b \le cx + d$$

$$ax + b > cx + d$$

$$ax + b \ge cx + d$$

Such inequalities can be solved without multiplying or dividing by expressions containing the variable; that is, all multiplication and division of both sides uses only constants and therefore we know which of principles B and C we are applying. Later on we will see situations where we must consider both possibilities.

Frequently we have two or more inequalities and we wish to find all numbers that satisfy both of them or either of them.

Example 5

Find all numbers x such that

$2x + 3 < 8$ *and* $3x - 1 > 2$

First we solve both inequalities:

$2x + 3 < 8$	$3x - 1 > 2$
$2x < 5$	$3x > 3$
$x < \frac{5}{2}$	$x > 1$

Thus the set of all solutions to both inequalities is the set of all numbers x such that $1 < x$ *and* $x < \frac{5}{2}$. A simpler way of denoting this is by

$1 < x < \frac{5}{2}$

Notice that the solution set for the above system is the *intersection* of the solution sets for the two inequalities. The graph of the solution is given in Figure 2–3. □

Figure 2–3

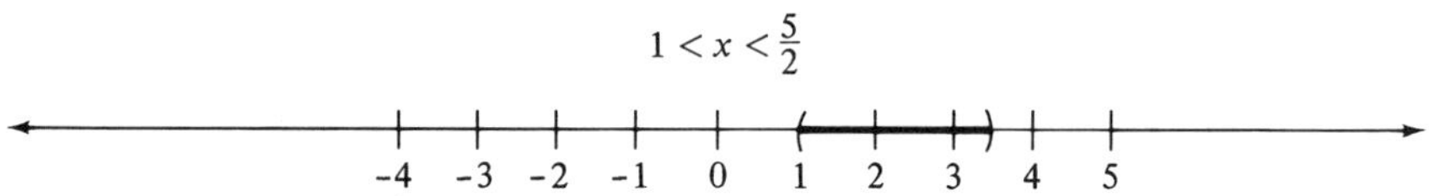

Example 6

Find the solution set of

$5x + 2 < 12$ *and* $2x - 4 > 2$

Solving both inequalities yields the two solution sets

$5x + 2 < 12$	$2x - 4 > 2$
$5x < 10$	$2x > 6$
$x < 2$	$x > 3$

In order for a number to satisfy both inequalities, it would have to be less than 2 *and* greater than 3 and this is impossible. Thus this system has no solutions; that is, it has an empty solution set. □

Example 7

Find all numbers satisfying

$4x + 2 < 14$ *and* $3x - 1 < 14$

Solving both inequalities yields the two solution sets

$4x + 2 < 14$	$3x - 1 < 14$
$4x < 12$	$3x < 15$
$x < 3$	$x < 5$

The solution set for the system is all numbers that are less than 3 and less than 5. These are just the numbers less than 3. Thus $\{x: x < 3\}$ is the solution set for the system. □

Example 8

Find all numbers satisfying

$5 < 2x + 4$ *or* $2x + 4 < -5$

Solving both inequalities yields

$1 < 2x$	$2x < -9$
$\frac{1}{2} < x$	$x < -\frac{9}{2}$

The desired solution set is

$\{x: \frac{1}{2} < x \text{ or } x < -\frac{9}{2}\}$

The graph of this solution set is given by Figure 2–4. □

Figure 2–4

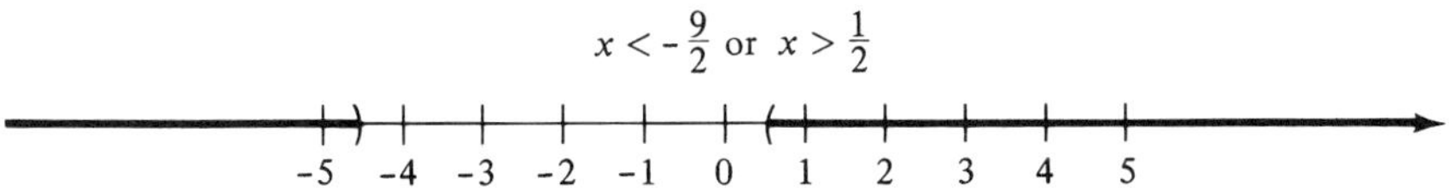

Frequently we need to solve inequalities involving the absolute value of numbers. Such inequalities are equivalent to a system of two inequalities and can be solved by solving the system.

If a is a positive number, we know from Chapter 1 that:

(i) The inequality $|y| < a$ is equivalent to the system of inequalities

$$-a < y \quad \textit{and} \quad y < a$$

that is,

$$-a < y < a$$

(ii) The inequality $|y| > a$ is equivalent to the system of inequalities

$$y < -a \quad \textit{or} \quad y > a$$

Example 9

Solve the inequality

$$|3x - 1| < 6$$

This inequality is equivalent to the system

$$-6 < 3x - 1 \quad \textit{and} \quad 3x - 1 < 6$$

Solving yields

$$-5 < 3x \quad \textit{and} \quad 3x < 7$$
$$-\tfrac{5}{3} < x \quad \textit{and} \quad x < \tfrac{7}{3}$$

Thus the solution set is

$$\{x: -\tfrac{5}{3} < x \quad \textit{and} \quad x < \tfrac{7}{3}\}$$

or just

$$\{x: -\tfrac{5}{3} < x < \tfrac{7}{3}\} \ \square$$

Example 10

Solve

$$|2x + 3| > 7$$

This inequality is equivalent to the system

$$2x + 3 < -7 \quad \textit{or} \quad 2x + 3 > 7$$

Solving yields

$$2x < -10 \quad \textit{or} \quad 2x > 4$$
$$x < -5 \quad \textit{or} \quad x > 2$$

The solution set is

$$\{x: x < -5 \textit{ or } x > 2\} \ \square$$

Notice that in Example 9 the solution set is the intersection of the two solution sets, whereas in Example 10 the solution set is the corresponding union.

Example 11

Solve $|5x + 3| \leq 2$.

$$-2 \leq 5x + 3 \quad \textit{and} \quad 5x + 3 \leq 2$$
$$-5 \leq 5x \quad \textit{and} \quad 5x \leq -1$$
$$-1 \leq x \quad \textit{and} \quad x \leq -\tfrac{1}{5}$$

The solution set is

$\{x: -1 \leq x \leq -\frac{1}{5}\}$ □

Example 12

Solve $|5x - 1| \geq 11$.

$$5x - 1 \geq 11 \quad \textit{or} \quad 5x - 1 \leq -11$$
$$5x \geq 12 \quad \textit{or} \quad 5x \leq -10$$
$$x \geq \tfrac{12}{5} \quad \textit{or} \quad x \leq -2$$

The solution set is

$\{x: x \geq \frac{12}{5} \quad \textit{or} \quad x \leq -2\}$ □

The graphs of the solutions of Examples 9–12 are given in Figure 2–5.

Figure 2–5

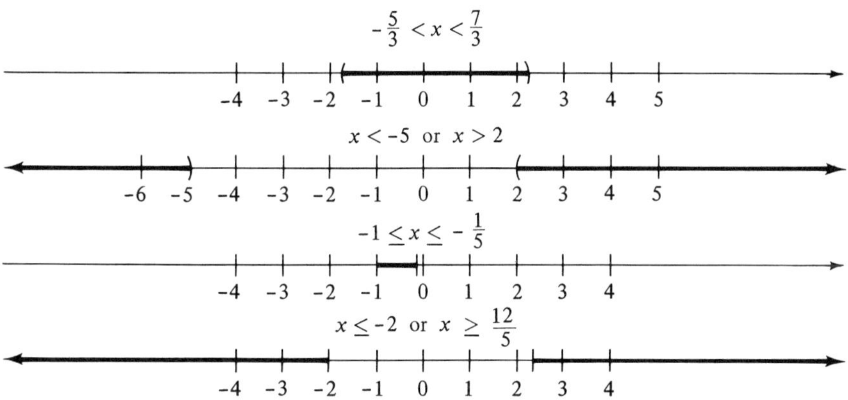

EXERCISES

1. Solve the following inequalities. Write out reasons for each step.
   (a) $-2x + 5 < 8$ (b) $4x - 7 < 9$
   (c) $.3x + 2 < 6$ (d) $3x + 2 \geq 4x - 7$
   (e) $5x - 2 \leq 13x + 9$ (f) $\frac{1}{3}x + \frac{1}{6} > \frac{4}{5}x + \frac{9}{7}$
   (g) $\frac{4}{3}y - \frac{2}{3} \leq \frac{1}{6}y + \frac{5}{3}$ (h) $4(x - 5) < 3(x + 2)$
   (i) $7(6 - 5x) \geq 4(3x + 2)$ (j) $9(4x - 3) < 8(4 - 3x)$
2. Solve the following inequalities and draw the graphs of the solutions.
   (a) $\frac{1}{3}(x + 1) < \frac{2}{3}(x + 2)$ (b) $\frac{4}{5}(2x - 3) \geq 4(\frac{1}{3}x - 2)$
   (c) $.9(.4x - .5) < .6(.8x - 7.1) + 4.3$ (d) $-\frac{2}{3}(4x - \frac{1}{9}) \leq \frac{4}{5}(3x - 7)$
   (e) $\frac{2}{3}(\frac{4}{5}x - \frac{2}{19}) < -\frac{4}{7}(\frac{3}{5}x + \frac{4}{19})$
3. Solve the following systems of inequalities and draw the graphs of the solutions.
   (a) $-2x + 5 < 8$ and $4x - 7 < 9$ (b) $3x + 2 \geq 4x - 7$ and $5x - 2 \leq 13x + 9$
   (c) $\sqrt{2}x + 3 < 7$ and $\sqrt{3}x + 1 > 2$ (d) $7x + 3 \leq 19$ and $3x - 1 < 4$
   (e) $6x - 5 < 3$ and $-3 < 6x - 5$ (f) $4x + 1 > 2$ or $3x - 2 < 1$
   (g) $13x - 19 \geq 5$ or $13x - 19 \leq -5$ (h) $1.5x - 7 \leq 3$ or $9.2x + 4 \geq 9$
4. Solve the following inequalities and draw the graphs of the solutions.
   (a) $|3x - 1| \leq 8$ (b) $|4x + 3| \leq 2$ (c) $|3 - 7x| < 5$
   (d) $|2x - 1| > 6$ (e) $|\sqrt{3}x - 2| \leq \sqrt{14}$ (f) $|.3x + .04| < .62$
   (g) $|.8x - .7| \geq .4$ (h) $|91x + 93| > 41$

3. Systems of Linear Equations

Thus far we have considered only equations and inequalities containing just one variable and solutions have been sets of real numbers. Consider now the equation

$$2x + 3y = 8$$

A solution for this equation would be an *ordered pair of numbers* which yield a true statement when the first is substituted for x and the second is substituted for y.

For example, if we let $x = 4$ and $y = 0$ we have a solution. Some other solutions are given by:

$x = 2$ and $y = \frac{4}{3}$ or just $(2, \frac{4}{3})$
$x = \frac{5}{2}$ and $y = 1$ or just $(\frac{5}{2}, 1)$
$x = 0$ and $y = \frac{8}{3}$ or just $(0, \frac{8}{3})$

In fact, each number we substitute for x will produce a simple equation in the one variable y and this equation will have a solution. Thus the equation

$$2x + 3y = 8$$

actually has infinitely many solutions. Realize, however, that not all ordered pairs of numbers are solutions. For example, (5, 10) is not a solution to this equation since $2 \cdot 5 + 3 \cdot 10 \neq 8$.

Frequently we have two equations in the same two variables. In such a case we wish to find all ordered pairs which are solutions to both equations, not just one of them. Thus, for example, suppose we have the above equation along with a second equation:

$$2x + 3y = 8$$
$$x + 4y = -1$$

The pair of numbers $(7, -2)$ is a solution to both of these equations since

$$2(7) + 3(-2) = 8$$

and

$$7 + 4(-2) = -1$$

A pair of equations in two variables is usually called a *system* of two equations in two unknowns or two variables. *To solve a system is to determine all ordered pairs that are solutions of both equations.* Our substitution principle for equality enables us to add, subtract, or multiply corresponding sides of equations and obtain new equations, since if

$$A = B \quad \text{and} \quad P = Q$$

it follows that

$$A + P = B + Q, \qquad A - P = B - Q, \quad \text{and} \quad AP = BQ$$

Furthermore,

$$\text{if } P \neq 0, \quad \text{then } \frac{A}{P} = \frac{B}{Q}$$

We seldom use the division technique because we must restrict ourselves to the case where the numbers are nonzero, and dividing by expressions involving variables may lead to division by zero.

Example 1

$$2x + 3y = 9$$
$$2x + 2y = 4$$

If we subtract the left side of the second equation from the left side of the first and the right side of the second from the right side of the first we have the following:

$$(2x + 3y) - (2x + 2y) = 9 - 4$$

and by arithmetic, we have

$$y = 5$$

Thus the only possible value for y is 5. When we substitute 5 for y in the first equation we obtain

$$\begin{aligned} 2x + 15 &= 9 \\ 2x &= -6 \\ x &= -3 \end{aligned}$$

Checking in both equations shows that $(-3, 5)$ is, in fact, a solution:

$$2(-3) + 3(5) = 9$$
$$2(-3) + 2(5) = 4 \quad \square$$

We were able to solve this system quite simply because $2x$ was a term in both equations and subtracting yielded an equation in the one variable y. We say that we "eliminated the x term" by subtracting.

Example 2

$$2x + 5y = 5$$
$$x + 2y = 3$$

In this case, subtraction would not yield an equation in one variable. Suppose, however, that we first multiply both sides of the second equation by 2. We now have the following system:

$$2x + 5y = 5$$
$$2x + 4y = 6$$

Subtracting yields

$$y = -1$$

Substituting in the second of the original equations yields

$$\begin{aligned} x + (-2) &= 3 \\ x &= 5 \end{aligned}$$

Checking shows that $(5, -1)$ is a solution. We leave the verification to the reader. $\square$

Example 3

$$2x + 5y = 9$$
$$3x + 2y = 4$$

In this case we could multiply both sides of the second equation by $\frac{2}{3}$ and then subtract and proceed as before. It would be computationally simpler, however, to multiply both sides of the first equation by 3, both sides of the second by 2, and then subtract.

$$\begin{aligned} 6x + 15y &= 27 \\ 6x + 4y &= 8 \\ 11y &= 19 \\ y &= \tfrac{19}{11} \end{aligned}$$

We could now substitute this value for y in one of the equations and then solve for x. An alternative method is to now choose multipliers so that subtracting yields an equation containing x but not y. If we use 2 as a multiplier in the first case and 5 in the second we obtain the following:

$$\begin{aligned} 2(2x + 5y) &= 2 \cdot 9 \\ 5(3x + 2y) &= 5 \cdot 4 \end{aligned}$$

or

$$\begin{aligned} 4x + 10y &= 18 \\ 15x + 10y &= 20 \end{aligned}$$

Subtracting yields

$$-11x = -2 \quad \text{and} \quad x = \tfrac{2}{11}$$

Checking, we see that

$$2\left(\frac{2}{11}\right) + 5\left(\frac{19}{11}\right) = \frac{4 + 95}{11} = \frac{99}{11} = 9$$

and

$$3\left(\frac{2}{11}\right) + 2\left(\frac{19}{11}\right) = \frac{6 + 38}{11} = \frac{44}{11} = 4$$

Thus $x = \frac{2}{11}$, $y = \frac{19}{11}$ is a solution to the system. □

A third possibility is to "solve" one equation for one variable in terms of the other, that is, to express one variable in terms of the other. Then we substitute this expression in the second equation obtaining an equation in one variable which we proceed to solve.

Example 4

$$\begin{aligned} x + 2y &= 5 \\ 2x + 3y &= 9 \end{aligned}$$

Solving the first equation for x in terms of y yields

$$x = 5 - 2y$$

Substituting $5 - 2y$ for x in the second yields

$$2(5 - 2y) + 3y = 9$$
$$10 - 4y + 3y = 9$$
$$10 - y = 9$$
$$y = 1 \qquad x = 5 - 2 = 3$$

Checking verifies that (3, 1) is a solution. □

The process of choosing multipliers and subtracting in order to obtain an equation containing only one variable is the most automatic technique and can be programmed for a computer. In general practice the most common technique probably involves using substitution to find the second number if the computation is not too messy. If the computation looks messy, use multiplication and subtraction again.

All of the equations shown thus far had solutions and all the systems of equations did also. The student should realize, however, that it is possible to write down equations with no solutions just as one can write English sentences that are never true. The equation $x = x + 1$ is such an equation. Similarly the system of equations

$$2x + 3y = 5$$
$$2x + 3y = 8$$

can have no solution because the number $2x + 3y$ cannot be both 5 and 8. If we subtract with this system, we obtain $0 = -3$ which we know is false. The system

$$2x + 3y = 4$$
$$4x + 6y = 8$$

has more than one solution because we have, essentially, only one equation. That is, although the sentences are different we can obtain the second by multiplying both sides of the first by 2. We could also obtain the first by multiplying both sides of the second by $\frac{1}{2}$. Any solution of one is a solution of the other and vice versa. Thus with a system of two equations we have the following possibilities:

no solution
exactly one solution
an infinite number of solutions

We will see later that systems of the type

$$ax + by = c$$
$$px + qy = r$$

have more than one solution only when every solution of one is a solution of the other. Thus a system cannot have more than one solution unless it has infinitely many.

The techniques involved in solving systems of 2 equations in 2 unknowns generalize to 3 equations in 3 unknowns and, in fact, to n equations in n unknowns.

Example 5

$$\begin{aligned} x + 2y + 3z &= 12 \\ 2x - y + z &= 9 \\ 3x + 2y - 2z &= -4 \end{aligned}$$

If we multiply both sides of the first equation by 2, then the first and second equations yield

$$\begin{aligned} 2x + 4y + 6z &= 24 \\ 2x - y + z &= 9 \end{aligned}$$

Subtracting yields

$$5y + 5z = 15$$

Dividing both sides by 5 we get

$$y + z = 3$$

If we use 3 as a multiplier, the first and third equations yield

$$\begin{aligned} 3x + 6y + 9z &= 36 \\ 3x + 2y - 2z &= -4 \end{aligned}$$

Subtracting yields

$$4y + 11z = 40$$

Now we have two equations in two unknowns, y and z, and we can proceed as before.

$$\begin{aligned} y + z &= 3 \\ 4y + 11z &= 40 \end{aligned}$$

Using 4 as a multiplier we have

$$\begin{aligned} 4y + 4z &= 12 \\ 4y + 11z &= 40 \end{aligned}$$

Subtracting yields

$$-7z = -28 \quad \text{and} \quad z = 4$$

Substituting into

$$y + z = 3$$

we have

$$y + 4 = 3$$
$$y = -1$$

Now substituting into the first equation we have

$$x + 2(-1) + 3(4) = 12$$
$$x - 2 + 12 = 12$$
$$x + 10 = 12$$
$$x = 2$$

Thus $x = 2$, $y = -1$, $z = 4$ is a solution to the system. We leave the verification for the student. □

As with systems of 2 equations, one could solve for each unknown by eliminating the other variables, but in most cases, the computation is simpler if we use substitution after a value has been determined for one or two variables. Systems of more than 3 equations are treated in the same manner; we eliminate one variable in order to obtain a smaller system of equations in fewer variables. For example, with a system of 4 equations in 4 unknowns we use the same techniques to obtain a system of 3 equations in 3 unknowns and then reduce this system to one with 2 unknowns.

EXERCISES

Solve the following systems of equations (if they have solutions).

1. $3x - 2y = 5$
   $x + y = 4$
2. $7x + 5y = 32$
   $3x - 2y = 6$
3. $1.3x - 5.2y = 11.1$
   $7x + 2y = 4$
4. $16x - 5y = 3$
   $6x + 7y = 1$
5. $9x + 6y = 5$
   $2x - y = 19$
6. $.43x - 10.2y = 9.3$
   $8.4x - 16y = 5.2$
7. $\frac{1}{2}x - \frac{1}{3}y = \frac{5}{4}$
   $\frac{2}{3}x - \frac{7}{2}y = \frac{1}{8}$
8. $\frac{6}{7}x + \frac{7}{6}y = 1$
   $\frac{4}{5}x - \frac{2}{3}y = 4$
9. $\frac{9}{2}x + \frac{4}{7}y = \frac{11}{3}$
   $\frac{6}{7}x + \frac{5}{4}y = \frac{2}{3}$
10. $2x - 3y = 5$
    $4x - 6y = 3$
11. $\sqrt{3}x - y = \sqrt{2}$
    $\sqrt{5}x + y = \sqrt{7}$
12. $\sqrt{2}x + y = 4$
    $x + \sqrt{3}y = 2$
13. $ax + by = p$
    $cx + dy = q$
    (Assume $ad - bc \neq 0$.)
    Use your results from Exercise 13 to solve the following three systems:
14. $23x + 42y = 79$
    $10x + 11y = 193$
15. $.4x + 1.6y = 2.7$
    $55x - 6y = 137$
16. $\sqrt{11}x + \sqrt{7}y = \sqrt{3}$
    $\sqrt{6}x + \sqrt{10}y = \sqrt{5}$

17. $x + 2y - 3z = 9$
$2x - y + 5z = -5$
$4x + 3y - 2z = 16$

18. $5x - 3y + 2z = 8$
$6x + y - z = 4$
$2x + 3y + z = 7$

19. $x + 2y + 3z = 4$
$2x - y + 2z = 5$
$4x + 3y + 8z = 15$

20. $x + 2y - 4z = 6$
$\sqrt{3}x + 5y - \sqrt{2}z = 4$
$x + y - z = 3$

*21. $ax + by = p$
$cy + fz = q$
$gx + kz = r$
(Assume $ack + gbf \neq 0$.)

*22. $ax + by + cz = p$
$dx + ey + fz = q$
$gx + hy + kz = r$
(Make the needed assumption as to certain numbers being nonzero.)

*23. $\sqrt{5}x + \sqrt{3}y + z = 4$
$\sqrt{3}x - \sqrt{2}y - z = 8$
$\sqrt{6}x + \sqrt{7}y - \sqrt{11}z = 3$
[*Hint*: Use your results from Exercise 22.]

24. Suppose (4, 2) and (7, 8) are solutions to $y = mx + b$. Compute b and m.

25. Suppose (p, q) and (h, k) are solutions to $y = mx + b$. Compute b and m in terms of p, q, h, and k.

4. Solving Polynomial Equations by Factoring

The technique of solving polynomial equations by factoring is based on one fundamental idea:

If A and B are numbers such that $AB = 0$, then either $A = 0$, $B = 0$, or both A and B are 0.

This is equivalent to our axiom which states that the product of two nonzero numbers is nonzero.

Consider now the equation

$$(x + 3)(2x - 1) = 0$$

In order for this statement to be true, we must have

$$x + 3 = 0 \quad \text{or} \quad 2x - 1 = 0$$

If $x + 3 = 0$, then $x = -3$.
If $2x - 1 = 0$, then $x = \frac{1}{2}$.

Checking we see that both -3 and $\frac{1}{2}$ are solutions of the equation.

The principle generalizes to more than two numbers; that is, if $A_1 A_2 \cdots A_n = 0$, then

$$A_1 = 0, \qquad A_2 = 0, \quad \ldots, \quad \text{or} \quad A_n = 0$$

Thus the solutions of the equation

$$7(2x + 3)(x - 1)(x + 4) = 0$$

are numbers that are solutions of

$$2x + 3 = 0, \qquad x - 1 = 0, \quad \text{or} \quad x + 4 = 0$$

Thus $x = -\frac{3}{2}$, $x = 1$, or $x = -4$.

The solution set for the equation is then

$$\{-\tfrac{3}{2}, 1, -4\}$$

In order to use this principle, we first must have an equation in a form where one side is just the number 0. *It is not applicable otherwise.* For example, if $(x - 5)(x + 2) = 1$, *there is no reason to expect* $x - 5 = 1$ *or* $x + 2 = 1$.

By suitable additions and subtractions, any equation can be transformed to one in which one side is 0. Once we have such an equation, we attempt to factor the polynomial.

Example 1

$$x^2 - 5x + 5 = 1$$
$$x^2 - 5x + 4 = 0$$
$$(x - 1)(x - 4) = 0$$
$$x - 1 = 0 \quad \text{or} \quad x - 4 = 0$$
$$x = 1 \quad \text{or} \quad x = 4$$

Checking: For $x = 1$,

$$1^2 - 5(1) + 5 = 1$$

For $x = 4$,

$$4^2 - 5(4) + 5 = 1 \ \square$$

Example 2

$$x^2 + 3x - 7 = 2x - 1$$
$$x^2 + x - 6 = 0$$
$$(x + 3)(x - 2) = 0$$
$$x + 3 = 0 \quad \text{or} \quad x - 2 = 0$$
$$x = -3 \quad \text{or} \quad x = 2$$

Checking: For $x = -3$,

$$(-3)^2 + 3(-3) - 7 = -7 \quad \text{and} \quad 2(-3) - 1 = -7$$

For $x = 2$,

$$(2)^2 + 3(2) - 7 = 3 \quad \text{and} \quad 2(2) - 1 = 3 \ \square$$

Example 3

$$x^4 - 12x^2 = x^2 - 36$$
$$x^4 - 13x^2 + 36 = 0$$
$$(x^2 - 4)(x^2 - 9) = 0$$
$$(x + 2)(x - 2)(x + 3)(x - 3) = 0$$
$$x + 2 = 0, \quad x - 2 = 0, \quad x + 3 = 0, \quad \text{or} \quad x - 3 = 0$$

Thus

$$x = -2, \quad x = 2, \quad x = -3, \quad \text{or} \quad x = 3$$

We leave the checking to the student. □

Example 4

$$40x^3 = 36x^2 - 8x$$
$$40x^3 - 36x^2 + 8x = 0$$
$$4x(10x^2 - 9x + 2) = 0$$
$$4x(2x - 1)(5x - 2) = 0$$

Thus

$$x = 0, \quad 2x - 1 = 0, \quad \text{or} \quad 5x - 2 = 0$$

(since $4 \neq 0$, only three factors must be considered) and

$$x = 0, \quad x = \tfrac{1}{2}, \quad \text{or} \quad x = \tfrac{2}{5}$$

We leave the checking to the student. □

EXERCISES

1. Solve the following equations.
   (a) $8x^2 + 10x = 3$ (b) $10x^2 = 33x + 7$ (c) $x^2 - 6x = -9$
   (d) $6x^2 - 97x + 18 = 2$ (e) $15x^2 - 14x = 8$ (f) $42x^2 = 23x + 5$
   (g) $2x^4 - 17x^3 + 21x^2 = 0$ (h) $x^3 - \frac{5}{6}x^2 = \frac{1}{6}x$ (i) $t^2 - 16 = 0$
   (j) $w^4 - 16 = 0$
2. Solve the following equations.
   (a) $x^4 - 29x^2 + 100 = 0$ (b) $y^3 + y^2 - 4y - 4 = 0$
   (c) $x^3 + 2x^2 - 9x - 18 = 0$ (d) $x^3 - 8 = 0$
   (e) $x^3 + 27 = 0$ (f) $x^4 - 5x^2 + 4 = 0$
   (g) $x^4 - 13x^2 + 36 = 0$ (h) $x^5 - x^3 - 8x^2 + 8 = 0$
3. Solve the following equations.
   (a) $x^2 - 9 = 0$ (b) $y^2 = 25$
   (c) $y^2 = 49$ (d) $(x + 1)^2 = 25$
   (e) $(x + 1)^2 = 49$ (f) $(x + 2)^2 = 49$
   (g) $(x + 3)^2 = 49$ (h) $(x + 2)^2 = 9$
   (i) $(x + 4)^2 = 25$ (j) $x^2 + 6x + 9 = 25$

(k) $x^2 + 2x + 1 = 9$ *(l) $(x + p)^2 = q^2$
*(m) $(x - p)^2 = q^2$ *(n) $x^2 + 2px + p^2 = q^2$

5. Completing the Square—The Quadratic Formula

A polynomial equation of the form

$$Ax^2 + Bx + C = 0 \qquad \text{with } A \neq 0$$

is called a *quadratic* equation. In this section we develop methods for solving all equations of this form.

If a and b are numbers such that $a^2 = b^2$, then $a^2 - b^2 = 0$. Thus $(a - b)(a + b) = 0$, $a - b = 0$ or $a + b = 0$, and $a = b$ or $a = -b$. Thus if $a^2 = b^2$, either $a = b$ or $a = -b$. We can use this principle to solve certain equations.

Example 1

$$(x + 2)^2 = 9$$
$$(x + 2)^2 = 3^2$$

Thus

$$x + 2 = 3 \quad \text{or} \quad x + 2 = -3$$
$$x = 1 \quad \text{or} \quad x = -5$$

Checking we see that both 1 and -5 are solutions:

$$(1 + 2)^2 = 3^2 = 9$$
$$(-5 + 2)^2 = (-3)^2 = 9 \ \square$$

Example 2

$$(x - 5)^2 = 7$$
$$(x - 5)^2 = (\sqrt{7})^2$$

Thus

$$x - 5 = \sqrt{7} \quad \text{or} \quad x - 5 = -\sqrt{7}$$
$$x = 5 + \sqrt{7} \quad \text{or} \quad x = 5 - \sqrt{7}$$

Checking:

$$[(5 + \sqrt{7}) - 5]^2 = (\sqrt{7})^2 = 7$$
$$[(5 - \sqrt{7}) - 5]^2 = (-\sqrt{7})^2 = 7 \ \square$$

In general, the equation

$$(x + p)^2 = q^2$$

is equivalent to the system of equations

$$x + p = q \quad \text{or} \quad x + p = -q$$

Thus $x = -p + q$ and $x = -p - q$ are the only possible solutions. [See Exercise 3(n) in Section 4.] It is easily verified that they are solutions. Thus we can solve a quadratic equation if we can express it in the form

$$(x + p)^2 = q^2$$

Example 3

Solve

$$x^2 + 4x = 7$$

The left side is not a perfect square but adding 4 to both sides yields

$$x^2 + 4x + 4 = 11$$

Then

$$(x + 2)^2 = 11$$
$$x + 2 = \sqrt{11} \quad \text{or} \quad x + 2 = -\sqrt{11}$$

and

$$x = -2 + \sqrt{11} \quad \text{or} \quad x = -2 - \sqrt{11}$$

Checking:

$$(-2 + \sqrt{11})^2 + 4(-2 + \sqrt{11}) = 4 - 4\sqrt{11} + 11 - 8 + 4\sqrt{11} = 7$$
$$(-2 - \sqrt{11})^2 + 4(-2 - \sqrt{11}) = 4 + 4\sqrt{11} + 11 - 8 - 4\sqrt{11} = 7 \; \square$$

Example 4

In a similar fashion we can solve

$$x^2 - 6x = 40$$

We need to choose a suitable number to add to both sides to make the left side a perfect square. The resultant equation is of the form

$$(x - a)^2 = 40 + \underline{\hspace{2cm}}$$

In order for

$$(x - a)^2 = x^2 - 2ax + a^2 = x^2 - 6x + \underline{\hspace{2cm}}$$

a must be $\frac{1}{2}$ of 6; that is, $a = 3$. Thus we must add $(3)^2$ to both sides.

$$x^2 - 6x = 40$$
$$x^2 - 6x + (3)^2 = 40 + 9$$
$$(x - 3)^2 = 49$$
$$x - 3 = 7 \quad \text{or} \quad x - 3 = -7$$
$$x = 10 \quad \text{or} \quad x = -4$$

Checking:

$$(10)^2 - 6(10) = 100 - 60 = 40$$
$$(-4)^2 - 6(-4) = 16 + 24 = 40 \quad \square$$

Example 5

$$x^2 + 5x = 8$$

We need to add the square of $\frac{1}{2}$ of 5 to both sides.

$$x^2 + 5x + (\tfrac{5}{2})^2 = 8 + (\tfrac{5}{2})^2$$
$$(x + \tfrac{5}{2})^2 = 8 + \tfrac{25}{4}$$
$$(x + \tfrac{5}{2})^2 = \tfrac{57}{4}$$

$$x + \frac{5}{2} = \frac{\sqrt{57}}{2} \quad \text{or} \quad x + \frac{5}{2} = -\frac{\sqrt{57}}{2}$$

$$x = -\frac{5}{2} + \frac{\sqrt{57}}{2} \quad \text{or} \quad x = -\frac{5}{2} - \frac{\sqrt{57}}{2}$$

$$= \frac{-5 + \sqrt{57}}{2} \quad \text{or} \quad = \frac{-5 - \sqrt{57}}{2}$$

We leave the checking to the student. □

Example 6

$$x^2 - 12x = 4$$
$$x^2 - 12x + 36 = 4 + 36 \qquad \text{Adding } 6^2 \text{ to both sides.}$$
$$(x - 6)^2 = 40$$
$$x - 6 = \sqrt{40} \quad \text{or} \quad x - 6 = -\sqrt{40}$$
$$x = 6 + \sqrt{40} \quad \text{or} \quad x = 6 - \sqrt{40}$$
$$= 6 + 2\sqrt{10} \quad \text{or} \quad = 6 - 2\sqrt{10}$$

We leave the checking as an exercise. □

Our technique suffices for equations of the form

$$x^2 + px = q$$

providing q is sufficiently large. We just add $(p/2)^2$ to both sides and solutions exist providing

$$q + \left(\frac{p}{2}\right)^2 \geq 0$$

that is, providing

$$q \geq -\left(\frac{p}{q}\right)^2$$

Example 7

The equation

$$2x^2 + 5x = 6$$

can be transformed to one of the previous form by dividing both sides by 2. Thus we have

$$x^2 + \tfrac{5}{2}x = 3$$

$$x^2 + \tfrac{5}{2}x + (\tfrac{5}{4})^2 = 3 + (\tfrac{5}{4})^2$$

$$(x + \tfrac{5}{4})^2 = \tfrac{73}{16}$$

$$x + \frac{5}{4} = \frac{\sqrt{73}}{16} \quad \text{or} \quad x + \frac{5}{4} = -\frac{\sqrt{73}}{16}$$

$$x = \frac{-5 + \sqrt{73}}{4} \quad \text{or} \quad x = \frac{-5 - \sqrt{73}}{4}$$

Checking:

$$2\left(\frac{-5 + \sqrt{73}}{4}\right)^2 + 5\left(\frac{-5 + \sqrt{73}}{4}\right) = \frac{25 - 10\sqrt{73} + 73}{8} + \frac{-25 + 5\sqrt{73}}{4}$$

$$= \frac{49 - 5\sqrt{73}}{4} + \frac{-25 + 5\sqrt{73}}{4}$$

$$= \frac{24}{4} = 6$$

$$2\left(\frac{-5 - \sqrt{73}}{4}\right)^2 + 5\left(\frac{-5 - \sqrt{73}}{4}\right) = \frac{49 + 5\sqrt{73}}{4} + \frac{-25 - 5\sqrt{73}}{4}$$

$$= \frac{24}{4} = 6 \; \square$$

Example 8

If we try our technique with the equation

$$x^2 + 4x + 7 = 2$$

we get

$$x^2 + 4x + 4 = -1$$

$$(x + 2)^2 = -1$$

There can be no solution because for any x, $(x + 2)^2 \geq 0$. $\square$

The method demonstrated in these examples is called *completing the square*. It is used, as in the examples, for solving quadratic equations and we will see later how a variation of the technique is very useful in graphing.

Consider now the general quadratic equation

$$Ax^2 + Bx + C = 0 \qquad \text{with } A \neq 0$$

If we can solve this general equation, we will have developed a formula that can be used with all solvable quadratic equations. We proceed as before:

Solve the equation

$$Ax^2 + Bx + C = 0 \qquad \text{where } A \neq 0$$

$$Ax^2 + Bx = -C \qquad \text{subtracting } C \text{ from both sides}$$

$$x^2 + \frac{B}{A}x = -\frac{C}{A} \qquad \text{dividing both sides by } A$$

We now add $(B/2A)^2$ to both sides to complete the square.

$$x^2 + \frac{B}{A}x + \left(\frac{B}{2A}\right)^2 = -\frac{C}{A} + \left(\frac{B}{2A}\right)^2$$

$$\left(x + \frac{B}{2A}\right)^2 = \frac{B^2}{4A^2} - \frac{C}{A}$$

$$\left(x + \frac{B}{2A}\right)^2 = \frac{B^2 - 4AC}{4A^2}$$

Thus

$$x + \frac{B}{2A} = \sqrt{\frac{B^2 - 4AC}{4A^2}} \qquad \text{or} \qquad x + \frac{B}{2A} = -\sqrt{\frac{B^2 - 4AC}{4A^2}}$$

that is,

$$x = -\frac{B}{2A} + \frac{\sqrt{B^2 - 4AC}}{2A} \qquad \text{or} \qquad x = -\frac{B}{2A} - \frac{\sqrt{B^2 - 4AC}}{2A}$$

$$x = \frac{-B + \sqrt{B^2 - 4AC}}{2A} \qquad \text{or} \qquad x = \frac{-B - \sqrt{B^2 - 4AC}}{2A}$$

This formula is usually condensed to

$$x = \frac{-B \pm \sqrt{B^2 - 4AC}}{2A}$$

where the $\pm$ sign indicates that we have combined the two formulas into one statement. This formula is known as the *quadratic formula* and can be used to find all solutions to quadratic equations if such solutions exist. It also will indicate when no solution exists. *The student should memorize this formula.*

Example 9

Solve

$x^2 + 2x - 5 = 0$

Using the formula yields

$$x = \frac{-2 \pm \sqrt{4 + 20}}{2}$$

since $A = 1$, $B = 2$, and $C = -5$.

$$\begin{aligned} x &= \frac{-2 \pm \sqrt{24}}{2} \\ &= \frac{-2 \pm 2\sqrt{6}}{2} \\ &= -1 \pm \sqrt{6} \end{aligned}$$

$$x = -1 + \sqrt{6} \quad \text{or} \quad x = -1 - \sqrt{6}$$

Checking:

$$(-1 + \sqrt{6})^2 + 2(-1 + \sqrt{6}) - 5 = 1 - 2\sqrt{6} + 6 - 2 + 2\sqrt{6} - 5 = 0$$
$$(-1 - \sqrt{6})^2 + 2(-1 - \sqrt{6}) - 5 = 1 + 2\sqrt{6} + 6 - 2 - 2\sqrt{6} - 5 = 0 \ \square$$

Example 10

Solve

$3x^2 - 12x + 7 = 0$

Applying the formula yields

$$\begin{aligned} x &= \frac{12 \pm \sqrt{144 - 84}}{6} \\ &= \frac{12 \pm \sqrt{60}}{6} \\ &= \frac{12 \pm 2\sqrt{15}}{6} \\ &= 2 \pm \frac{\sqrt{15}}{3} \end{aligned}$$

$$x = 2 + \frac{\sqrt{15}}{3} \quad \text{or} \quad x = 2 - \frac{\sqrt{15}}{3}$$

We leave the checking to the student. $\square$

Example 11

If we try to apply the formula to

$$x^2 + 3x + 5 = 0$$

we obtain

$$x = \frac{-3 \pm \sqrt{9 - 20}}{2}$$

$$= \frac{-3 \pm \sqrt{-11}}{2}$$

This is impossible because there is no real number whose square is -11. □

The equation

$$Ax^2 + Bx + C = 0$$

has no solution in the real number system if $B^2 - 4AC$ is negative because $\sqrt{B^2 - 4AC}$ does not exist.

EXERCISES

1. Solve the following by completing the square.
   (a) $x^2 + 4x = 8$ (b) $x^2 - 7x = 3$ (c) $x^2 + x = 2$
   (d) $x^2 + 12x = 5$ (e) $2x^2 + 3x = 5$ (f) $3x^2 + 2x + 1 = 6$
   (g) $4x^2 + x + 2 = 3$ (h) $3x^2 + 2x = 1$
2. Solve the equations from Exercise 1 using the quadratic formula.
3. Use the quadratic formula to solve the following equations (if they have solutions).
   (a) $2x^2 + 3x - 4 = 0$ (b) $x^2 + 5x - 2 = 0$
   (c) $7x^2 + x - 9 = 0$ (d) $x^2 - 4x - 5 = 0$
   (e) $5x^2 + 7x + 1 = 0$ (f) $\sqrt{7}x^2 + \sqrt{5}x - \sqrt{2} = 0$
   (g) $14x^2 + 17x - 32 = 0$ (h) $16x^2 - 15x + 3 = 0$
   (i) $11x^2 + 12x + 13 = 0$ (j) $1.3x^2 - 2.4x + 1.2 = 0$
4. Verify that

$$Ax^2 + Bx + C = A\left[x + \frac{B - \sqrt{B^2 - 4AC}}{2A}\right]\left[x + \frac{B + \sqrt{B^2 - 4AC}}{2A}\right]$$

   [*Hint*: All you need do is "multiply out" the right side to see if you get the left side.]
5. Use the assertion from Exercise 4 to factor the following expressions.
   (a) $x^2 + 5x - 10$ (b) $3x^2 + 2x - 7$ (c) $5x^2 - 3x - 1$
   (d) $16x^2 - 17x + 2$ (e) $7x^2 + 3x - 19$ (f) $4x^3 + 7x^2 + x$
   (g) $5x^2 - 3x + 2$ (h) $.4x^2 + 5x - 1.2$
6. Determine the average of the solutions to each of the equations in Exercise 1; for example, if the solutions are 3 and 11, the average is 7.
7. Use the quadratic formula to determine the average of the solutions of the general quadratic equation

$$Ax^2 + Bx + C = 0$$

8. Use your result from Exercise 7 to determine the average of the solutions to each of the equations in Exercise 3.
9. What is the sum of the two roots of the general quadratic equation?
10. What is the product of the two roots of the general quadratic equation?

6. Polynomial Inequalities

In Section 3 we studied linear inequalities and systems of linear inequalities. We now consider inequalities involving polynomials. The basic technique is to reduce a polynomial inequality to a system of linear inequalities.

Example 1

Solve the inequality

$$(x - 1)(x - 3) > 0$$

In order for the product $(x - 1)(x - 3)$ to be positive, either both factors are positive or both are negative; that is, either Case I or Case II below must hold.

Case I

$$\begin{aligned} x - 1 &> 0 \quad \text{and} \quad x - 3 > 0 \\ x &> 1 \quad \text{and} \quad x > 3 \end{aligned}$$

In this case $x > 3$ satisfies both conditions and makes both factors positive.

Case II

$$\begin{aligned} x - 1 &< 0 \quad \text{and} \quad x - 3 < 0 \\ x &< 1 \quad \text{and} \quad x < 3 \end{aligned}$$

In this case $x < 1$ satisfies both conditions and makes both factors negative.

Thus the solution is

$$\{x: x < 1 \text{ or } x > 3\}$$

The graph of this set is given in Figure 2–6. □

Figure 2–6

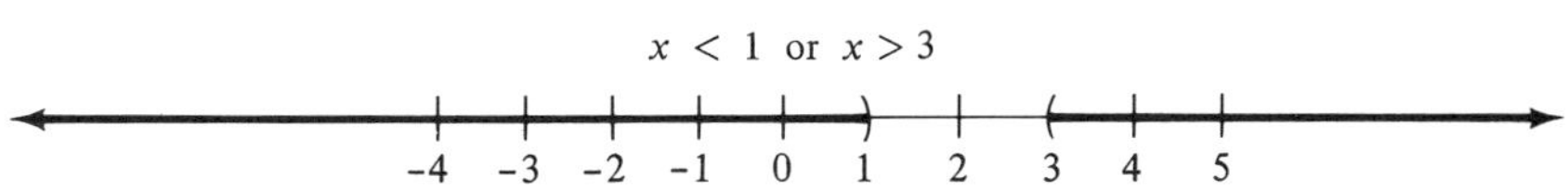

Example 2

Solve

$$(x - 5)(2x + 1)(3x - 2) > 0$$

This product is positive if all factors are positive or one is positive and the other two negative. Thus there are four possible cases:

Case I

$$x - 5 > 0, \quad 2x + 1 > 0, \quad \text{and} \quad 3x - 2 > 0 \quad \text{(all factors positive)}$$
$$x > 5, \quad x > -\tfrac{1}{2}, \quad \text{and} \quad x > \tfrac{2}{3}$$

Thus the solution set for Case I is $x > 5$.

Case II

$$x - 5 < 0, \quad 2x + 1 < 0, \quad \text{and} \quad 3x - 2 > 0 \quad \text{(first two factors}$$
$$x < 5, \quad x < -\tfrac{1}{2}, \quad \text{and} \quad x > \tfrac{2}{2} \quad \text{negative)}$$

Since $x < -\frac{1}{2}$ and $x > \frac{2}{3}$ is impossible, Case II produces no solutions.

Case III

$$x - 5 < 0, \quad 2x + 1 > 0, \quad \text{and} \quad 3x - 2 < 0 \quad \text{(first and third}$$
$$x < 5, \quad x > -\tfrac{1}{2}, \quad \text{and} \quad x < \tfrac{2}{3} \quad \text{factors negative)}$$

Thus Case III yields $\{x: -\frac{1}{2} < x < \frac{2}{3}\}$.

Case IV

$$x - 5 > 0, \quad 2x + 1 < 0, \quad \text{and} \quad 3x - 2 < 0 \quad \text{(last two factors}$$
$$x > 5, \quad x < -\tfrac{1}{2}, \quad \text{and} \quad x < \tfrac{2}{3} \quad \text{negative)}$$

Since $x > 5$ and $x < \frac{2}{3}$ is impossible, Case IV produces no solutions. Thus

$$\{x: -\tfrac{1}{2} < x < \tfrac{2}{3} \text{ or } 5 < x\}$$

is the solution set. The graph of the solution set is given in Figure 2–7. □

Figure 2–7

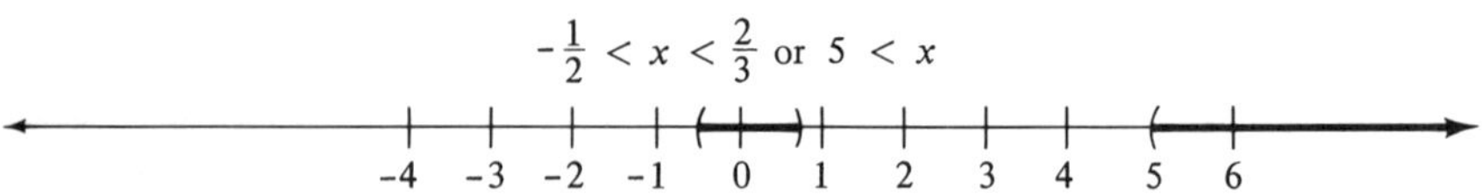

Example 3

Solve

$x^2 - 2x - 3 < 0$

Factoring yields

$(x - 3)(x + 1) < 0$

In order for this product to be negative, we must have one factor positive and the other negative and there are two possible cases:

Case I

$$x - 3 < 0 \quad \text{and} \quad x + 1 > 0$$
$$x < 3 \quad \text{and} \quad x > -1$$
$$-1 < x < 3$$

Case II

$$x - 3 > 0 \quad \text{and} \quad x + 1 < 0$$
$$x > 3 \quad \text{and} \quad x < -1$$

This is impossible.

The solution set is

$\{x: -1 < x < 3\}$

The graph is given in Figure 2–8. □

Figure 2–8

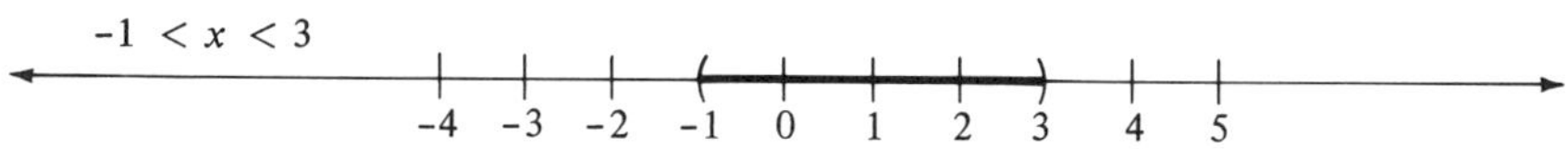

Example 4

Solve

$x^2 - 5x + 6 > 0$

$(x - 3)(x - 2) > 0$

Case I

$$x - 3 > 0 \quad \text{and} \quad x - 2 > 0$$
$$x > 3 \quad \text{and} \quad x > 2$$
$$x > 3$$

Case II

$$x - 3 < 0 \quad \text{and} \quad x - 2 < 0$$
$$x < 3 \quad \text{and} \quad x < 2$$
$$x < 2$$

The solution set is

$\{x: x < 2 \text{ or } x > 3\}$

The graph is given in Figure 2–9. □

Figure 2–9

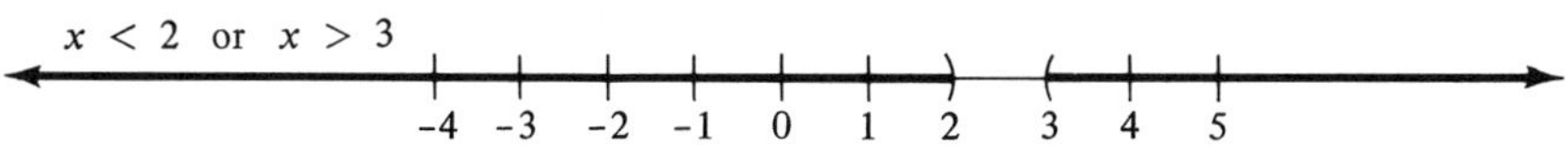

Example 5

Solve

$x^2 - 7x > 8$

$x^2 - 7x - 8 > 0$ Subtracting 8 from both sides

$(x - 8)(x + 1) > 0$

Case I

$$x - 8 > 0 \quad \text{and} \quad x + 1 > 0$$
$$x > 8 \quad \text{and} \quad x > -1$$
$$x > 8$$

Case II

$$x - 8 < 0 \quad \text{and} \quad x + 1 < 0$$
$$x < 8 \quad \text{and} \quad x < -1$$
$$x < -1$$

The solution set is

$\{x: x < -1 \text{ or } x > 8\}$

The graph is given in Figure 2–10. □

Figure 2–10

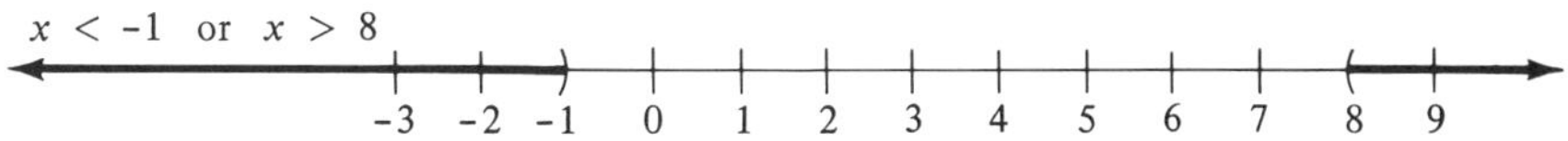

The general technique illustrated above is as follows:

A. By suitable additions or subtractions transform the original inequality into one in which one side is 0.
B. Factor the polynomial as much as possible.
C. Determine *all systems of inequalities* that are compatible with the original inequality.
D. Solve the resulting systems.
E. The solution set is the union of the solution sets of the systems.

Other methods of solving polynomial inequalities will be discussed after we develop the basic techniques of graphing polynomial functions.

EXERCISES

1. Solve the following inequalities and graph their solution sets.
   (a) $(x - 4)(x + 5) < 0$
   (b) $(2x - 1)(5x - 7) > 0$
   (c) $x^2 - 9x + 20 < 0$
   (d) $x^3 - 3x^2 - 4x > 0$
   (e) $30x^2 - 17x + 2 < 0$
   (f) $30x^3 - 17x^2 + 2x > 0$
   (g) $x^2 - 4 > 0$
   (h) $x^2 - 9 < 0$
   (i) $12x^4 - 17x^3 + 6x^2 > 0$
   (j) $x^3 - x^2 - x + 1 > 0$
2. Let q be a positive number. Show that

   $a^2 > q$ if and only if $a > \sqrt{q}$ or $a < -\sqrt{q}$

   and that

   $a^2 < q$ if and only if $-\sqrt{q} < a < \sqrt{q}$

3. Use the principle established in Exercise 2 to solve the following inequalities. Graph the solution sets.
   (a) $x^2 > 3$
   (b) $y^2 < 5$
   (c) $(x + 2)^2 > 9$
   (d) $(x - 3)^2 < 100$
   (e) $(2x + 1)^2 > 25$
   (f) $(x - \frac{1}{2})^2 < 4$
   (g) $(x + 4)^2 > .5$
   (h) $(x - 3)^2 < 7$
   (i) $(x - \frac{1}{2})^2 < \frac{4}{9}$
   (j) $(x + .2)^2 < 10$
4. Use the completing the square technique and the technique of Exercise 3 to solve the following inequalities. Graph the solution sets.
   (a) $x^2 + 4x < 7$
   (b) $x^2 - 6x > 4$
   (c) $x^2 + 10x < 6$
   (d) $x^2 - 5x < 2$
   (e) $2x^2 + 6x < 11$
   (f) $x^2 - 2x < -3$
   (g) $x^2 - 6x > -10$
   (h) $12x^2 - 5x > 7$

5. Assume $A > 0$. Use the factoring of $Ax^2 + Bx + C$ from the exercises in Section 5 or the completing the square technique to solve

$$Ax^2 + Bx + C > 0 \quad \text{and} \quad Ax^2 + Bx + C < 0$$

Draw the general graphs of the two solution sets.

6. If $A > 0$ and p is the average of the solutions of $Ax^2 + Bx + C = 0$, is $Ap^2 + Bp + C \leq 0$ or is $Ap^2 + Bp + C \geq 0$? Justify your answer using Exercise 5.

7. More General Equations and Systems of Equations

Frequently equations arise that involve algebraic fractions. The most general method for solving such equations is to multiply both sides by a suitable expression and thereby reduce the problem to solving a polynomial equation.

Example 1

Solve the equation

$$\frac{1}{x} + \frac{4}{x+3} = 1$$

If we multiply both sides by $x(x + 3)$ we have

$$(x)(x+3)\left[\frac{1}{x} + \frac{4}{x+3}\right] = (x)(x+3)$$

$$\begin{aligned} (x+3) + 4x &= x^2 + 3x \\ 5x + 3 &= x^2 + 3x \\ x^2 - 2x - 3 &= 0 \\ (x-3)(x+1) &= 0 \\ (x-3) &= 0 \quad \text{or} \quad (x+1) = 0 \\ x &= 3 \quad \text{or} \quad x = -1 \end{aligned}$$

Checking: For $x = 3$,

$\frac{1}{3} + \frac{4}{6} = 1$

For $x = -1$,

$(-1) + 2 = 1$ □

Example 2

Solve the equation

$$\frac{x+1}{x-1} + \frac{x-4}{x+2} = \frac{2x+11}{x+4}$$

The common denominator of the three fractions is $(x - 1)(x + 2)(x + 4)$. Multiplying both sides by this expression yields the equation

$$(x + 2)(x + 4)(x + 1) + (x - 4)(x - 1)(x + 4) = (2x + 11)(x - 1)(x + 2)$$

Simplifying we have

$$\begin{aligned}
(x^3 + 7x^2 + 14x + 8) + (x^3 - x^2 - 16x + 16) &= 2x^3 + 13x^2 + 7x - 22 \\
2x^3 + 6x^2 - 2x + 24 &= 2x^3 + 13x^2 + 7x - 22 \\
6x^2 - 2x + 24 &= 13x^2 + 7x - 22 \\
7x^2 + 9x - 46 &= 0
\end{aligned}$$

Using the quadratic formula

$$\begin{aligned}
x &= \frac{-9 \pm \sqrt{81 + 1288}}{14} \\
&= \frac{-9 \pm \sqrt{1369}}{14} \\
&= \frac{-9 \pm 37}{14}
\end{aligned}$$

$$x = 2 \quad \text{or} \quad x = \frac{-46}{14} = \frac{-23}{7}$$

Checking: For $x = 2$,

$$\frac{3}{1} + \frac{-2}{4} = \frac{15}{6}$$

For $x = \dfrac{-23}{7}$,

$$\begin{aligned}
\frac{\frac{-23}{7} + 1}{\frac{-23}{7} - 1} + \frac{\frac{-23}{7} - 4}{\frac{-23}{7} + 2} &= \frac{-16}{-30} + \frac{-51}{-9} \\
&= \frac{16}{30} + \frac{51}{9} = \frac{31}{5}
\end{aligned}$$

and

$$\frac{2(-23/7) + 11}{-23/7 + 4} = \frac{-46/7 + 11}{-23/7 + 4} = \frac{31}{5}$$

Thus, $\{2, \frac{-23}{7}\}$ is the solution set. □

In dealing with equations involving algebraic fractions, checking is more important than ever. When we multiply both sides of an equation by an expression involving the variable, we may introduce *extraneous* roots, that is, the new equation may have roots that are not roots of the original equation. We must test all possible solutions in the original equation.

Example 3

If we start with the equation $x = 7$ and multiply both sides by x, we obtain $x^2 = 7x$, which has 7 and 0 as roots, although 0 is not a root of the original equation. □

The systems of equations studied in Section 3 were *linear* systems, that is, all the equations in the system were linear. No variables appeared in denominators, no variables were multiplied together, and no variables were raised to powers other than 1. Our general technique was to eliminate one or more variables until we had reduced the system to a linear equation in one variable. In more general systems of equations, the variables may appear in many ways, but our general technique is still the same. We use methods that enable us to reduce the system to a simpler system and, hopefully, to an equation in one variable which we then attempt to solve using techniques developed in earlier sections.

Example 4

Solve the system

$$\begin{aligned} 2x + y &= 1 \\ x^2 + 3x + 5y &= 13 \end{aligned}$$

Solving the first equation for y in terms of x we have

$$y = 1 - 2x$$

Substituting in the second equation yields

$$x^2 + 3x + 5(1 - 2x) = 13$$

We now need to solve the resulting quadratic equation.

$$\begin{aligned} x^2 + 3x + 5 - 10x &= 13 \\ x^2 - 7x - 8 &= 0 \\ (x - 8)(x + 1) &= 0 \end{aligned}$$

$$\begin{aligned} x - 8 = 0 \quad &\text{or} \quad x + 1 = 0 \\ x = 8 \quad &\text{or} \quad x = -1 \end{aligned}$$

If $x = 8$, then $y = 1 - 2(8) = -15$.
If $x = -1$, then $y = 1 - 2(-1) = 3$.

Checking: If $x = 8$ and $y = -15$, then

$$2(8) + (-15) = 1 \quad \text{and} \quad (8)^2 + 3(8) + 5(-15) = 13$$

because

$$16 + (-15) = 1 \quad \text{and} \quad 64 + 24 - 75 = 13$$

If $x = -1$ and $y = 3$, then

$$2(-1) + 3 = 1 \quad \text{and} \quad (-1)^2 + 3(-1) + 5(3) = 13$$

since

$$(-2) + 3 = 1 \quad \text{and} \quad 1 - 3 + 15 = 13 \ \square$$

Example 5

Solve the system

$$x^2 + 3x + y^2 + 4y = 1$$
$$x^2 + 2x + y^2 + 3y = 3$$

Subtracting the left side of the second equation from the left side of the first and the right side of the second from the right side of the first yields the equation

$$x + y = -2$$

At this point all we know is that all solutions satisfy this equation. Solving for y in terms of x we have

$$y = -x - 2$$

Substituting in the first equation yields

$$x^2 + 3x + (-x - 2)^2 + 4(-x - 2) = 1$$

Simplifying yields

$$x^2 + 3x + x^2 + 4x + 4 - 4x - 8 = 1$$
$$2x^2 + 3x - 5 = 0$$

Using the quadratic formula,

$$x = \frac{-3 \pm \sqrt{9 + 40}}{4}$$
$$= \frac{-3 \pm \sqrt{49}}{4}$$
$$= \frac{-3 \pm 7}{4}$$

$$x = 1 \quad \text{or} \quad x = -\tfrac{5}{2}$$

If $x = 1$, $y = -3$.
If $x = -\frac{5}{2}$, $y = \frac{1}{2}$.

Checking the solution $(1, -3)$:

$$1 + 3 + 9 - 12 = 1$$
$$1 + 2 + 9 - 9 = 3$$

Checking the solution $(-\frac{5}{2}, \frac{1}{2})$:

$$\frac{25}{4} - \frac{15}{2} + \frac{1}{4} + 2 = 1$$

$$\frac{25}{4} - 5 + \frac{1}{4} + \frac{3}{2} = 3 \; \Box$$

Example 6

Solve the system

$$xy = 8$$
$$2x + 3y = 16$$

Solving the second equation for y in terms of x we have

$$y = \frac{-2}{3}x + \frac{16}{3}$$

Substituting in the first we have

$$x\left(-\frac{2}{3}x + \frac{16}{3}\right) = 8$$

$$-\frac{2}{3}(x^2) + \frac{16x}{3} = 8$$

$$-2x^2 + 16x = 24$$
$$x^2 - 8x = -12$$
$$x^2 - 8x + 12 = 0$$
$$(x - 6)(x - 2) = 0$$
$$x = 6 \quad \text{or} \quad x = 2$$

If $x = 2$, $y = 4$; if $x = 6$, $y = \frac{4}{3}$.

Checking: For $x = 2$, $y = 4$,

$$2 \cdot 4 = 8$$
$$2(2) + 3(4) = 16$$

For $x = 6$, $y = \frac{4}{3}$,

$$6(\tfrac{4}{3}) = 8$$
$$2(6) + 3(\tfrac{4}{3}) = 16 \; \Box$$

EXERCISES

Solve the following equations and systems of equations (if they have solutions).

1. $\frac{2}{x} + \frac{3}{x+1} = \frac{9}{2x}$
2. $\frac{4}{x+1} + \frac{5}{x-1} = \frac{22}{3x-4}$
3. $\frac{3}{x+1} + \frac{4}{x+2} = \frac{11}{x+7}$
4. $\frac{3}{x} + \frac{2}{3x} - \frac{5}{x+1} = \frac{7}{2x}$
5. $\frac{x}{x+1} = \frac{3x-2}{x+6}$
6. $\frac{2x}{x-2} = \frac{5x+1}{3x-2}$
7. $\frac{2x+3}{x+2} = \frac{3x-1}{x-2}$
8. $\frac{3x+8}{6x-1} = \frac{2x+5}{4x-3}$
9. $y = 2x$ and $x^2 + y^2 = 5$
10. $x + 2y = 1$ and $y = x^2 + 2x - 5$
11. $3x + 4y = 8$ and $x^2 + 4y^2 = 16$
12. $5x + y = 9$ and $x^2 - 2xy + y^2 = 10$
13. $x^2 + 9y^2 = 1$ and $4x^2 + y^2 = 1$
14. $y = x^2 + 4x - 8$ and $y = 5 + 2x - x^2$
15. $y = x^2 - 4x + 5$ and $y = \frac{1}{4}x^2 + 2x + 2$

8. Determinants

In Exercises 13 and 22 in Section 3 you were asked to solve the general systems

$$\begin{aligned} ax + by &= p \\ cx + dy &= q \end{aligned} \quad \text{and} \quad \begin{aligned} ax + by + cz &= p \\ dx + cy + fz &= q \\ gx + hy + kz &= r \end{aligned}$$

You were allowed to assume that $ad - bc \neq 0$ for the first system and were instructed to make the necessary assumptions in order to assure a solution for the second system. The solutions for the first system are

$$x = \frac{pd - bq}{ad - bc} \quad \text{and} \quad y = \frac{aq - pc}{ad - bc}$$

and the solutions for the second system are

$$x = \frac{pek - phf - qbk + qhc + rbf - rec}{D}$$

$$y = \frac{aqk - arf - dpk + drc + gpf - gqc}{D}$$

$$z = \frac{aer - ahq - dbr + dhp + gbq - gep}{D}$$

where

$$D = aek - ahf - dbk + dhc + gbf - gec$$

and we assume $D \neq 0$.

These solutions are obtained by the methods discussed and illustrated in Section 3. If one has a general method for solving certain types of equations or types of systems, then it can be applied to find a general formula for solving all such equations or systems. In these exercises you found general formulas for solving systems of two or three linear equations just as we developed, in Section 5, a general formula for solving equations of the form $Ax^2 + Bx + C = 0$, with $A \neq 0$. The formulas developed in these exercises are complicated and difficult to remember. The notions of a *matrix* and its *determinant* will help one to remember these formulas and simplify the computation of the solutions. A *matrix* is just a rectangular array of numbers, each *row* having the same number of numbers (rows go across) and each *column* with the same number of numbers (columns go down). At present, we need consider only square matrices, that is, those in which the number of rows and the number of columns are the same. These arrays of numbers are usually set off by brackets.

Examples of matrices:

$$\begin{bmatrix} 2 & 3 & 4 & 7 \\ 1 & -2 & 3 & 8 \end{bmatrix} \quad \begin{bmatrix} 4 & 6 & 9 \\ 3 & 1 & 2 \\ 7 & 1 & 3 \end{bmatrix} \quad \begin{bmatrix} 0 & 1 & 8 \\ 6 & 4 & 4 \\ 8 & 0 & 9 \end{bmatrix} \quad \begin{bmatrix} 2 & 3 \\ 1 & -9 \\ 4 & 7 \end{bmatrix} \quad \begin{bmatrix} 1 & 9 \\ 2 & 8 \end{bmatrix} \quad \begin{bmatrix} 0 & 2 & -3 \\ 5 & 4 & 0 \\ -6 & 0 & 8 \\ 1 & 5 & 3 \end{bmatrix}$$

A matrix with p rows and q columns is said to be a $p \times q$ matrix. The above matrices are, respectively, 2×4, 3×3, 3×3, 3×2, 2×2, and 4×3.

The *matrix of coefficients* of a system of equations is the matrix obtained by writing the coefficients of the first variable as the first column, the coefficients of the second variable as the second column, and so on.

Example 1

The matrix of coefficients of the system

$$2x + 7y = 8$$
$$3x + 5y = 2$$

is

$$\begin{bmatrix} 2 & 7 \\ 3 & 5 \end{bmatrix} \square$$

Example 2

The matrix of coefficients of the system

$$\begin{aligned} 7x + 3y - 2z &= 9 \\ 6x - 19y + 8z &= 15 \\ 4x + 5y - z &= 3 \end{aligned}$$

is

$$\begin{bmatrix} 7 & 3 & -2 \\ 6 & -19 & 8 \\ 4 & 5 & -1 \end{bmatrix} \square$$

The matrix of coefficients of

$$\begin{aligned} ax + by &= p \\ cx + dy &= q \end{aligned}$$

is

$$\begin{bmatrix} a & b \\ c & d \end{bmatrix}$$

The matrix of coefficients of

$$\begin{aligned} ax + by + cz &= p \\ dx + ey + fz &= q \\ gx + hy + kz &= r \end{aligned}$$

is

$$\begin{bmatrix} a & b & c \\ d & e & f \\ g & h & k \end{bmatrix}$$

Notice that in solving these systems, the denominator of each of the solutions depends only on the coefficients of the variables and therefore is dependent on the matrix of coefficients. The number D obtained from these matrices is called the *determinant* of the matrix. More precisely, the determinant of

$$\begin{bmatrix} a & b \\ c & d \end{bmatrix}$$

is denoted by

$$\begin{vmatrix} a & b \\ c & d \end{vmatrix}$$

and defined by

$$\begin{vmatrix} a & b \\ c & d \end{vmatrix} = ad - bc$$

The determinant of

$$\begin{bmatrix} a & b & c \\ d & e & f \\ g & h & k \end{bmatrix}$$

is denoted similarly and defined by

$$\begin{vmatrix} a & b & c \\ d & e & f \\ g & h & k \end{vmatrix} = aek - ahf - dbk + dhc + gbf - gec$$

The formula for

$$\begin{vmatrix} a & b \\ c & d \end{vmatrix}$$

is easy to remember, just multiply diagonal elements and subtract:

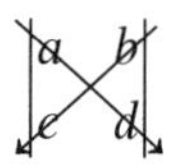

The second formula can be remembered by another graphic device. Rewrite the matrix and write the first and second columns to the right of the matrix. Compute the products following the diagonal lines to the right and sum them. Compute the products following the diagonal lines to the left and sum them. Subtract the second sum from the first sum:

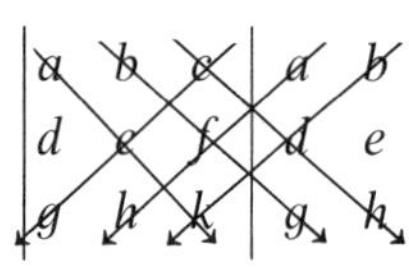

Example 3

Evaluate

$$\begin{vmatrix} 2 & 4 & 5 \\ 3 & -2 & 1 \\ 6 & 3 & 4 \end{vmatrix}$$

First we rewrite the matrix repeating the first and second columns:

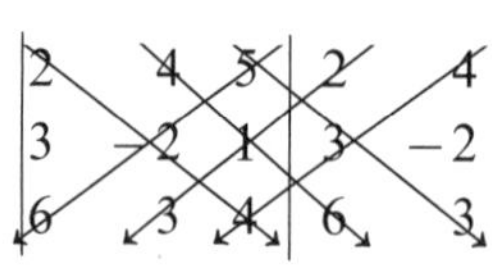

The three products following diagonal lines to the right are -16, 24, and 45. Their sum is 53.

The three products following diagonal lines to the left are 48, 6, and -60. Their sum is -6.

Subtracting -6 from 53 yields 59. Thus the determinant is 59. □

This graphic device is useful as a memory aid but *does not* generalize for larger systems. The second formula can also be generalized by using determinants of 2×2 matrices obtained from the original matrix. Notice that

$$\begin{aligned} aek - ahf - dbk + dhc + gbf - gec &= a(ek - hf) - d(bk - hc) + g(bf - ec) \\ &= a\begin{vmatrix} e & f \\ h & k \end{vmatrix} - d\begin{vmatrix} b & c \\ h & k \end{vmatrix} + g\begin{vmatrix} b & c \\ e & f \end{vmatrix} \end{aligned}$$

The three smaller matrices are obtained from the original matrix by crossing out the row and columns containing, respectively, a, d, and g. You need to remember that the second expression has a minus sign.

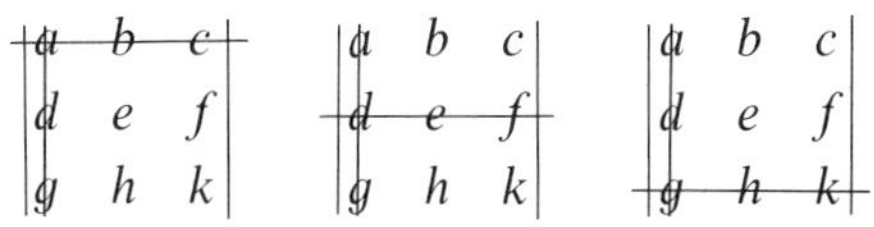

This process of evaluating a determinant is called *expanding by minors*, the *minors* being the determinants of the smaller matrices obtained by crossing out one row and one column.

By straightforward computation, it is simple to show that

$$\begin{vmatrix} a & b & c \\ d & e & f \\ g & h & k \end{vmatrix} = a\begin{vmatrix} e & f \\ h & k \end{vmatrix} - b\begin{vmatrix} d & f \\ g & k \end{vmatrix} + c\begin{vmatrix} d & e \\ g & h \end{vmatrix}$$

$$\begin{vmatrix} a & b & c \\ d & e & f \\ g & h & k \end{vmatrix} = -b\begin{vmatrix} d & f \\ g & k \end{vmatrix} + e\begin{vmatrix} a & c \\ g & k \end{vmatrix} - h\begin{vmatrix} a & c \\ d & f \end{vmatrix}$$

and

$$\begin{vmatrix} a & b & c \\ d & e & f \\ g & h & k \end{vmatrix} = g\begin{vmatrix} b & c \\ e & f \end{vmatrix} - h\begin{vmatrix} a & c \\ d & f \end{vmatrix} + k\begin{vmatrix} a & b \\ d & e \end{vmatrix}$$

In fact, the determinant can be expressed in many more ways using the entries from any row or any column and the corresponding minors. We will not write them all down but will, in fact, leave some of these verifications to the student.

In order to discuss more complicated matrices and determinants, it is customary to use double subscripts instead of several different letters in denoting the entries of the matrix. The first subscript denotes the row and the second the column. (Note that a_{11} means a-one-one, not a-eleven.)

$$\begin{bmatrix} a_{11} & a_{12} & a_{13} \\ a_{21} & a_{22} & a_{23} \\ a_{31} & a_{32} & a_{33} \end{bmatrix}$$

By our results above we have

$$\begin{vmatrix} a_{11} & a_{12} & a_{13} \\ a_{21} & a_{22} & a_{23} \\ a_{31} & a_{32} & a_{33} \end{vmatrix} = a_{11}\begin{vmatrix} a_{22} & a_{23} \\ a_{32} & a_{33} \end{vmatrix} - a_{21}\begin{vmatrix} a_{12} & a_{13} \\ a_{32} & a_{33} \end{vmatrix} + a_{31}\begin{vmatrix} a_{12} & a_{13} \\ a_{22} & a_{23} \end{vmatrix}$$

and

$$\begin{vmatrix} a_{11} & a_{12} & a_{13} \\ a_{21} & a_{22} & a_{23} \\ a_{31} & a_{32} & a_{33} \end{vmatrix} = -a_{12}\begin{vmatrix} a_{21} & a_{23} \\ a_{31} & a_{33} \end{vmatrix} + a_{22}\begin{vmatrix} a_{11} & a_{13} \\ a_{31} & a_{33} \end{vmatrix} - a_{32}\begin{vmatrix} a_{11} & a_{13} \\ a_{21} & a_{23} \end{vmatrix}$$

Notice that when we expand by minors, if the sum of the subscripts on a_{ij} is even, we have a plus sign and if it is odd we have a minus sign.

Now let us go back to our original problems that gave rise to these concepts and formulas. Notice that in solving

$$ax + by = p$$
$$cx + dy = q$$

we have

$$x = \frac{\begin{vmatrix} p & b \\ q & d \end{vmatrix}}{\begin{vmatrix} a & b \\ c & d \end{vmatrix}} \quad \text{and} \quad y = \frac{\begin{vmatrix} a & p \\ c & q \end{vmatrix}}{\begin{vmatrix} a & b \\ c & d \end{vmatrix}}$$

Furthermore, in solving

$$ax + by + cz = p$$
$$dx + ey + fz = q$$
$$gx + hy + kz = r$$

we have

$$x = \frac{\begin{vmatrix} p & b & c \\ q & e & f \\ r & h & k \end{vmatrix}}{D}$$

$$y = \frac{\begin{vmatrix} q & p & c \\ d & q & f \\ g & r & k \end{vmatrix}}{D}$$

and

$$z = \frac{\begin{vmatrix} a & b & p \\ d & e & q \\ g & h & r \end{vmatrix}}{D}$$

where, as before,

$$D = \begin{vmatrix} a & b & c \\ d & e & f \\ g & h & k \end{vmatrix}$$

In both systems, the solutions can be found by:

1. Computing the determinant D of the matrix of coefficients.
2. Replacing the coefficients of x (or y or z) in the matrix of coefficients by the constants, computing the resulting determinant, and dividing by D.

Example 4

The solutions of

$$2x + 3y = 5$$
$$4x - 2y = 8$$

are given by

$$x = \frac{\begin{vmatrix} 5 & 3 \\ 8 & -2 \end{vmatrix}}{\begin{vmatrix} 2 & 3 \\ 4 & -2 \end{vmatrix}} \quad \text{and} \quad y = \frac{\begin{vmatrix} 2 & 5 \\ 4 & 8 \end{vmatrix}}{\begin{vmatrix} 2 & 3 \\ 4 & -2 \end{vmatrix}}$$

$$x = \frac{-34}{-16} = \frac{17}{8} \qquad y = \frac{-4}{-16} = \frac{1}{4} \square$$

Example 5

The solutions of

$$2x + 3y + 4z = 5$$
$$x - 2y + 3z = 9$$
$$3x + 4y + 7z = 2$$

are given by

$$x = \frac{\begin{vmatrix} 5 & 3 & 4 \\ 9 & -2 & 3 \\ 2 & 4 & 7 \end{vmatrix}}{D}, \qquad y = \frac{\begin{vmatrix} 2 & 5 & 4 \\ 1 & 9 & 3 \\ 3 & 2 & 7 \end{vmatrix}}{D}, \qquad z = \frac{\begin{vmatrix} 2 & 3 & 5 \\ 1 & -2 & 9 \\ 3 & 4 & 2 \end{vmatrix}}{D}$$

where

$$D = \begin{vmatrix} 2 & 3 & 4 \\ 1 & -2 & 3 \\ 3 & 4 & 7 \end{vmatrix}$$

Computing yields $D = -6$ and

$$x = \frac{-141}{-6} = \frac{47}{2}, \qquad y = \frac{24}{-6} = -4, \qquad z = \frac{45}{-6} = \frac{-15}{2} \qquad \square$$

This general method or formula is known as *Cramer's rule* and can be extended for larger systems of equations once the notion of determinant is extended. The general definition of a determinant is an example of a *recursive* definition in that we define the determinant of an $(n + 1) \times (n + 1)$ matrix in terms of the determinants of $n \times n$ matrices obtained by eliminating a row and a column of the original matrix. If

$$\begin{bmatrix} a_{11} & \cdots & a_{1n} \\ \vdots & & \\ a_{n1} & \cdots & a_{nn} \end{bmatrix}$$

is an $n \times n$ matrix, we will define M_{ij} to be the ijth minor which is the determinant of the matrix obtained from the original by eliminating the ith row and the jth column. Our definition of the determinant of

$$\begin{bmatrix} a_{11} & a_{12} & a_{13} \\ a_{21} & a_{22} & a_{23} \\ a_{31} & a_{32} & a_{33} \end{bmatrix}$$

then becomes

$$\begin{vmatrix} a_{11} & a_{12} & a_{13} \\ a_{21} & a_{23} & a_{23} \\ a_{31} & a_{32} & a_{33} \end{vmatrix} = a_{11}M_{11} - a_{21}M_{21} + a_{31}M_{31}$$

We define the determinant of a 4×4 matrix in a similar manner:

$$\begin{vmatrix} a_{11} & a_{12} & a_{13} & a_{14} \\ a_{21} & a_{22} & a_{23} & a_{24} \\ a_{31} & a_{32} & a_{33} & a_{34} \\ a_{41} & a_{42} & a_{43} & a_{44} \end{vmatrix} = a_{11}M_{11} - a_{21}M_{21} + a_{31}M_{31} - a_{41}M_{41}$$

The determinant of a 5×5 matrix is defined similarly using the definition for 4×4 matrices:

$$\begin{vmatrix} a_{11} & a_{12} & a_{13} & a_{14} & a_{15} \\ a_{21} & a_{22} & a_{23} & a_{24} & a_{25} \\ a_{31} & a_{32} & a_{33} & a_{34} & a_{35} \\ a_{41} & a_{42} & a_{43} & a_{44} & a_{45} \\ a_{51} & a_{52} & a_{53} & a_{54} & a_{55} \end{vmatrix} = a_{11}M_{11} - a_{21}M_{21} + a_{31}M_{31} - a_{41}M_{41} + a_{51}M_{51}$$

In general, once the determinant of an $n \times n$ matrix has been defined, we define the determinant of an $(n + 1) \times (n + 1)$ matrix by

$$\begin{vmatrix} a_{11} & \cdots & a_{1\,n+1} \\ \vdots & & \\ a_{n+1\,1} & \cdots & a_{n+1\,n+1} \end{vmatrix} = a_{11}M_{11} - a_{21}M_{21} + \cdots + a_{n+1\,n+1}M_{n+1\,n+1}$$

where the sign preceding a_{ij} is + if $i + j$ is even and − if $i + j$ is odd. Thus the general term in this sum is

$$(-1)^{i+1}a_{i1}M_{i1}$$

The general form of Cramer's rule is stated as follows:

A system of equations in n variables $x_1, \ldots, x_n$

$$\begin{aligned} a_{11}x_1 + a_{12}x_2 + \cdots + a_{1n}x_n &= p_1 \\ a_{21}x_1 + a_{22}x_2 + \cdots + a_{2n}x_n &= p_2 \\ &\vdots \\ a_{n1}x_1 + a_{n2}x_2 + \cdots + a_{nn}x_n &= p_n \end{aligned}$$

with

$$\begin{vmatrix} a_{11} & \cdots & a_{1n} \\ \vdots & & \\ a_{n1} & \cdots & a_{nn} \end{vmatrix} = D \neq 0$$

has a unique solution

$$x_1 = \frac{k_1}{D}, \qquad x_2 = \frac{k_2}{D}, \qquad \ldots, \qquad x_n = \frac{k_n}{D}$$

where

$$k_i = \begin{vmatrix} a_{11} & \cdots & a_{1\,i-1} & p_1 & a_{1\,i+1} & \cdots & a_{1n} \\ a_{21} & \cdots & a_{2\,i-1} & p_2 & a_{2\,i+1} & \cdots & a_{2n} \\ \vdots & & & & & & \\ a_{n1} & \cdots & a_{n\,i-1} & p_n & a_{n\,i+1} & \cdots & a_{nn} \end{vmatrix}$$

that is, k_i is the determinant of the matrix obtained by replacing the ith column of the matrix of coefficients by the constants, $p_1, \ldots, p_n$. We will not provide a proof of this general form; you have already done or seen the proof for systems of 2 or 3 equations in 2 or 3 unknowns.

In using Cramer's rule, we can evaluate determinants according to the definition, but once the matrix is larger than 4×4, the computation becomes very time-consuming. The following theorems provide some basic facts and some methods for simplifying computations with determinants. We will not provide general proofs but will ask the student to verify them for the 3×3 case using the general matrix

$$\begin{bmatrix} a_{11} & a_{12} & a_{13} \\ a_{21} & a_{22} & a_{23} \\ a_{31} & a_{32} & a_{33} \end{bmatrix}$$

THEOREM 1

Let

$$\begin{bmatrix} a_{11} & \cdots & a_{1n} \\ \vdots & & \\ a_{n1} & \cdots & a_{nn} \end{bmatrix}$$

be an $n \times n$ matrix. Then

(i) $$\begin{vmatrix} a_{11} & \cdots & a_{1n} \\ \vdots & & \\ a_{n1} & \cdots & a_{nn} \end{vmatrix} = (-1)^{1+j}a_{1j}M_{1j} + (-1)^{2+j}a_{2j}M_{2j} + \cdots + (-1)^{n+j}a_{nj}M_{nj}, \qquad \text{for any } j,\ 1 \le j \le n$$

(ii) $$\begin{vmatrix} a_{11} & \cdots & a_{1n} \\ \vdots & & \\ a_{n1} & \cdots & a_{nn} \end{vmatrix} = (-1)^{i+1}a_{i1}M_{i1} + (-1)^{i+2}a_{i2i2} + \cdots + (-1)^{i+n}a_{in}M_{in}, \qquad \text{for any } i,\ 1 \le i \le n$$

(iii) $$\begin{vmatrix} a_{11} & \cdots & a_{1n} \\ \vdots & & \\ a_{n1} & \cdots & a_{nn} \end{vmatrix} = \begin{vmatrix} a_{11} & \cdots & a_{n1} \\ \vdots & & \\ a_{1n} & \cdots & a_{nn} \end{vmatrix}$$

Part (i) indicates that we can expand by minors using the entries for any column as coefficients of the minors and part (ii) indicates that we can expand by minors using the entries from any row as coefficients of the minors. The term is assigned a plus sign if the sum of the subscripts is even and a minus sign if it is odd. This is easily remembered by using a "checkerboard pattern" of alternating "+" and "−" signs in which the upper left corner is a "+" sign.

$$\begin{bmatrix} + & - & + & - & + \\ - & + & - & + & - \\ + & - & + & - & + \\ - & + & - & + & - \\ + & - & + & - & + \end{bmatrix}$$

Part (iii) states that transposing a matrix or "flipping" it around the diagonal from upper left to lower right does not change its determinant.

Example 6

$$\begin{vmatrix} 2 & 5 \\ 3 & 4 \end{vmatrix} = \begin{vmatrix} 2 & 3 \\ 5 & 4 \end{vmatrix} \qquad \begin{vmatrix} 1 & 2 & 3 \\ 4 & 5 & 6 \\ 7 & 8 & 9 \end{vmatrix} = \begin{vmatrix} 1 & 4 & 7 \\ 2 & 5 & 8 \\ 3 & 6 & 9 \end{vmatrix} \ \square$$

Note that if you prove (i) and (iii), you can easily prove (ii) using them.

THEOREM 2

Let

$$\begin{bmatrix} a_{11} & \cdots & a_{1n} \\ \vdots & & \\ a_{n1} & \cdots & a_{nn} \end{bmatrix}$$

be an $n \times n$ matrix, let D denote its determinant, and let $b \in R$.

(i) If

$$\begin{bmatrix} c_{11} & \cdots & c_{1n} \\ \vdots & & \\ c_{n1} & \cdots & c_{nn} \end{bmatrix}$$

is the matrix obtained by interchanging two rows or two columns of the original matrix, then its determinant is $-D$.

(ii) $$\begin{vmatrix} a_{11} & \cdots & ba_{1j} & \cdots & a_{1n} \\ a_{21} & \cdots & ba_{2j} & \cdots & a_{2n} \\ \vdots & & & & \\ a_{n1} & \cdots & ba_{nj} & \cdots & a_{nn} \end{vmatrix} = bD$$

for any j, $1 \leq j \leq n$; that is, multiplying a column of the matrix by b multiplies the determinant by b.

(iii) $$\begin{vmatrix} a_{11} & \cdots & a_{1n} \\ \vdots & & \\ ba_{i1} & \cdots & ba_{in} \\ \vdots & & \\ a_{n1} & \cdots & a_{nn} \end{vmatrix} = bD$$

for any i, $1 \leq i \leq n$, that is multiplying a row of the matrix by b multiplies the determinant by b.

(iv) $$\begin{vmatrix} a_{11} & a_{12} & \cdots & a_{1n} \\ \vdots & & & \\ a_{i1} + ba_{k1} & a_{i2} + ba_{k2} & & a_{in} + ba_{kn} \\ \vdots & & & \\ a_{n1} & a_{n2} & \cdots & a_{nn} \end{vmatrix} = D$$

if $i \neq k$; that is, multiplying a row by b and adding it to another row does not change the value of the determinant.

(v) $$\begin{vmatrix} a_{11} & \cdots & a_{1j} + ba_{1k} & \cdots & a_{n1} \\ a_{21} & & a_{2j} + ba_{2k} & \cdots & a_{n2} \\ \vdots & & & & \\ a_{n1} & & a_{nj} + ba_{nk} & \cdots & a_{nk} \end{vmatrix} = D$$

if $j \neq k$; that is, multiplying a column by b and adding it to another column does not change the value of the determinant.

Notice that part (i) indicates that if two rows or columns of a matrix are the same, its determinant is 0 since exchanging them would change the sign of the determinant but would not change the matrix. Thus we would have $D = -D$ and D must be 0.

From the definition of determinant and by examining the general case for $n = 2$, 3, and 4 we see that the determinant consists of a sum of products, each product containing exactly one factor from each row and each column of the matrix. Thus:

(a) If a matrix has a row or column of zeros, its determinant is 0.
(b) If all entries above the diagonal from upper left to lower right are zeros, then the determinant is just the product of the elements on this diagonal. Similarly, if all entries below this diagonal are zeros, the determinant is the product of the elements on the diagonal.

A matrix in which all entries below the diagonal are 0 or in which all entries above the diagonal are 0 is called a *triangular matrix*. One of the best ways to evaluate a determinant is to apply the above theorem, transform the original matrix into a triangular matrix, and then compute.

Example 7

$$\begin{vmatrix} 1 & 4 & 3 \\ 2 & 1 & 1 \\ 1 & 5 & -2 \end{vmatrix} = \begin{vmatrix} 1 & 4 & 3 \\ 0 & -7 & -5 \\ 0 & 1 & -5 \end{vmatrix}$$

Adding (-2) times the first row to the second and (-1) times the first row to the third.

$$= \begin{vmatrix} 1 & 4 & 3 \\ 0 & 0 & -40 \\ 0 & 1 & -5 \end{vmatrix}$$

Adding 7 times the third row to the second.

$$= -\begin{vmatrix} 1 & 4 & 3 \\ 0 & 1 & -5 \\ 0 & 0 & -40 \end{vmatrix}$$

Exchanging second and third rows.

$$= -(-40)$$
$$= 40 \ \square$$

Example 8

$$\begin{vmatrix} 1 & 5 & 6 & 2 \\ 2 & 8 & 4 & 10 \\ 3 & -1 & 2 & -3 \\ -2 & 1 & 6 & 4 \end{vmatrix}$$

$$= \begin{vmatrix} 1 & 5 & 6 & 2 \\ 0 & -2 & -8 & 6 \\ 0 & -16 & -16 & -9 \\ 0 & 11 & 18 & 8 \end{vmatrix}$$ Adding suitable multiples of the first row to the others.

$$= (-2)\begin{vmatrix} 1 & 5 & 6 & 2 \\ 0 & 1 & 4 & -3 \\ 0 & -16 & -16 & -9 \\ 0 & 11 & 18 & 8 \end{vmatrix}$$ By part (iii) of Theorem 2.

$$= (-2)\begin{vmatrix} 1 & 5 & 6 & 2 \\ 0 & 1 & 4 & -3 \\ 0 & 0 & 48 & -57 \\ 0 & 0 & -26 & 42 \end{vmatrix}$$ Adding suitable multiples of the second row to the third and fourth.

$$= (-2)(3)(-2)\begin{vmatrix} 1 & 5 & 6 & 2 \\ 0 & 1 & 4 & -3 \\ 0 & 0 & 16 & -19 \\ 0 & 0 & 13 & -21 \end{vmatrix}$$ By part (iii) of Theorem 2.

$$= 12\begin{vmatrix} 1 & 5 & 6 & 2 \\ 0 & 1 & 4 & -3 \\ 0 & 0 & 16 & -19 \\ 0 & 0 & 0 & -69/16 \end{vmatrix}$$ Adding $-13/16$ times the third row to the fourth.

$$= 12(-69) = -828 \quad \square$$

EXERCISES

1. Compute the following determinants.

(a) $\begin{vmatrix} 2 & 4 \\ -5 & 6 \end{vmatrix}$ (b) $\begin{vmatrix} \sqrt{6} & \sqrt{10} \\ \sqrt{3} & \sqrt{5} \end{vmatrix}$ (c) $\begin{vmatrix} 1 & 4 & 7 \\ 0 & 5 & 9 \\ 3 & 12 & 17 \end{vmatrix}$

(d) $\begin{vmatrix} 2 & 1 & 7 \\ 3 & -4 & 6 \\ 5 & 1 & -7 \end{vmatrix}$ (e) $\begin{vmatrix} \frac{1}{2} & \frac{4}{7} & \frac{2}{3} \\ 5 & 1 & 2 \\ 1 & 1 & 3 \end{vmatrix}$ (f) $\begin{vmatrix} 1 & 3 & 4 & 8 \\ 4 & 1 & 4 & -2 \\ 2 & 6 & 1 & 1 \\ 1 & -2 & 3 & 2 \end{vmatrix}$

2. Use Cramer's rule to solve the following systems of equations.

(a) $$\begin{aligned} 2x + y + 7z &= 8 \\ 3x - 4y + 6z &= 4 \\ 5x + y - 7z &= 2 \end{aligned}$$

(b) $$\begin{aligned} \tfrac{1}{2}x + \tfrac{4}{7}y + \tfrac{2}{3}z &= 2 \\ 5x + y + 2z &= 3 \\ x + y + 3z &= 2 \end{aligned}$$

(c) $$\begin{aligned} x + 3y + 4z + 8w &= 2 \\ 4x + y + 4z - 2w &= 0 \\ 2x + 6y + z + w &= 1 \\ x - 2y + 3z + 2w &= 5 \end{aligned}$$

3. Use Cramer's rule to solve the following systems.

(a) $$\begin{aligned} \tfrac{2}{3}x + \tfrac{4}{7}y &= \tfrac{8}{15} \\ \tfrac{3}{5}x - \tfrac{2}{11}y &= \tfrac{6}{13} \end{aligned}$$

(b) $$\begin{aligned} 2.7x + 1.32y &= 8.41 \\ 3.9x + 5.2y &= 9.37 \end{aligned}$$

(c) $$\begin{aligned} \sqrt{2}x + \sqrt{3}y &= \sqrt{6} \\ \sqrt{5}x - \sqrt{10}y &= \sqrt{7} \end{aligned}$$

(d) $$\begin{aligned} .004x - .03y &= .009 \\ 6x + 5y &= 11 \end{aligned}$$

4. Solve the following systems.

(a) $$\begin{aligned} 3x + 4y + 5z &= 2 \\ 2x - y + 3z &= 8 \\ 5x - 6y + 7z &= 11 \end{aligned}$$

(b) $$\begin{aligned} \tfrac{1}{2}x + \tfrac{2}{3}y - \tfrac{4}{5}z &= \tfrac{8}{3} \\ \tfrac{1}{4}x - \tfrac{1}{6}y + \tfrac{2}{3}z &= 1 \\ \tfrac{1}{5}x - \tfrac{1}{5}y + \tfrac{4}{3}z &= \tfrac{4}{15} \end{aligned}$$

(c) $$\begin{aligned} x + 2y + 3z + 4w &= 5 \\ 2x - y - z + 2w &= 4 \\ 3x - y - z &= 1 \\ 2x + 3y + 4z - 2 &= 2 \end{aligned}$$

(d) $$\begin{aligned} x + 3y + 4z + 5w &= 10 \\ 2x + y + z - w &= 8 \\ 4x + 3y + 2z - 2w &= 5 \\ x - y + z + w &= 11 \end{aligned}$$

9. Variation

The *ratio* between two numbers is just their quotient. This quotient, however, is usually expressed as a fraction. Thus, for example, the ratio between 2 and 5 or the ratio of 2 to 5 is $\frac{2}{5}$. Many quantities are related by a constant ratio. For example, when you buy gasoline, the ratio of the amount you spend to the amount purchased is just the price per gallon:

$$\frac{\text{Amount spent}}{\text{Amount purchased}} = \text{Price per gallon}$$

Two quantities or variables are said to *vary directly* if the ratio of one to another is a constant. In the example above, the amount spent varies directly with the amount purchased. To determine this ratio, we need to know just one set of values.

Example 1

If you spend \$9.12 and buy 16.8 gallons of gasoline, then the ratio is

$$\frac{\$9.12}{16.8}$$

which is approximately \$.519 per gallon.

If you purchase 14 gallons and let x denote the amount spent, then

$$\frac{x}{14} = \$.519 \quad \text{and} \quad x = \$7.27 \ \square$$

To say that x varies directly with y means that x/y is a constant. Thus, for some constant c, $x/y = c$.

Example 2

If x varies directly with y and $x = 4.2$ when $y = 6$, then

$$\frac{x}{y} = \frac{4.2}{6} = .7 \quad \text{and} \quad x = .7y$$

Notice that the equation $x/y = c$ for some constant c means the same as

$$x = cy$$

for some constant c.

We could also say that y varies directly with x and then

$$\frac{y}{x} = k$$

for some constant k, or

$$y = kx \ \square$$

Example 3

If x varies directly with y, and $x/y = 4$, then

$$\frac{y}{x} = \frac{1}{4}$$

and y varies directly with x.

In general, if x and y are related by an equation of the form

$$\frac{y}{x} = k, \qquad \text{with } k \neq 0$$

then

$$\frac{x}{y} = \frac{1}{k}$$

and each varies directly with the other. $\square$

Notice that an equation of the form $y = cx$ with $c \neq 0$ assures us that each variable varies directly with the other.

Example 4

If $y = 5x$, then

$$\frac{y}{x} = 5 \quad \text{and} \quad \frac{x}{y} = \frac{1}{5}$$

and each varies directly with the other. □

The notion of direct variation can also be applied to quantities represented by algebraic expressions.

Example 5

Suppose y varies directly with $x + 2$ and $y = 5$ when $x = 8$. Determine the relationship between x and y.

Since y varies directly with $x + 2$,

$$\frac{y}{x + 2} = c$$

for some constant c. Since $y = 5$ when $x = 8$,

$$\frac{5}{8 + 2} = c$$

$$\tfrac{5}{10} = c \quad \text{and} \quad c = \tfrac{1}{2}$$

Thus,

$$\frac{y}{x + 2} = \frac{1}{2} \quad \text{and} \quad 2y = x + 2$$ □

Example 6

If $y - 7$ varies directly with $2x - 1$, and $y = 10$ when $x = 3$, then

$$\frac{y - 7}{2x - 1} = k$$

for some constant k. Since $y = 10$ when $x = 3$,

$$\frac{10 - 7}{2 \cdot 3 - 1} = k$$

$$\frac{3}{5} = k$$

Thus,

$$\frac{y - 7}{2x - 1} = \frac{3}{5}$$

and

$$5(y - 7) = 3(2x - 1)$$
$$5y - 35 = 6x - 3$$
$$5y - 6x = 32 \quad \square$$

Example 7

If $y + 2$ varies directly with $2x + 4$, and $y = 6$ when $x = 2$, then what is the value of y when $x = 4$?

First

$$\frac{y + 2}{2x + 4} = c$$

because $y + 2$ varies directly with $2x + 4$.

$$\frac{6 + 2}{8 + 2} = c$$

$$\frac{8}{10} = c \quad \text{and} \quad c = \frac{4}{5}$$

Thus,

$$\frac{y + 2}{2x + 4} = \frac{4}{5}$$

and, when $x = 4$,

$$\frac{y + 2}{12} = \frac{4}{5}$$

$$5y + 10 = 48$$
$$5y = 38$$
$$y = \frac{38}{5} \quad \square$$

Example 8

If y varies directly with x^2 and $y = 4$ when $x = 3$, then what is the value of y when $x = 5$?

First

$$y = kx^2$$

for some constant k. Also,

$$4 = k(9) \quad \text{and} \quad k = \frac{4}{9}$$

Thus,

$$y = \frac{4}{9}x^2$$

and when $x = 5$,

$$y = \frac{4}{9}(5)^2 = \frac{4}{9}(25) = \frac{100}{9} \quad \square$$

Two quantities are said to *vary inversely* with each other if their product is a nonzero constant. Thus y varies inversely with x if $yx = k$ for some nonzero constant k. Notice that this equation is equivalent to the equation $y = k/x$. The latter form may be more convenient since we can compute values of y just by substituting values for x and dividing. Notice that if k is positive, then y increases as x decreases and vice versa. This fact is the reason for using the word "inversely" to describe the variation. As with direct variation, one set of values for the quantities will determine the constant and, therefore, the relationship.

Example 9

If y varies inversely with x and $y = 8$ when $x = 3$, then, since $yx =$ all x and y,

$$8 \cdot 3 = k \quad \text{and} \quad k = 24$$

Thus

$$yx = 24 \quad \text{or} \quad y = \frac{24}{x}$$

When $x = 6$, $y = 4$; when $x = -2$, $y = -12$. $\square$

Example 10

If y varies inversely with x and $y = 40$ when $x = \frac{1}{2}$, then

$$yx = k$$

for some k and

$$(40)(\tfrac{1}{2}) = k$$

Thus

$$yx = 20 \quad \text{or} \quad y = \frac{20}{x}$$

When $x = 5$, $y = 4$; when $x = 100$, $y = \frac{1}{5}$. $\square$

Example 11

If the distance d is fixed, then rate r varies inversely with time t since $rt = d$. If we go a fixed distance in 4 hours at 55 mph, then we know that $(55)(4) = d$ and $d = 220$. Thus $rt = 220$, or $r = 220/t$. If we make the same trip in 4.4 hours, then $r = 220/4.4 = 50$ mph. □

As with direct variation, quantities that vary inversely with respect to each other may be represented by algebraic expressions more complicated than just one variable.

Example 12

If $y(2x + 1) = 10$, then y varies inversely with $(2x + 1)$. Notice that we can write this as $y = 10/(2x + 1)$. □

As before, one set of associated values will determine the relationship.

Example 13

Suppose y varies inversely with $(4x + 3)$ and $y = 5$ when $x = 1$. Since

$$y(4x + 3) = k$$

for some constant k,

$$5(7) = k \quad \text{and} \quad k = 35$$

Thus

$$y(4x + 3) = 35 \quad \text{or} \quad y = \frac{35}{4x + 3}$$

When $x = -1$, $y = -35$; when $x = 3$, $y = 35/15 = 7/3$. □

Example 14

Suppose y varies inversely with $x^2 + 1$ and $y = 5$ when $x = 1$. Since

$$y(x^2 + 1) = k$$

for some constant k,

$$5(2) = k \quad \text{and} \quad k = 10$$

Thus

$$y = \frac{10}{x^2 + 1}$$

When $x = 0$, $y = 10$; when $x = 2$, $y = 2$. □

EXERCISES

In each of the exercises below, determine an equation that gives the relationship between x and y and compute y when $x = 10$.

1. y varies directly with x and $y = 5$ when $x = 2$.
2. y varies directly with x and $y = 9$ when $x = -5$.
3. y varies inversely with x and $y = 5$ when $x = 11$.
4. y varies directly with $3x + 7$ and $y = 10$ when $x = 5$.
5. $y - 5$ varies directly with $x^2 + 1$ and $y = 4$ when $x = 1$.
6. $y + 6$ varies inversely with $3x - 7$ and $y = 3$ when $x = 5$.
7. y^2 varies inversely with x^2 and $y = 2$ when $x = 7$.
8. y varies directly with $x + 2$ and $y = 5$ when $x = 4$.
9. y^3 varies directly with $5x^2 + 3$ and $y = 10$ when $x = 3$.
10. y^2 varies inversely with $5x + 2$ and $y = 2$ when $x = 2$.

10. Problem Solving

Problem solving in mathematics frequently involves a translation process. The problem usually arises in a real world situation but we work in the idealized world of mathematics. For example, if our problem involves the length of a board, we translate the problem to one involving line segments (which are abstract ideas in the mathematical world), solve an analogous problem involving segments, and then translate our answer back to the physical world. Usually we are not aware of this process but every time we draw a picture and apply our geometric knowledge we are involved in such a translation. In general, then, to solve a problem using mathematics, we need to formulate a mathematical problem that will provide an answer to the real world problem. This process of building or using mathematical models is the essence of applied mathematics. In this section we will be concerned with many problems that can be solved using the algebraic techniques covered in the first nine sections.

Example 1

What percent of 92 is 11?

If we translate this to symbols we might have

$$\underline{\qquad\qquad}\% \text{ of } 92 = 11$$

We know, for instance, that 5 percent of a number means 5/100 times the number. If we let x become the number we wish to find, we have

$$\frac{x}{100} \cdot 92 = 11$$

If we now solve this equation for x we have

$$\frac{92x}{100} = 11$$

$$\begin{aligned} 92x &= 11(100) \\ &= 1100 \end{aligned}$$

$$x = \frac{1100}{92}$$

and x is approximated by 11.9. (Divide 1100 by 92 and round off to the nearest tenth.) Thus 11 is approximately 11.9 percent of 92. □

When many students read this example their reaction is "What a lot of bother! I learned in fifth grade to determine what percent one number is of another and I didn't need algebra!"

If this were all we were going to do, they would be right and you could close the book right now. The point of this example is that any percentage problem involves the basic equation among three numbers x, y, and z.

$$x\% \text{ of } y = z$$

or

$$\frac{x}{100} \cdot y = z$$

The fundamental percentage relationship is given by this equation and to solve elementary percentage problems, we only need to know which two numbers are given in order to solve for the third.

Example 2

Suppose you own a home that you wish to sell and you need to receive net proceeds of at least \$31,500 in order to buy the bigger home you desire. If the legal fees are \$200, the real estate commission is 7 percent of the selling price, and the transfer tax is 1 percent of the selling price, what must your selling price be in order to net \$31,500? Does the problem sound difficult? It isn't if you translate it into algebraic form.

Let x be the selling price of the house.
Let y be the total cost of selling the house.
You require \$31,500 as the net proceeds.

Since

(selling price) − (costs) = net proceeds

the first equation is

$$x - y = \$31{,}500$$

Furthermore the costs are:

$.07x$	for real estate commission
$.01x$	for transfer tax
\$200	for legal fees

Therefore the total cost is

$$.07x + .01x + 200$$

that is,

$$y = .07x + .01x + 200$$

and

$$y = .08x + 200$$

is the second equation.

Now to solve the problem, we need only to solve the system of equations.

$$\begin{aligned} x - y &= 31{,}500 \\ y &= .08x + 200 \end{aligned}$$

Substituting $.08x + 200$ for y in the first equation yields

$$\begin{aligned} x - (.08x + 200) &= 31{,}500 \\ .92x - 200 &= 31{,}500 \\ .92x &= 31{,}700 \\ x &= \frac{31{,}700}{.92} \end{aligned}$$

x is approximately 34,456.52.

If you price the house at \$34,500, you would have what you need. Realize that solving this problem would not, however, determine your market price. You would price the home at or above the appraised value but would not take less than \$34,456.52 because any lower price would not satisfy your needs. If you could not sell for at least this much, you would not sell. □

Example 3

Suppose you pay \$240.35 for a stereo set including 4.5 percent sales tax. What was the actual price of the set and how much tax did you pay? The basic equation is

(selling price) + (tax) = total cost

If we let x denote the selling price, then $.045x$ is the tax and the equation becomes

$$x + .045x = 240.35$$

Then

$$1.045x = 240.35$$

$$x = \frac{240.35}{1.045}$$

$$= 230$$

Thus the actual price of the stereo was \$230. □

The last two problems involved percentage and it is doubtful that your methods from elementary arithmetic would have solved the problem for you.

An effective technique for solving problems by using algebra is:

A. Determine the relationships among the quantities under discussion and state them as verbal equations or inequalities.
B. Express the quantities using algebraic representations.
C. State the relationships using the algebraic representations.
D. Solve the resulting equation, inequality, or system.
E. Translate those solutions into answers to your original problem.

Example 4

Your automobile runs smoothly on both leaded and nonleaded gasoline. If leaded gasoline costs 64.9¢ per gallon and yields 16 miles per gallon, and nonleaded gasoline costs 61.9¢ per gallon, how many miles per gallon must you get in order for you to benefit financially by using nonleaded gasoline?

The question is, what gas mileage on nonleaded gas assures that:

cost per mile on nonleaded gas < cost per mile on leaded gas

If we let x denote the number of miles per gallon you get on nonleaded gasoline, then $61.9/x$ represents the cost per mile on nonleaded gas and $64.9/16$ is the cost per mile on leaded gas.

Thus to solve the problem, we need to solve the inequality

$$\frac{61.9}{x} < \frac{64.9}{16}$$

Since the cost per mile is positive, $x > 0$ and we can multiply both sides by $16x$ and preserve the direction of the inequality. Thus

$$(61.9)(16) < 64.9x$$

and

$$\frac{(61.9)(16)}{64.9} < x$$

$$\frac{9904}{649} < x$$

For practical purposes we would approximate the quotient by 15.3:

$$15.3 < x$$

Thus if the nonleaded gasoline yields 15.3 or more miles per gallon, you save money. □

Example 5

Suppose you are offered two equally pleasant positions for summer vacation. The first pays \$2.50 an hour with a 10¢ raise every week and the second pays \$2.75 an hour with a 5¢ raise every week. Assuming a 40-hour work week, after how many weeks do they pay the same?

If we let w denote the number of weeks, y the rate on the first job, and z the rate on the second job, we have

$$y = 2.50 + .1w$$
$$z = 2.75 + .05w$$

The question is, for what value of w is $y = z$? If $y = z$, then

$$\begin{aligned} 2.50 + .1w &= 2.75 + .05w \\ .05w &= .25 \\ 5w &= 25 \\ w &= 5 \end{aligned}$$

Thus at the end of 5 weeks, the two jobs pay the same. Notice that the fact of a 40-hour work week was redundant, that is, it does not enter into the problem. (Can you decide which job is better for a 12-week summer vacation?) □

When we translate the real problem situation into the world of mathematics, solve our mathematical problem, and then translate back, sometimes not all of our mathematical answers make sense in the practical situation.

Example 6

A rectangular flower garden is 5 times as long as it is wide. If the total area is 720 sq ft, what are the dimensions of the garden?

We know $A = lw$, where A denotes area, l denotes length, and w

denotes width. We are given that $A = 720$ and that $l = 5w$. Thus we have the equation

$$720 = (5w)w$$
$$144 = w^2$$
$$w = 12 \quad \text{or} \quad w = -12$$

and

$$l = 60 \quad \text{or} \quad l = -60$$

Since the negative dimensions for a flower garden make no sense, the width is 12 and the length is 60.

Example 7

Using physics and calculus, it can be shown that if an object is thrown upward from the ground with a velocity of v_0 at time 0, then its distance above the ground at time t is given approximately by

$$d(t) = -16t^2 + v_0t$$

This formula ignores the effect of air friction. The velocity is given in feet per second, the time t in seconds, and $d(t)$ is measured in feet. Note that this equation makes sense only for certain values of t, that is, only from the time that the object is thrown (when $t = 0$) until it returns to the ground [the first positive value of t such that $d(t) = 0$].

If a ball is thrown upward with an initial velocity of 64 feet per second, at what time will it hit the ground?

We wish to know for what values of t is $d(t) = 0$. In this case

$$v_0 = 64$$

Thus

$$d(t) = -16t^2 + 64t$$

We need to solve

$$0 = -16t^2 + 64t$$

This can be rewritten as

$$16t^2 - 64t = 0$$

Factoring yields

$$16t(t - 4) = 0$$

Either

$$t = 0 \quad \text{or} \quad t - 4 = 0$$

Thus

$t = 0 \quad \text{or} \quad t = 4$

Since the object is thrown at time $t = 0$, it must return to the ground after 4 seconds. □

Example 8

Suppose Mr. Thurston has two quarts of 3:2 martinis (by this we mean 3 parts gin to 2 parts vermouth) and he decides that a 7:3 mixture would be more to his taste. How much gin must he add to the two quarts?

Since he is adding only gin, the following relations hold:

Amount of vermouth in original mixture = amount in final mixture.
Amount of vermouth in original mixture is $\frac{2}{5}$ of two quarts.
Amount of vermouth in the final mixture is $\frac{3}{10}$ of final mixture.

If we let x be the amount added, then $2 + x$ is the amount of the final mixture. Thus the first equation becomes

$$\tfrac{2}{5}(2) = \tfrac{3}{10}(2 + x)$$

or

$$\tfrac{4}{5} = \tfrac{3}{10}(2 + x)$$

Thus

$$8 = 3(2 + x)$$
$$8 = 6 + 3x$$
$$2 = 3x$$

and

$$x = \tfrac{2}{3}$$

Thus he must add $\frac{2}{3}$ quart of gin.

Checking in the original situation, recall that $\frac{3}{5}$ of the original 2 quarts was gin. Thus he began with $\frac{3}{5}(2) = \frac{6}{5}$ quarts of gin and ends up with $\frac{6}{5} + \frac{2}{3} = \frac{28}{15}$ quarts of gin. The final amount of martinis is $2\frac{2}{3}$ or $\frac{8}{3}$ quart. Since he wishes $\frac{7}{10}$ of the final mixture to be gin, he wants $(\frac{7}{10})(\frac{8}{3}) = \frac{56}{30} = \frac{28}{15}$ which is exactly what he has. Cheers! □

Example 9

Suppose that a fullback can run 100 yd in 10.9 sec in full football equipment. If he breaks into the clear at the 20-yd line and is 5 yd ahead of you, what would your average time for 100 yd have to be in order for you to catch him before he gets to the 5-yd line. (If you get him after that, the momentum will carry him over.)

Let x be your time for 100 yd. You are both going to run for the same time and we have the equation

$$\text{your time} = \text{his time}$$

Since you are going to run 80 yd, your time is $\frac{80}{100}x$ or $.8x$. Since he is going to run 75 yd, his time is $\frac{75}{100}(10.9)$ or $.75(10.9)$. The equation becomes

$$.8x = .75(10.9)$$
$$.8x = 8.175$$
$$8x = 81.75$$
$$x = \frac{81.75}{8}$$

or x is approximately 10.22. Thus if your time is 10.22 or better, you have a good chance to catch him. □

In natural science, social science, and business, many quantities are related by direct or inverse variation. Using this fact and one observation, one may completely describe the relation between such quantities.

Example 10

In business, there are two basic types of costs, fixed costs and variable costs. Examples of fixed costs are rent and utilities. The cost of raw materials would be a variable cost since it would change with the amount of goods produced. The total cost is the sum of fixed costs and variable costs. Frequently the variable costs vary directly with the amount of goods purchased. Suppose you own a car wash in which the fixed daily costs are \$15 and general experience tells you that the variable costs vary directly with the number of cars washed. Let C denote the total costs, s be the number of cars washed, and suppose on a day in which you wash 85 cars the total costs are \$42.20. Can you express the total costs in terms of the number of cars washed per day?

The fixed costs are \$15 and C is the total costs, so

$$C - 15 = \text{variable costs}$$

Since the variable costs vary directly with the number of cars washed, we have

$$\frac{C - 15}{s} = k$$

for some constant k. When $s = 85$, $C = 42.20$; thus

$$\frac{42.20 - 15}{85} = k$$

$$\frac{27.20}{85} = k$$

and

$$k = .32$$

Then

$$\frac{C - 15}{s} = .32$$

$$C - 15 = .32s$$

and

$$C = .32s + 15$$

Thus, if you wash 120 cars, the costs will be

$$C = .32(120) + 15 = 38.40 + 15 = 53.40$$ □

Example 11

Usually in business, the number of sales depends on the price charged. Usually the higher the price, the lower the number of sales. Suppose s denotes the number of sales and p denotes the price. Again, let us consider the example of the car wash. Suppose that if $p = 0$, that is, if car washes are free, then you would wash at maximum capacity, and let us assume that 300 is the maximum number of cars per day. Thus, when $p = 0$, $s = 300$. Suppose, further, that at a price of \$4, the price is too high; no one will patronize you and $s = 0$. Thus the price p is between 0 and 4, and the number of washes s is between 0 and 300. Suppose now that the sales vary directly with the difference between \$4 and the price p. Then

$$\frac{s}{4 - p} = k$$

for some constant k. When $p = 0$, $s = 300$,

$$\frac{300}{4} = k$$

$$k = 75$$

Thus,

$$\frac{s}{4 - p} = 75$$

$$s = 75(4 - p)$$
$$s = 300 - 75p \qquad \text{for } 0 \le p \le 4$$

If you establish a price of \$2, you will wash 150 cars per day. At a price of \$2.50, you will wash only about 112 cars. □

Example 12

In business one is usually interested in profits. Profit is the difference between gross income and costs. If we let P denote profit, G denote gross income, and C denote costs, then

$$P = G - C$$

In the above examples, $G = sp$, where s is the number of cars washed and p is the price. Furthermore,

$$C = .32s + 15$$

Thus,

$$P = sp - (.32s + 15)$$

but we know that

$$s = 300 - 75p$$

Hence,

$$\begin{aligned} P &= (300 - 75p)p - [.32(300 - 75p) + 15] \\ &= -75p^2 + 300p - 96 + 24p - 15 \\ &= -75p^2 + 324p - 111 \end{aligned}$$

where p denotes the price charged.

We now have expressed the profit in terms of the price charged, and we can estimate the daily profit at any particular price p. Thus, for example, if we charge \$3, then

$$\begin{aligned} P &= -75(3^2) + 324(3) - 111 \\ &= -75(9) + 972 - 111 \\ &= -675 + 972 - 111 \\ &= 186 \end{aligned}$$ □

The following exercise set contains many different problems. We have purposely avoided classifying them as to "types" of problems. It is not our objective to have you learn "how to do" certain "types" of problems but rather to have you develop the ability to analyze a problem, determine the relationships involved, and then use your knowledge of algebra to solve it. A few of the exercises are mathematician's puzzles but most of the problems are similar to real applications. We have included the puzzles because they are of interest to some students and, in fact, are frequently easier than "real" problems.

EXERCISES

1. The sum of two numbers is 12 and their product is 30. What are the numbers?
2. If good quality corporate bonds are paying $8\frac{1}{2}$ percent interest, all of which is taxable, and you are in a bracket in which you pay 71 percent of all additional income as taxes, what rate of interest would you be willing to accept on good, dependable municipal bonds on which the interest is not taxable?
3. Mr. Thurston has just mixed 2 quarts of martinis according to his new favorite formula of 10:1; that is, 10 parts gin to 1 part vermouth. His wife contends that this is almost equivalent to waving the vermouth bottle over the glass and that these are not suitable for company. In order to make her happy, he decides to mix a quart of 7:3 as usual but, to his horror, he finds that he used the last of the gin when he made his 10:1 mixture. The only way to make the quart of 7:3 is to add vermouth to some of the 10:1 mixture. How much vermouth and how much 10:1 mixture should he use to obtain the desired quart of 7:3 mixture?
4. The price of rugs of the same material should vary directly with the area of the rug. (This is at least a reasonable approximation.) If a circular nylon woven rug 5 ft in diameter sells for \$60, what would you expect to pay for a circular rug of the same material but with a diameter of 9 ft?
5. Simple interest is computed using the formula $i = Prt$ (P represents the principal, r represents the rate, and t represents the units of time) and is paid at the end of the period of time, that is, on the maturity date of the loan. If the amount you pay, including principal and 8 percent interest for one year, is \$1350, what was the amount of the principal you borrowed?
6. Discount interest is paid at the time you borrow the money; that is, the interest charged is subtracted from the amount you borrow and at the end of the period of the loan, the amount you pay is just the principal. Thus, for example, if you borrowed \$1000 for one year at 6 percent interest per year, you would receive \$940 and you pay your creditor \$1000 at the end of one year. If you borrow \$1000 at 8 percent interest per year on a discount basis, what rate is this equivalent to at simple interest? Answer the same question if you borrow \$3000.
7. Generalize Exercise 6. If you borrow K dollars at rate r on a discount basis, what rate is this equivalent to at simple interest?
8. The sum of two numbers is 3 and the difference in their squares is 4. What are the numbers?
9. Two private planes leave an airport, one flying south at 150 mph and the other flying east at 210 mph.

(a) How far apart will they be in $\frac{1}{2}$ hour?
(b) How long will it take them to be at least 540 miles apart?

10. If a gun is fired upward from ground level and a bullet leaves the gun with a velocity of 1120 ft/sec, in how many seconds will it return to the ground?
11. If two numbers are such that the sum of their squares is 8 and the difference in their squares is 2, what can the numbers be?
12. Suppose you have two numbers such that the absolute value of their sum is 10 and the absolute value of their difference is 2. What could the numbers be?
13. A nut shop manager wishes to mix peanuts, cashews, almonds, and pecans in order to sell a mixture of mixed nuts. If peanuts are 59¢ a pound, cashews \$1.49 a pound, almonds \$1.59 a pound, and pecans \$1.69 a pound, and if the amounts of cashews, almonds, and pecans used in the mixture are the same, what portions of each kind of nut should he use to produce a mixture selling at \$.99 a pound?
14. If a truck leaves a turnpike service station at 10:15 traveling eastward at 55 mph and a car leaves the same station at 10:25 and travels after it at 70 mph, at what time will the car overtake the truck?
15. The strength of rope varies directly with the square of its diameter. If the breaking strength of nylon rope is 800 lb for $\frac{1}{4}$-in. rope, what is the strength of $\frac{1}{2}$-in. nylon rope?
16. If an object is thrown upward from ground level with velocity v_0, its velocity at time t is given by $v(t) = -32t + v_0$ (until it returns to the ground). Note that velocity is a directed quantity in that upward velocity is positive and downward velocity is negative. How high is the object when the velocity is 0? Is this as high as the object will go? Explain. What is the velocity at the instant the object hits the ground?
17. Suppose an automobile manufacturing company has announced a new engine option for their line of cars for next year. Consumer tests show that this new engine increases gasoline mileage by 15 percent for highway driving and by 10 percent for city driving. The new engine option will cost \$400 extra per car. Suppose you average 20,000 miles per year with 55 percent of your driving on highways and the balance in cities. Furthermore, suppose that you expect to keep a car for 4 years and that the standard model you are considering buying can be expected to yield 13 mpg on the highway and 9 mpg in the city. What must the average price of gasoline be for the next 4 years for you to profit from buying the new engine?
18. Suppose you are a singing star about to negotiate a new recording contract. One company offers you 10 percent on all sales whereas another offers you 8 percent on the first 50,000 records sold, 9 percent on the next 100,000, and 12 percent on all records sold over 150,000. (These rates apply to each record, that is, 50,000 copies of the same record, and so on.) How many records must your average release sell for the second contract to be better?
19. Suppose that the cost of color TV tubes varies directly with the area of the screen. Assume that the sides and diagonal of the screen are always in the proportions of 3, 4, 5. If a 17-in (diagonal measure) picture tube costs \$114, what does a 19-in tube cost?
20. Suppose you are boating on a stream and wish to know your speed in still water. If you know that a particular bridge is just 2 miles from your starting place and you travel downstream to the bridge in 8 min but it takes you 11 min to return, what is your speed in still water? What is the rate of the current?

21. In a particular city, the labor department finds an overall unemployment rate of 5.8 percent with an unemployment rate of 3.1 percent among whites. If nonwhites make up 34 percent of the population of the city, what is the unemployment rate among nonwhites?
22. Show that a circular region enclosed by p ft of fence is larger than a square region enclosed by p ft of fence.
23. Suppose you have an average of 73 percent before the final exam in a course. If the final exam counts $\frac{1}{5}$ of the grade, what must you receive on the final in order to have at least a 75 percent average for the course?
24. The distance d that a car with speed s will travel after the breaks are applied is given by the equation $d = .02s^2 + .7s$ where speed is given in feet per second and d is in feet.
    (a) If an automobile is going 60 mph, how far will it travel after the breaks are applied?
    (b) If skid marks measure 300 ft, how fast was the car going when the breaks were slammed on?
    (c) What speed provides a stopping distance of no more than 50 ft after the breaks are applied?
25. Suppose that experience shows that the average shipping cost C per unit for a particular machinery business varies directly with the square of the average distance from the factory to the customer. If the average shipping cost in one year is $42 and the average distance from the factory to the customer is 100 miles, then what can you expect the average shipping cost to be if the business is expanded so that the average customer is 120 miles from the factory?
26. An aluminum siding contractor assures you that if he puts siding on your house, it will not need any painting or maintenance for at least 20 years. Suppose that previous experience tells you that in order to keep the house in good condition, it must be painted at least every 4 years. Furthermore, assume that the house needs painting now and that it will cost $925 to do the job.
    (a) If the siding job costs $4500, which will cost you more over the long run, painting or siding?
    (b) Perhaps the question in part (a) was too simple. If you paint the house now and invest the balance of $3575 (the difference between painting and siding) in a savings account at 5 percent per year and use the interest in painting the house, then which costs more? (Ignore the compounding of interest.)
    (c) What would the costs of painting have to be for the costs of both to be the same over a 20-year period?
    *(d) Answer the same questions as in (a)–(c) taking into consideration an estimated inflation of 4.5 percent per year.
27. The grade-point *average* (A) of a student is given by the formula $A = P/C$, where P indicates the quality *points* earned and C is the number of *credits* earned. The number of quality points earned in a course is the product of the number of credits in a course and the number which corresponds to the grade. At most colleges and universities A corresponds to 4, B to 3, C to 2, D to 1, and F to 0. The total number of quality points, P, is determined by summing those earned for each course. Suppose you have an average of 2.36, have earned 45 credits, and are taking 16 credits this term. What must be your average for the term in order to bring your average up to 2.50?

28. Suppose 180 credits are required for graduation and you have earned a total of 164 credits with an average of 2.42. If you are taking 16 credits this term and an average of 2.25 is required for graduation, what is the minimal average you can earn this term and still graduate?
29. Two boys are to mow a yard 100×120. Their procedure is to mow a swath completely around the rectangle and keep going until the job is finished. How wide a swath should the first boy mow so that they mow the same amount?
30. From experience it is found that the monthly profit P of a certain television store is given by the equation $P = 4S - .02S^2 - 100$, for $0 \leq S \leq 800$, where S denotes the number of units sold.
    (a) What are the profits on sales of 300 units?
    (b) What must the sales be in order to avoid a loss?
    (c) What sales are necessary for a profit of at least \$3000 a month?
31. By testing your car at 55 mph and 70 mph under similar driving conditions, you determine that it yields 14.2 mpg at 55 mph and 12 mpg at 70 mph.
    (a) If gasoline is 54¢ per gallon, how much money do you save on a 600-mile trip by driving 55 mph instead of 70 mph?
    (b) How much longer does the trip take at 55 mph?
    (c) How much per hour are you earning? (Your savings are your earnings.)
    (d) What would the price of gasoline have to be in order for you to earn at least \$2 per hour?
    (e) Does the 600-mile distance enter into your answers for parts (c) and (d)? Explain.
32. The intensity of light varies inversely with the square of the distance from the source. If the intensity at a distance of 12 cm is 24 units, what is the intensity at a distance of 18 cm? At what distance is the intensity 100 units?
33. Mr. Pinchpenny operates a hamburger stand and wishes to spend as little as possible on ketchup, mustard, and so on. It is his practice to add water to the ketchup to make it go farther and he has found that if he adds 1 pint of water to each gallon of ketchup, no one can tell the difference. On April 16 he feels that he needs to save even more money and decides to add a quart of water to each gallon of ketchup and mixes up 10 gal of his new mixture using 2 gal of water and 8 gal of ketchup. Usually an 8-gal supply would last him for a week but he notices that 2 gal of the new mixture are used in only one day with no appreciable change in the number of sandwiches sold. The mixture is so thin that the customers are pouring more than they expect to. He realizes that he should go back to his original mixture. How much pure ketchup should he add to the remaining 8 gal?
34. A sportscaster suggests that the penalty for a false start in the 440-yd run should be a 5-yd set back, that is, the false starter must give the others a 5-yd head start. Suppose you are penalized and still win with a time of 58 sec. What was your average time for the actual 440? (Does this seem like a reasonable penalty?)
35. Suppose two companies sell batteries for your car, one with a 4-year prorated guarantee and one with a 5-year prorated guarantee. Suppose, also, that the battery with the longer guarantee sells for a firm price of \$31 and you are dickering with the dealer over the price of the one with the shorter guarantee. Assuming that nothing lasts beyond its guarantee, what would the price have to be in order for the battery with the 4-year guarantee to be the better buy for you?
36. In geometry it is shown that any triangle constructed with all vertices on a circle and with a diameter of the circle as one side is a right triangle, the right angle being

opposite the diameter. Furthermore, any rectangle constructed with all vertices on the circle consists of two such triangles. Suppose you have a circle with diameter d. What is the largest rectangle that can be so constructed and what is its shape?

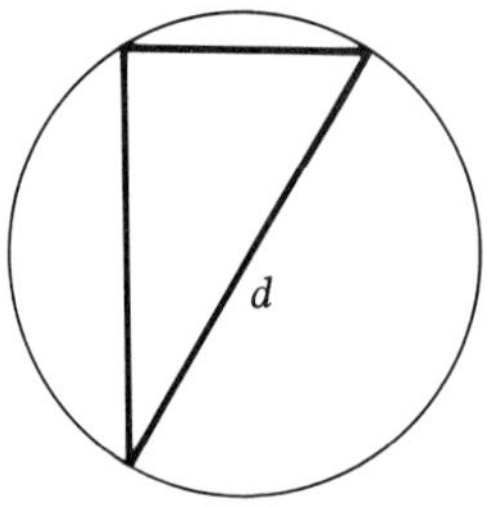

37. Suppose you have two outboard motors, the older of which uses 1 part oil to 20 parts gasoline and the newer of which uses 1 part oil to 50 parts of gasoline. If you have an abundant supply of gasoline for the newer motor, how much oil would you have to add to 3 gal of the 1:50 mixture to make it suitable to burn in the old motor?
38. A tank contains 12 gal of 6 percent salt brine. How much water must be evaporated to yield a solution of 7 percent salt? How much water must be added to yield a solution of 4 percent salt?
39. The volume of a sphere is given by $V = \frac{4}{3}\pi r^3$, where r denotes the radius. Suppose you have two spheres, one with radius 2 and the other with unknown radius but suppose you know that the one with unknown radius has twice the volume of the first. What is its radius? If we double the radius of a sphere, how does that affect the volume?
40. Suppose you hire two typists to type a paper for you. One types 45 words per minute and the other 65 words per minute. If the paper contains about 15,000 words, how long will it take them to type the paper assuming that they work the same amount of time?
41. Refer to Exercise 40. Suppose you pay the slower of your typists \$2 per hour. What should you pay the second if you wish to pay according to the amount of work done? What will the typing cost you at these prices?
42. Automotive engineers have determined that current automobile engines can easily be adapted to operate on a mixture of 80 percent gasoline and 20 percent methol alcohol. If the cost of gasoline delivered to the station is 39¢ per gallon, the retail price of gasoline is 57¢ per gallon, and the cost of producing and delivering the methol alcohol to the station is 15¢ per gallon, what would be the price of the 80—20 mixture given the same percentage of mark up?
43. A salesman is offered two equally attractive positions in terms of products and working conditions. The first company offers him \$10,000 a year salary plus 3 percent commission on all sales. The second offers him commissions of 7 percent on all sales with a yearly guarantee of \$14,000 in earnings. What yearly sales on his part would make the second position more profitable for him?
44. The distance from your cold, windswept campus to Fort Lauderdale is 1200 miles and the round trip air fare is \$140. Assume that your car averages 14 mpg for highway driving and that, besides the cost of gasoline, you estimate that it will cost about 1.1¢

more per mile to operate the car and that you will need to rent a room at $9 per night at the halfway point on each way of your auto trip. At what price per gallon for gasoline does it become more economical to fly?

45. Suppose Mr. Bunker wishes to make some green paint by mixing yellow and blue together. His perfect memory and judgment tell him that he should use 8 parts of yellow to 3 parts of blue and so he mixes a gallon this way. To his amazement he discovers that it is too light. Amid gales of laughter (from others) he leaves for the corner bar and turns the job over to his son-in-law (the one leading the laughter). The son-in-law relies on his equally perfect memory and judgment and, with his usual disdain for money, starts with new paint and mixes a gallon using 8 parts yellow and 5 parts blue. He is amazed to find that his mixture is too dark and sheepishly leaves to join his father-in-law at the saloon, leaving the job to his wife and mother-in-law. The mother-in-law, by first mixing very small portions, determines that the proper mixture is 2 parts yellow and 1 part blue. Instead of starting again, however, she wishes to use the light and dark mixtures to make one that is just right. How much of each should she use to obtain a gallon of the right shade?
46. Suppose you have 120 ft of fence and wish to enclose a rectangular garden with area of about 600 sq ft. To the nearest foot, what should be the dimensions of the garden?
47. One orthodontist makes no initial charge but charges $60 a month for the period of time that the patient wears the braces. Another charges $350 at the time the braces are installed and $30 a month for the period of time that the patient wears the braces. How many months would a patient have to wear the braces in order for the charges to be less from the second orthodontist?
48. Suppose the sum of three numbers is 24. If the sum of the first two is 13 and the sum of the last two is 19, what are the three numbers?
49. The force of gravitational attraction between two objects with fixed masses varies inversely with the square of the distance between them. If the force of attraction of two objects 200 cm apart is 12 units, what is this force when they are 150 cm apart? What is it when they are 300 cm apart?
50. The circumference of one circle is 7 times that of another. If the smaller circle has a radius of 9, what is the radius of the larger? Can you generalize the principle involved in this problem?
51. Suppose a baseball player has a batting average of .245 after 350 times at bat. If he can expect to bat about 200 more times during the season, how many hits must he get in order to assure that he bats at least .275?
52. As with everything else, the price of paper keeps increasing. Suppose that you are a newspaper publisher and you find that by changing type face and the printing style of your newspaper, you can cut the amount of paper used by 7 percent. What percentage increase in paper price can you offset by making these changes?
53. Suppose you are in a federal income tax bracket in which the government takes 22 percent of the first $2000 of any annual increase in your income and the state tax takes 3 percent of any increase in your income. How much of a raise do you need in order to increase your monthly net income by $80?
54. The air fare from Detroit to Dallas is $170 but the special excursion price of $110 each is available if at least 10 people travel together. It is not really necessary for 10 people to go; it is only necessary for 10 fares to be paid at the excursion rate and for those going to travel at the same time on the same plane. How many people actually need to go in order to save money by buying the 10 tickets?

55. A publisher of a monthly magazine has determined, from experience, that the following relationships hold:
    (i) The number S of subscription customers is related to A, the annual subscription price (in dollars) by the equation $S = -30{,}000A + 240{,}000$.
    (ii) The monthly costs C of printing and mailing the magazine and operating the business are given by the equation $C = .17N + .03S + 15{,}000$, where N is the number of magazines printed per month.
    (iii) The advertising revenue is given by $R = .08N$.
    (iv) The newsstand sales income T is given by $T = pn$, where p is the net price and n is the number sold on newsstands.

    Let us assume, for this problem, that $n = 10{,}000$ and $N = S + 10{,}000$, that is, newsstand sales are 10,000 and they sell all they print. Assume also that $p = 35¢$.
    (a) If the annual subscription price is \$4, what is the net profit per month?
    (b) If postage rates are changed, the increased costs must be passed on to the subscription customers. (The $.03S$ term in the equation for C is the present mailing cost. If mailing costs change, this changes.) Determine the net profit per month if the cost of mailing is increased to 8¢ per copy and this is passed on to the subscribers by changing the subscription price.
    (c) What would be the net profit if mailing costs remain the same and the subscription cost is lowered to \$3 per year?
56. Suppose that a rocket is launched with an initial velocity of 102.4 ft/sec.
    (a) How long will it take to reach a height of 640 ft?
    (b) Will it ever reach a height of 20,000 ft? Explain your answer.
57. Suppose your experience in the car wash business tells you that the number of cars washed per day and the price charged per car are related by the equation $N = -100p + 300$, where N is the number washed, p is the price, and $0 < p < \$3$. Furthermore, suppose your costs are given by $C = \$30 + .25N$.
    (a) Express the profit P in terms of the price p.
    (b) What will be the profits at a price of \$1.25?
    (c) What should the price be in order to assure a profit of at least \$145 a day?
    *(d) Could you charge more and make more? How much and how much?
58. The strength of steel wire made from a particular alloy varies directly with the square of its diameter. If wire made from this alloy with a diameter of .03 has a breaking strength of 10 lb, what is the diameter of a wire with a breaking strength of 25 lb?
59. Show that the fact that the strength varies directly with the square of the diameter in Exercise 58 is equivalent to stating that the strength varies directly with the area of a cross section.
60. Suppose you have your choice of two fertilizers, type A and type B. To fertilize your fields with type A costs \$5 more per acre than with type B but type A brings a 7 percent higher yield than does type B. What must your gross receipts per acre be in order for type A to be a more profitable investment than type B?

3

Graphs and Functions

The French mathematician and philosopher Descartes invented analytic geometry and thus provided us with one of the most useful tools in mathematics. In this chapter we see how algebra and geometry can be blended so that methods and concepts from each area enlighten our understanding of the other.

1. The Coordinate System and the Distance Formula

When using number lines we have assumed that we can assign a real number to a point on a line so that the distance between two points will be the absolute value of the difference between the corresponding numbers or coordinates. For example, on a number line, the distance between the point with coordinate 5 and the point with coordinate 9 is just $9 - 5$ or $|5 - 9|$. In order to coordinatize a plane, we use two perpendicular lines. Usually we call one X, or the X-*axis*, and the other Y, or the Y-*axis*, and choose the coordinates so that the intersection of the lines has the coordinate 0 on both lines; that is, as a point on X, the intersection is named 0 and also as a point on Y it is named 0. This point is called the *origin* of the system. It is customary to orient the lines as in Figure 3–1. We could choose other orientations, but for

Figure 3–1

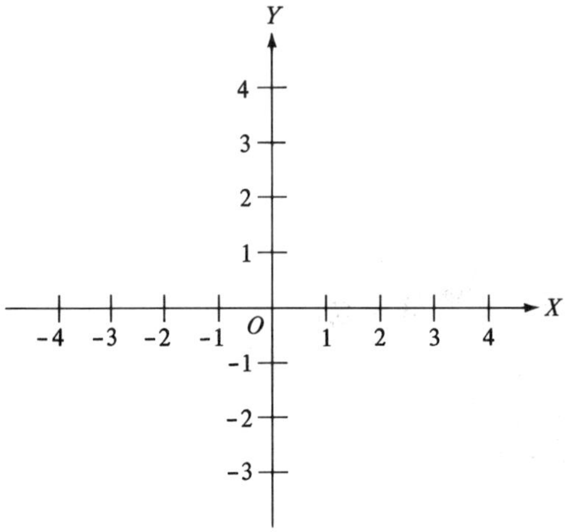

most purposes the one above is sufficient. Now let P be any point in the plane. If we construct the lines through P parallel to Y and X, that is, the perpendiculars from P to X and Y, then these lines will pass through X and Y. We assign to P the ordered pair of numbers (a, b), where a is the coordinate of the foot of the perpendicular from P to X and b is the coordinate of the foot of the perpendicular from P to Y. The numbers a and b are called the *coordinates* of P, and we use the ordered pair of numbers (a, b) to indicate that a is the "X-coordinate" and b the "Y-coordinate." In Figure 3–2 the coordinates of P are (2.1, 3.4). It is customary to speak or write of the "point (a, b)" rather

Figure 3–2

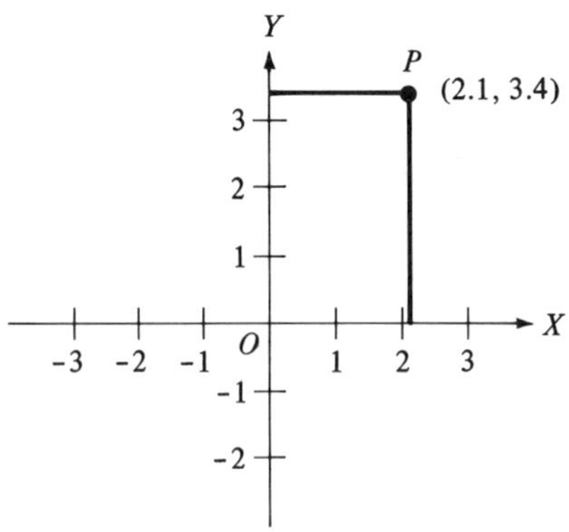

than the "point P with coordinates (a, b)." If we are given an ordered pair of numbers, we can locate a point by reversing the above process. For example, the point $(5, -2)$ is the intersection of the vertical line through the point "5" on X and the horizontal line through the point "-2" on Y. The fundamental assumption of analytic geometry is that this process determines a one-to-one correspondence between points in the plane and the set $R \times R$, the set of all ordered pairs of real numbers; that is, each point determines an ordered pair of numbers and each such pair determines a point. This correspondence is called a *coordinate system* for the plane.

Plotting points on a coordinate system is much easier if we use coordinate paper as in Figure 3–3.

Figure 3-3

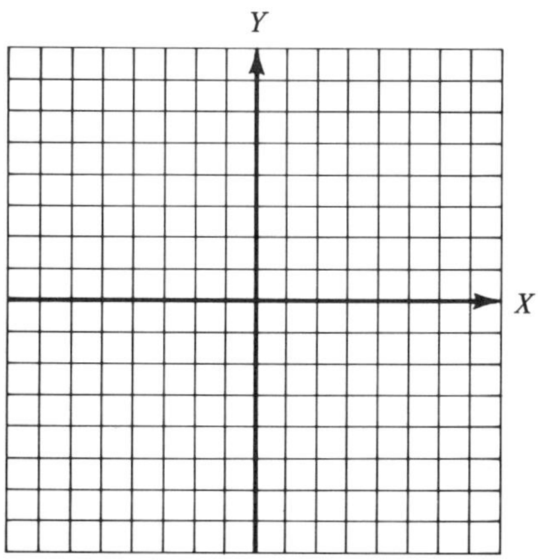

Since the coordinates of a point are determined by the distances from the point to the axes, we might expect the distance between two points to be determined by their coordinates and such is, in fact, the case. Before developing the general formula, let us consider a particular example.

Example 1

Determine the distance between (3, 2) and (7, 5). (See Figure 3–4.)

The points (3, 2), (7, 5), and (7, 2) form a right triangle with the hypotenuse being the line segment from (3, 2) to (7, 5). The two sides have lengths 3 and 4 and from the Pythagorean theorem we know that the length of the third side is given by

$$\sqrt{3^2 + 4^2} = \sqrt{9 + 16} = \sqrt{25} = 5 \quad \square$$

Figure 3–4

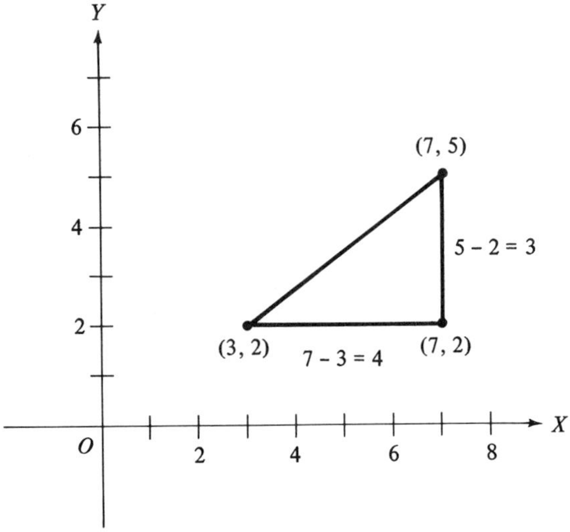

This example can be generalized to provide a simple formula for the distance between two points (x_1, y_1) and (x_2, y_2). Figure 3–5 shows $x_1 < x_2$ and $y_1 < y_2$ but the argument does not use this. The points (x_1, y_1), (x_2, y_2), and (x_2, y_1) form a right triangle with the right angle at (x_2, y_1). The sides have length $|y_2 - y_1|$ and $|x_2 - x_1|$ and, by the Pythagorean theorem, the

Figure 3–5

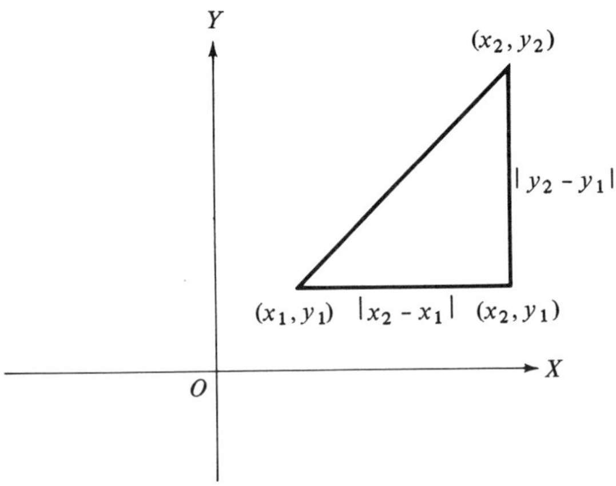

hypotenuse must have length given by

$$\sqrt{|x_2 - x_1|^2 + |y_2 - y_1|^2}$$

Since

$$|x_2 - x_1|^2 = (x_2 - x_1)^2$$

and

$$|y_2 - y_1|^2 = (y_2 - y_1)^2$$

the distance d between the points is given by the following *distance formula*:

$$d = \sqrt{(x_2 - x_1)^2 + (y_2 - y_1)^2}$$

Example 2

The distance from (5, 2) to (7, 9) is given by

$$d = \sqrt{(7 - 5)^2 + (9 - 2)^2} = \sqrt{2^2 + 7^2} = \sqrt{4 + 49} = \sqrt{53} \quad \square$$

Example 3

The distance from (4, 3) to (−5, 5) is given by

$$d = \sqrt{(-5 - 4)^2 + (5 - 3)^2} = \sqrt{(-9)^2 + 2^2} = \sqrt{81 + 4} = \sqrt{85} \quad \square$$

Example 4

The distance from (17, 1) to (13, 8) is given by

$$d = \sqrt{(17 - 13)^2 + (1 - 8)^2} = \sqrt{4^2 + 7^2} = \sqrt{16 + 49} = \sqrt{65} \quad \square$$

Notice that it makes no difference which of the two points is viewed as (x_1, y_1) and which one is viewed as (x_2, y_2).

EXERCISES

1. Compute the distances between each of the following pairs of points.
   (a) (6, 19), (11, 7) (b) (1, −4), (5, 9)
   (c) (−3, −7), (−5, 19) (d) (4.7, 3.9), (7.6, −4.2)
   (e) $(\sqrt{7}, \sqrt{3}), (\sqrt{3}, \sqrt{7})$ (f) $(\sqrt{11}, 2\sqrt{3}), (\sqrt{19}, \sqrt{5})$
   (g) $(a, b), (-a, -b)$ (h) $(a, b), (b, a)$
   (i) $(0, 0), (a, b)$ (j) $(0, 0), (a, a)$
2. (a) What is the set of all points having X-coordinate 5? Draw the set on a coordinate system.

(b) What is the set of all points having Y-coordinate 4? Draw the set on a coordinate system.

3. Use the distance formula to show that $\left(\frac{a+c}{2}, \frac{b+d}{2}\right)$ is the midpoint of the line segment from (a, b) to (c, d).
4. Compute the midpoints of the following segments.
   (a) the segment from (4, 7) to (16, 23)
   (b) the segment from (−2, 5) to (4, 11)
   (c) the segment from (3, −6) to (6, 10)
   (d) the segment from (−4, −9) to (−11, −23)
   (e) the segment from (6, 13) to (−2, −4)
5. (a) Find 5 points that are a distance of 5 units from (0, 0).
   (b) Draw the graph of all points that are 5 units from (0, 0).
   (c) Draw the graph of all points that are 5 units from (2, 3).
6. Use the distance formula to show that the triangle determined by (−1, 2), (2, 6), and (6, 9) is isosceles.
7. Show that the triangle determined by (5, 2), (3, 4), and (8, 9) is a right triangle.
8. Find the coordinates of the trisection points of the segment from (a, b) to (c, d). Prove your assertion.
9. Write an equation describing the set of all points (x, y) satisfying the condition that the distance from (x, y) to (5, 0) plus the distance from (x, y) to (−5, 0) is 20.
10. Write an equation describing the set of all points (x, y) satisfying the condition that the distance from (x, y) to (2, 3) plus the distance from (x, y) to (−4, 3) is 16.
11. Write an equation describing the set of all points (x, y) satisfying the condition that the distance from (x, y) to (2, 7) is the same as the distance from (x, y) to the line $y = 1$.
12. Write an equation describing the set of all points (x, y) satisfying the condition that the difference in the distance from (x, y) to (4, 0) and the distance from (x, y) to (−4, 0) is 1.

2. Sketching Graphs

The equation

$$y = x + 2$$

cannot be solved uniquely for x or for y. There are infinitely many pairs (x, y) which are solutions and, in fact, with each real number we substitute for x, we obtain a unique value for y. If we plot all such possible ordered pairs on a coordinate system, we get a *graph* of the equation. A graph of an equation is just a "picture" of the solution set of an equation. The graph of $y = x + 2$ is given by Figure 3–6.

The basic technique in graphing equations is to compute enough points to give us an idea of the shape of the graph. We then plot these and sketch the rest of the drawing.

Figure 3–6

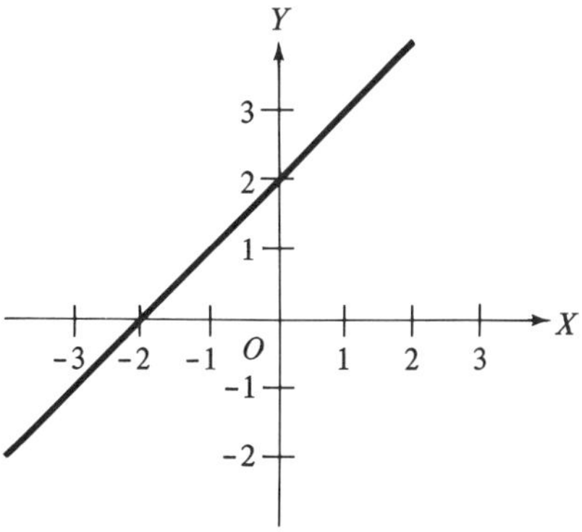

Example 1

Draw the graph of $y = x^2$.

We make a table of values of x with the corresponding values of y. After plotting these, we then sketch the rest of the graph. See Figure 3–7. □

Figure 3–7

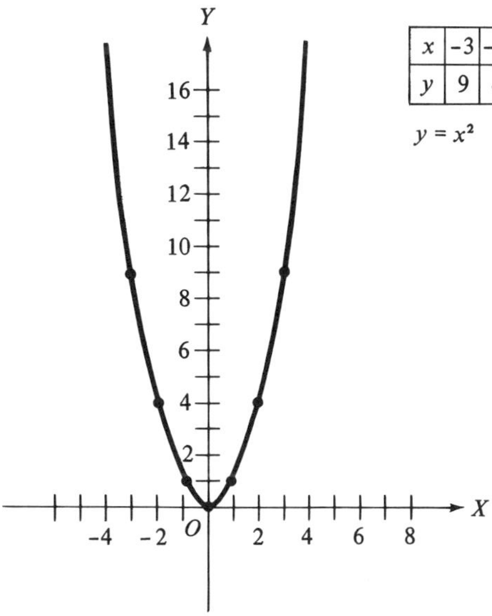

x	-3	-2	-1	0	1	2	3
y	9	4	1	0	1	4	9

$y = x^2$

In the exercises of this section you are asked to sketch several graphs by this method. Sections 3–7 will be concerned with some very special graphs which depend on our geometric and algebraic knowledge.

Example 2

$y = x^3 - x$

See Figure 3–8. □

Figure 3-8

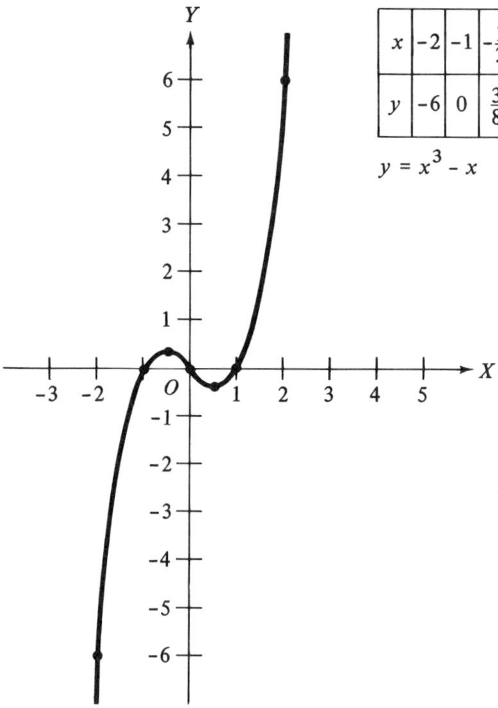

x	-2	-1	$-\frac{1}{2}$	0	$\frac{1}{2}$	1	2
y	-6	0	$\frac{3}{8}$	0	$-\frac{3}{8}$	0	6

$y = x^3 - x$

Example 3

$$y = \frac{1}{x}$$

See Figure 3–9. □

Figure 3-9

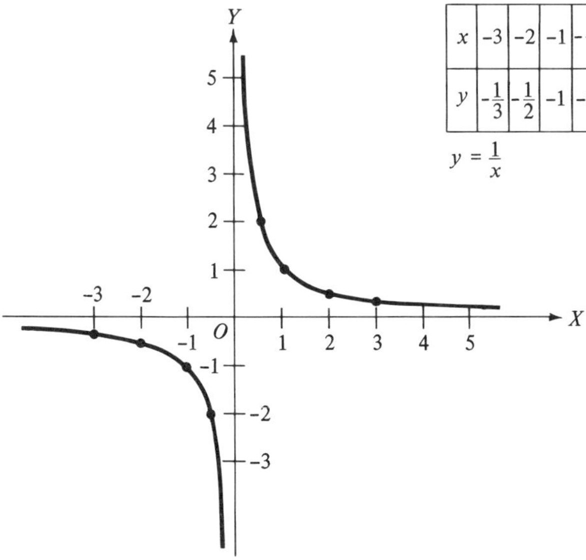

x	-3	-2	-1	$-\frac{1}{2}$	$\frac{1}{2}$	1	2	3
y	$-\frac{1}{3}$	$-\frac{1}{2}$	-1	-2	2	1	$\frac{1}{2}$	$\frac{1}{3}$

$y = \frac{1}{x}$

EXERCISES

1. Sketch the graphs of the following equations.
   (a) $y = 2x - 3$ (b) $y = -3x + 1$ (c) $y = 5x - 4$
   (d) $y = \frac{3}{4}x + 2$ (e) $y = 7x + 4$ (f) $y = -\frac{6}{5}x - 2$
   (g) What do you think is the shape of the graph of $y = mx + b$? Can you give two points on the graph?
2. Sketch the graphs of the following equations.
   (a) $2x + y = 2$ (b) $3x - y = 3$ (c) $4x + 3y = 12$
   (d) $4x + 5y = 20$ (e) $x/2 + y/3 = 1$ (f) $x/6 + y/5 = 1$
   (g) What do you think is the shape of the graph of $x/a + y/b = 1$? Can you give two points on the graph?
3. Sketch the graphs of the following equations.
   (a) $y = 2x^2$ (b) $y = x^2 + 1$ (c) $y = x^3$
   (d) $y = -x^3$ (e) $y = x^3 - 2$ (f) $y = (x - 1)^2$
   (g) $y = x^2 - 2x + 3$ (h) $y = x^3 - x$ (i) $x^2 - y = 4$
   (j) $x^2 + y^2 = 25$ (k) $y = 1/x^2$ (l) $y = 2/(x^2 + 3)$
   (m) $y = -x/(x^2 + 2)$ *(n) $y = 5/(x^2 - 4)$
4. Sketch the graphs of the following equations.
   (a) $y = 3x^2$ (b) $y = 2x^2 + 1$ (c) $y = -2x^3$
   (d) $y = -x^2$ (e) $y = (x + 2)^2$ (f) $y = (x + 2)^2 + 1$
   (g) $y = 2x^3 + 1$ (h) $y = 3(x - 1)^2$ (i) $y - x^2 = 5$
   (j) $y^2 + x^2 = 16$ (k) $y = 3/x^2$ (l) $y = 1/(x^2 + 2)$
   (m) $y = x/(x^2 + 1)$ *(n) $y = 4/(x^2 - 1)$

5. Plot the points (1, 2) and (7, 11) and draw the straight line through them. Now choose any other point (x, y) on the line and compute $(y - 2)/(x - 1)$, $(y - 11)/(x - 7)$. Choose another point (x, y) and make the same computations. What can you say about your results? What do you think will happen if you do the same thing with with a third point on the line?
6. If y varies directly with x and $y = 8$ when $x = 3$, draw the graph of the equation that gives the relationship between x and y.
7. If y varies inversely with x and $y = 4$ when $x = 2$, draw the graph of the equation that gives the relationship between x and y.
8. If y varies directly with $x^2 + 1$ and $y = 10$ when $x = 2$, draw the graph of the equation that gives the relationship between x and y.
9. If y varies inversely with x^2 and $y = 4$ when $x = 1$, draw the graph of the equation that gives the relationship between x and y.

3. Lines

Suppose we plot two points (x_1, y_1) and (x_2, y_2) and draw the line through them. Furthermore, let us assume that $x_1 < x_2$ and $y_1 \neq y_2$. Our drawing will be similar to Case I illustrated in Figure 3–10 if $y_1 < y_2$ and similar to Case II if $y_2 < y_1$.

Figure 3–10

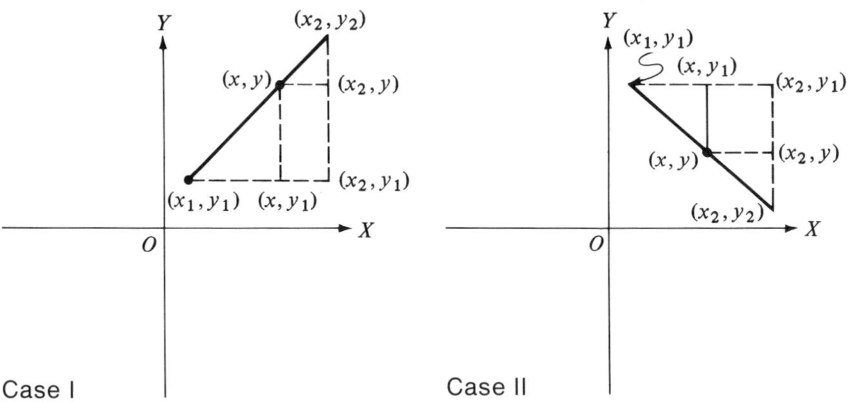

In either case, the points (x_1, y_1), (x_2, y_2), and (x_2, y_1) determine a right triangle. Now suppose we have any point (x, y) on the line through the points (x_1, y_1) and (x_2, y_2). The triangle determined by (x_1, y_1), (x, y), and (x, y_1) is similar to the triangle formed by (x_1, y_1), (x_2, y_2), and (x_2, y_1) and, therefore, the corresponding sides are proportional.

Thus, in Case I, we have

$$\frac{y - y_1}{x - x_1} = \frac{y_2 - y_1}{x_2 - x_1}$$

and in Case II, we have

$$\frac{y_1 - y}{x - x_1} = \frac{y_1 - y_2}{x_2 - x_1}$$

Multiplying both sides of this equation by -1 yields

$$\frac{y - y_1}{x - x_1} = \frac{y_2 - y_1}{x_2 - x_1}$$

The drawings in Figure 3–10 illustrate only the case when (x, y) is on the segment between (x_1, y_1) and (x_2, y_2). If (x, y) is on the line but not between the points, the same equation results and we leave the necessary illustrations and arguments to the student. Furthermore, if $y_1 = y_2$, then $y = y_1 = y_2$ and both sides of the equation are 0. If $x_2 < x_1$, similar arguments yield the same equation.

In summary, then, if (x_1, y_1) and (x_2, y_2) are two points with $x_1 \neq x_2$, then any other point (x, y) on the line through them must satisfy the equation

$$\text{I.} \quad \frac{y - y_1}{x - x_1} = \frac{y_2 - y_1}{x_2 - x_1}$$

Example 1

Every point [except (3, 4)] on the line through (3, 4) and (4, 7) must satisfy the equation

$$\frac{y - 4}{x - 3} = \frac{7 - 4}{4 - 3}$$

that is,

$$\frac{y - 4}{x - 3} = 3 \; \square$$

Example 2

Every point [except (11, 2)] on the line through (11, 2) and (9, 7) must satisfy

$$\frac{y - 2}{x - 11} = \frac{7 - 2}{9 - 11}$$

that is,

$$\frac{y - 2}{x - 11} = -\frac{5}{2}$$

See Figure 3–11. □

Figure 3-11

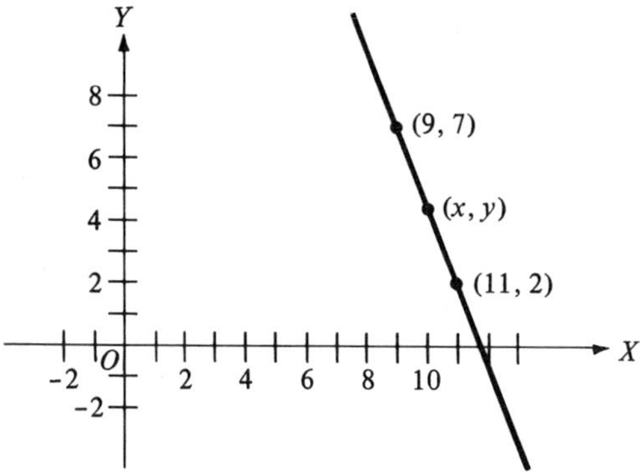

Suppose (x_3, y_3) and (x_4, y_4) are two other points on the line through (x_1, y_1) and (x_2, y_2). By our arguments above

$$\frac{y_3 - y_1}{x_3 - x_1} = \frac{y_2 - y_1}{x_2 - x_1}$$

because (x_3, y_3) is on the line through (x_1, y_1) and (x_2, y_2) and

$$\frac{y_1 - y_3}{x_1 - x_3} = \frac{y_4 - y_3}{x_4 - x_3}$$

because (x_1, y_1) is on the line through (x_3, y_3) and (x_4, y_4). Since the left sides of these equations are equal, the right sides are also and, therefore,

$$\frac{y_2 - y_1}{x_2 - x_1} = \frac{y_4 - y_3}{x_4 - x_3}$$

Thus the number

$$\frac{y_2 - y_1}{x_2 - x_1}$$

is determined not just by the points (x_1, y_1) and (x_2, y_2) but by any two points on the line. This number is called the *slope* of the line. See Figure 3–12.

Figure 3–12

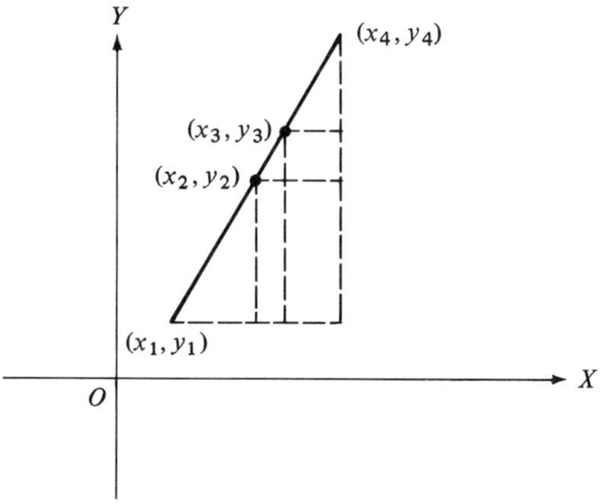

Example 3

The slope of the line through (4, 7) and (8, 19) is

$$\frac{19 - 7}{8 - 4} = \frac{12}{4} = 3 \quad \square$$

Example 4

The slope of the line through (14, 6) and (10, 3) is

$$\frac{6 - 3}{14 - 10} = \frac{3}{4} \quad \square$$

Example 5

The slope of the line through (−11, 2) and (4, 14) is

$$\frac{14 - 12}{4 - (-11)} = \frac{2}{15} \quad \square$$

Intuitively, the slope of a line can be viewed as "the rise divided by the run." The "rise" is the difference between the Y-coordinates and the "run" is the difference in X-coordinates. If the "run" is positive, then the "rise" is positive if the line "runs uphill to the right" and is negative if the line "runs downhill to the right." See Figure 3–13. It is customary to denote the slope

Figure 3-13

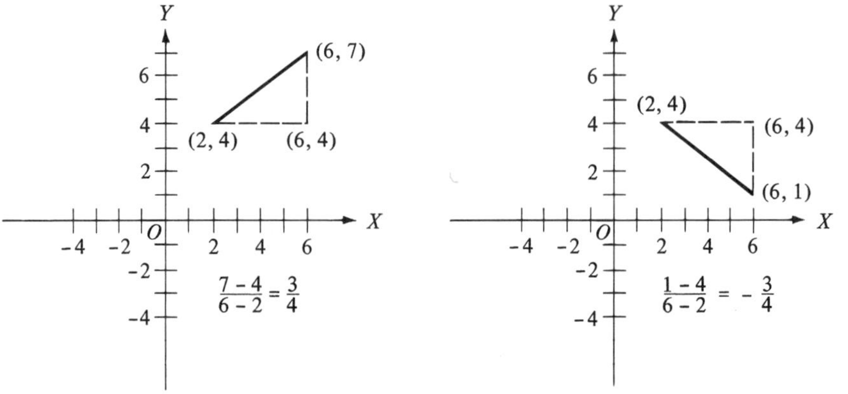

of a line by the letter m. Thus

$$\frac{y_2 - y_1}{x_2 - x_1} = m$$

Equation (I) becomes

$$\frac{y - y_1}{x - x_1} = m$$

Any point satisfying this equation satisfies the equation

$$\text{II.} \quad y - y_1 = m(x - x_1)$$

and the point (x_1, y_1) does also. Thus Equation (II) is satisfied by every point on the line through (x_1, y_1) with slope m. Conversely if (p, q) is any point satisfying Equation (II) and $p \neq x_1$, then

$$\frac{q - y_1}{p - x_1} = m$$

and (p, q) is on the line. Thus the set of all points satisfying Equation (II) is precisely the line through (x_1, y_1) with slope m. Equation (II) is called the *point-slope form* of the equation of a line.

Since two points determine a line, they determine an equation for the line. We know that for points (x_1, y_1) and (x_2, y_2) with $x_1 \neq x_2$, the slope is given by

$$m = \frac{y_2 - y_1}{x_2 - x_1}$$

Thus we have

$$\text{III.} \quad y - y_1 = \frac{y_2 - y_1}{x_2 - x_1}(x - x_1)$$

This is called the *two-point form* of the equation of a line.

Using the two forms above, if we know one point on a line and its slope or if we know two points on a line, we can write an equation for the line.

Example 6

An equation of the line through (2, 4) with slope 5 is

$y - 4 = 5(x - 2)$ □

Example 7

An equation for the line through (4, −6) with slope $\frac{2}{3}$ is

$y + 6 = \frac{2}{3}(x - 4)$ □

Example 8

An equation for the line through (14, 6) and (11, 16) is

$$y - 6 = \frac{16 - 6}{11 - 14}(x - 14)$$

or

$$y - 6 = \frac{-10}{3}(x - 14) \quad \square$$

Example 9

An equation for the line through (5, 7) and (−11, 14) is

$$y - 7 = \frac{7 - 14}{5 - (-11)}(x - 5)$$

or

$$y - 7 = \frac{-7}{16}(x - 5) \quad \square$$

Example 10

An equation for the line through (0, 6) with slope $\frac{1}{4}$ is given by

$$y - 6 = \tfrac{1}{4}(x - 0)$$

or

$$y = \tfrac{1}{4}x + 6 \ \square$$

Example 10 leads to a special case of the point-slope form of the equation. If we have a line with slope m that intersects the Y-axis at the point $(0, b)$, then an equation for this line is given by

$$y - b = m(x - 0)$$

or

$$y = mx + b$$

Conversely, the equation

$$y = mx + b$$

is the equation of the line through $(0, b)$ and $(1, m + b)$ and

$$\frac{(m + b) - b}{1 - 0} = m$$

An equation of the form

$$y = mx + b$$

is said to be the *slope-intercept form* of the equation for a line. The number b is called the Y-*intercept* since the line meets the Y-axis at $(0, b)$. This form is very convenient for several reasons. First of all, we can read the slope and one point directly from the equation.

Example 11

The graph of $y = 4x + 2$ is a line through (0, 2) with slope 4 because 2 is the Y-intercept and $m = 4$. □

Example 12

The graph of $y = -\frac{2}{3}x - 7$ is a line through $(0, -7)$ with slope $-\frac{2}{3}$. □

Notice that, since the Y-intercept and the slope determine a line, two equations of this form have the same graph if and only if the equations are the same; that is,

$$y = mx + b \quad \text{and} \quad y = px + q$$

can have the same graph if and only if $m = p$ and $b = q$.

In order to draw the graph of the equation of a line, we need only to plot two of its points and draw the line through them. If we have an equation of the form

$$y = mx + b$$

we immediately know that $(0, b)$ is on the graph. We can then use m to determine another point. For example, $(1, m + b)$ will also be on the graph. In drawing graphs of lines, our picture usually is better if we do not choose points that are close together. Thus if $|m|$ is small, the points $(0, b)$ and $(1, m + b)$ do not yield as good a drawing as, say, $(0, b)$ and $(10, 10m + b)$. Since the slope is the "rise divided by the run," if m is a rational number in fractional form, we can "change the X-coordinate by the denominator and change the Y-coordinate by the numerator."

Example 13

To draw the graph of

$$y = 4x + 2$$

draw a line through $(0, 2)$ and $(1, 6)$. See Figure 3–14. □

Example 14

To draw the graph of

$$y = -\tfrac{2}{3}x - 7$$

draw a line through

$$(0, -7) \quad \text{and} \quad (3, -9)$$

In this case we "change x by 3 and y by -2." See Figure 3–14. □

Suppose that we have two different lines with the same slope and the equations

$$y = mx + b \quad \text{and} \quad y = mx + d$$

If the two lines meet at some point, say where $x = a$, then

$$(a, ma + b) = (a, ma + d)$$

Figure 3–14

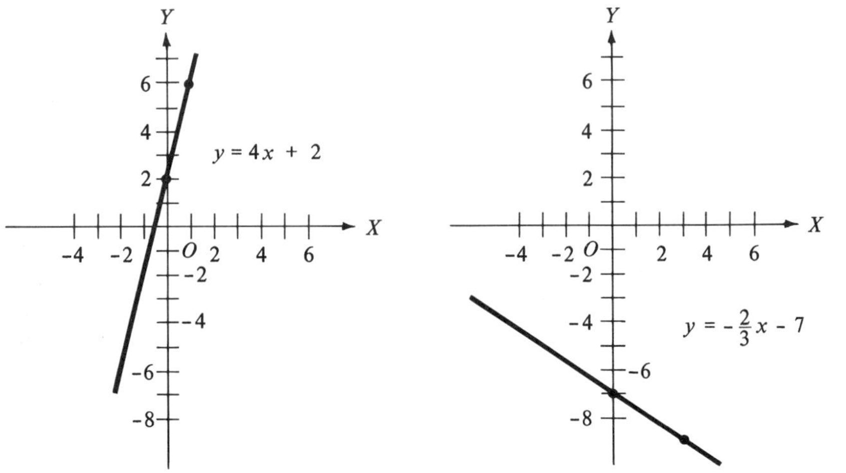

Thus

$$ma + b = ma + d$$
$$b = d$$

and the lines are the same, contrary to our assumption. Thus if two different lines have the same slope, they do not meet and, hence, are parallel.

Conversely, if

$$y = mx + b \quad \text{and} \quad y = nx + d$$

are equations of two lines and $m \neq n$, then they meet at the point

$$\left(\frac{d - b}{m - n}, \frac{md - nb}{m - n}\right)$$

(We leave it to the reader to verify that this point is on both lines.) Thus *two lines are parallel if and only if they have the same slope.*

Two lines with equations

$$y = mx + b \quad \text{and} \quad y = nx + d$$

are perpendicular if and only if $nm = -1$. We will not prove this in the text but will ask the reader to do so in an exercise.

Example 15

The graph of $y = 4x + 2$ is parallel to the graph of $y = 4x - 1$ because they have the same slope but different Y-intercepts. See Figure 3–15. □

Figure 3–15

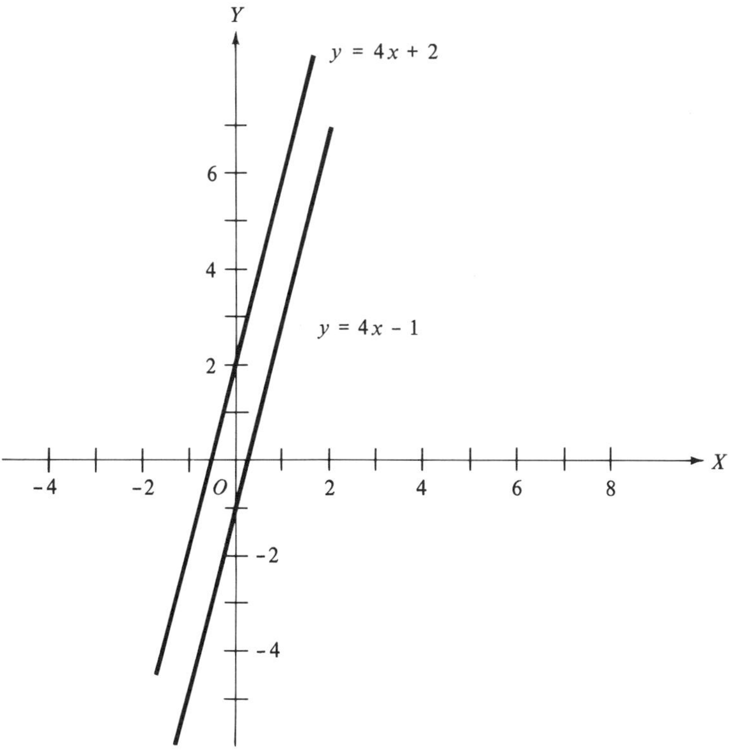

Example 16

Write the equation of a line through (3, 8) parallel to the graph of

$y = 5x - 1$

The slope is 5. Using the point-slope form we have

$y - 8 = 5(x - 3)$ □

Example 17

An equation for the line through (2, 7) perpendicular to the graph of $y = 6x + 1$ is

$y - 7 = -\frac{1}{6}(x - 2)$ □

A horizontal line has the property that all of its Y-coordinates are the same.

Example 18

The horizontal line through (4, 7) is just the set of all points with Y-coordinate 7 and, therefore, has the equation

$$y = 7$$

See Figure 3–16. □

In general, the equation of the horizontal line through (h, k) is

$$y = k$$

Notice that this can be rewritten

$$y = 0x + k$$

As we would expect, the slope is 0 since if (p, k) and (q, k) are any points on the line, the slope is given by $(k - k)/(p - q) = 0$. Conversely, if the slope of a line is 0, then two points must have the same Y-coordinate. Hence all points will have the same Y-coordinate and the line is horizontal. *Thus a line is horizontal if and only if it has 0 slope.*

Similarly, the vertical line through (h, k) is just the set of all points with X-coordinate h and its equation is

$$x = h$$

Example 19

The equation of the vertical line through (4, 7) is

$$x = 4$$

See Figure 3–16. □

Figure 3–16

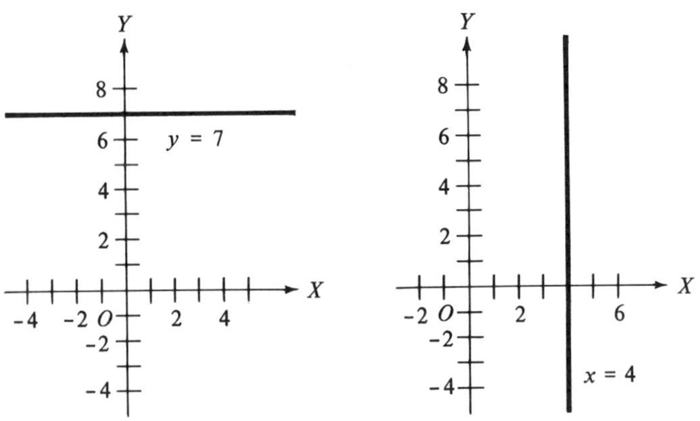

Since all X-coordinates of a vertical line are the same, if (x_1, y_1) and (x_2, y_2) are any two points on the line, then $x_1 = x_2$ and the expression

$$\frac{y_2 - y_1}{x_2 - x_1}$$

is meaningless because the denominator is 0.

Conversely, the slope can fail to exist only if all points have the same X-coordinate and the line is vertical. *Thus a line is vertical if and only if it has no slope.*

An equation of the form

$$px + qy = c$$

with $p \neq 0$ or $q \neq 0$ can be transformed into one of the form

$$y = mx + b \quad \text{or} \quad x = h$$

Thus equations of this form have graphs that are lines. We can draw the graphs directly from the original equation by computing two points or we can transform it to one of the standard forms and use the techniques previously established.

Example 20

Draw the graph of $4x + 6y = 12$.

If $x = 0$, $y = 2$.

If $y = 0$, $x = 3$.

Thus (0, 2) and (3, 0) are two points on the graph. We plot them and draw the line. If we solve this equation for y, we would obtain the following equation.

$$y = -\tfrac{2}{3}x + 2$$

This equation also easily yields the points (0, 2) and (3, 0). See Figure 3–17. □

An equation of the form

$$\frac{x}{a} + \frac{y}{b} = 1$$

is equivalent to the equation

$$bx + ay = ab$$

which is the equation of a line. For $x = 0$, we have

$$\frac{0}{a} + \frac{y}{b} = 1$$

Figure 3–17

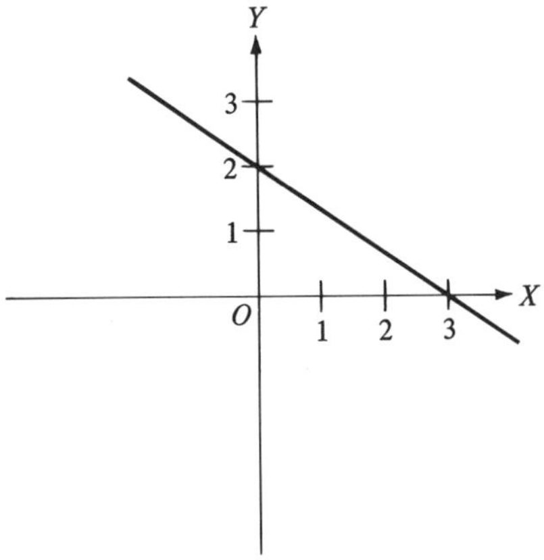

and $y = b$. Similarly, for $y = 0$,

$$\frac{x}{a} + \frac{0}{b} = 1$$

and $x = a$. Thus b is the Y-intercept and a is the X-intercept. The form

$$\frac{x}{a} + \frac{y}{b} = 1$$

is called the *intercept form* of the equation of a line.

Example 21

The graph of

$$\frac{x}{4} + \frac{y}{7} = 1$$

is the line through (4, 0) and (0, 7). See Figure 3–18. □

In summary:

(i) If (x_1, y_1) and (x_2, y_2) are two points on a line, the slope m of the line is given by

$$m = \frac{y_2 - y_1}{x_2 - x_1}$$

(ii) $y - y_1 = m(x - x_1)$ is the point-slope form of the equation of the line with slope m through the point (x_1, y_1).

(iii) $y = mx + b$ is the slope-intercept form of the equation of a line with slope m and Y-intercept b.

Figure 3-18

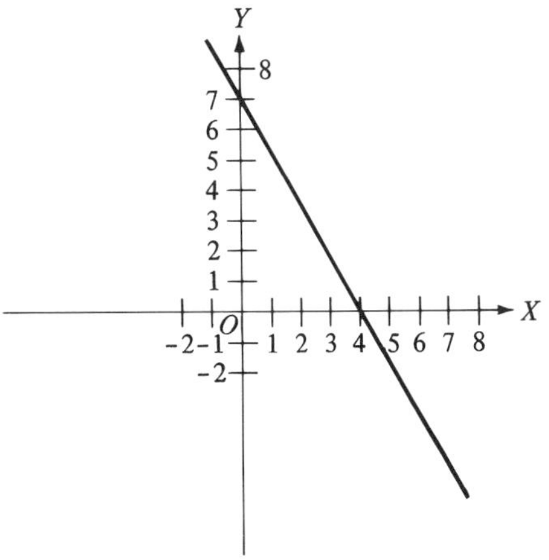

(iv) $x/a + y/b = 1$ is the intercept form of the equation of a line with X-intercept a and Y-intercept b.
(v) $y = k$ is the equation of the horizontal line through the point (h, k).
(vi) $x = h$ is the equation of the vertical line through (h, k).

EXERCISES

1. Draw graphs of the following equations.
   (a) $y = 2x + 3$ (b) $y = 2x - 1$ (c) $y = \frac{3}{4}x + 2$
   (d) $y = -\frac{4}{3}x - 1$ (e) $y = 5$ (f) $x = 2$
   (g) $x/3 + y/4 = 1$ (h) $y - 5 = 2(x - 3)$
2. Draw graphs of the following equations.
   (a) $y - 3x = 4$ (b) $2y - 3x = 4$ (c) $6y + 5x = 30$
   (d) $6y + 5x = 2$ (e) $16y + 32x = 96$ (f) $5x + 7y = 11$
   (g) $11x + 4y = 3$ (h) $\sqrt{2}x + 5y = 3$ (i) $\sqrt{5}x - \sqrt{3}y = \sqrt{7}$
   (j) $\pi x + \pi^2 y = \pi^3$
3. Write an equation for the line:
   (a) through (0, 3) with slope 5
   (b) through (0, −5) with slope −3
   (c) through (5, 2) with slope $-\frac{2}{3}$
   (d) through (1, 5) with slope 10
   (e) through (7, 1) and (2, 3)
   (f) through (6, 1) and (8, 5)
   (g) through (0, 4) and (1, 8)
   (h) through (4, 7) parallel to $x = 5$
   (i) through (1, 5) parallel to the line with equation $y = 5x - 4$
   (j) through (1, 5) perpendicular to the line with equation $y = 5x - 4$
   (k) which is horizontal through (7, 3)
   (l) which is vertical through (7, 3)

4. Write an equation for the line:
   (a) through (0, 7) with slope -3
   (b) through (0, -3) with slope 8
   (c) through (7, 9) with slope 17
   (d) through (2, 6) with slope $\frac{4}{5}$
   (e) through (4, 7) and (8, 13)
   (f) through (8, -3) and (13, 5)
   (g) through (0, -7) and (4, 10)
   (h) through (5, 8) parallel to the line $y = 2$
   (i) through (4, 9) parallel to the line with equation $y = 6x + 5$
   (j) through (7, -4) perpendicular to the line with equation $y = 6x + 2$
   (k) which is horizontal through (9, 11)
   (l) which is vertical through (9, 11)
5. (a) If $m \neq n$, verify that

$$\left(\frac{d-b}{m-n}, \frac{md-nb}{m-n}\right)$$

   is a point on the graphs of

$$y = mx + b \quad \text{and} \quad y = nx + d$$

   (b) Explain why the point

$$\left(\frac{d-b}{m-n}, \frac{md-nb}{m-n}\right)$$

   is the only point on both graphs.
6. Complete the argument that

$$\frac{y - y_1}{x - x_1} = \frac{y_2 - y_1}{x_2 - x_1}$$

   for the cases where (x, y) is a point on the line through the points (x_1, y_1), (x_2, y_2) but not on the segment between them.
7. If p is any number, then $(p, mp + b)$ and $(-p, -mp + b)$ are two points on the graph of $y = mx + b$ which are equidistant from $(0, b)$. To prove that the graph of $y = (-1/m)x + b$ is perpendicular to the graph of $y = mx + b$, it suffices to show that each point on the graph is equidistant from these two points. Do this by letting q be any real number and use the distance formula to show that $(q, (-1/m)q + b)$ is equidistant from the two points $(p, mp + b)$ and $(-p, -mp + b)$.

   Since any line perpendicular to the graph of $y = mx + b$ must be parallel to every other such line, all such lines are parallel to the graph of $y = (-1/m)x + b$ and, therefore, must have slope $-1/m$.

4. Linear Variation

Two quantities, say x and y, are said to *vary linearly* with respect to each other if the relationship between them can be expressed by a linear equation; that is, if the graph of the set of all corresponding values is a line. Since two points determine a line, two sets of values determine the relationship if we know that the variation is linear.

Example 1

If y varies linearly with x and we know that $y = 5$ when $x = 2$ and $y = 7$ when $x = 6$, then we can use the two-point form of the equation of a line to determine the equation that gives the relationship:

$$y - 5 = \frac{7 - 5}{6 - 2}(x - 2)$$

or

$$\begin{aligned} y - 5 &= \tfrac{1}{2}(x - 2) \\ y &= \tfrac{1}{2}x + 4 \quad \square \end{aligned}$$

Example 2

Suppose that experience shows that the sales of donuts in your shop vary linearly with the price, at least for prices between \$.50 and \$1.00. Determine this relationship if the sales at \$.50 are 500 dozen per day and the sales at \$1.00 are 200 dozen per day.

Again using the two-point form we have

$$\begin{aligned} S - 500 &= \frac{500 - 200}{.50 - 1.00}(P - .50) \\ &= -600(P - .50) \\ S &= -600P + 800 \end{aligned}$$

If we set the price at \$.75, we will sell $S = -600(.75) + 800 = 350$ dozen. $\square$

EXERCISES

1. Suppose y varies linearly with x and assume that $y = 2$ when $x = 7$ and $y = 8$ when $x = 4$. Determine an equation that gives the relationship between x and y.
2. Suppose y varies linearly with x and assume that $x = 9$ when $y = 4$ and $x = 13$ when $y = 8$. Determine an equation that gives the relationship between x and y.
3. Suppose y varies linearly with x and $y = 0$ when $x = 0$. Is this a variation of one of the types studied earlier?
4. The price of gasoline varies linearly with the octane rating. If 91 octane sells for 53.9¢ per gallon and 97 octane sells for 59.9¢ per gallon, what should be the price per gallon of 95 octane? (You may find it interesting to try this problem using the prices posted at one of the stations where you do business.)
5. The number of men enlisting in the army varies linearly with the unemployment rate in their city of residence. If 1200 volunteered from Gotham City when the unemployment rate was 4 percent, and 1350 enlisted when the rate was 5.2 percent, how many will enlist if the rate is 6 percent? How many would enlist if there was full employment, that is, 0 percent unemployed?

6. The rent charged some stores varies linearly with the amount of sales the store makes.
   (a) If you have such a store, determine the relationship between the rent R and sales S if $R = \$500$ when $S = \$15{,}000$ and $R = \$550$ when $S = \$17{,}000$.
   (b) What would the rent be if sales were \$20,000?
   (c) What is the minimum rent, that is, the rent on zero sales?
7. Air fares vary linearly with the distance traveled (approximately, anyway). If the fare from Detroit to Chicago is \$32 for a distance of 250 miles, and the fare from Detroit to New York is \$52 for a distance of 700 miles, determine the approximate fare from Detroit to Miami, a distance of 1300 miles.
8. Costs in business are of two basic types, fixed costs and variable costs. Fixed costs are those that do not change with the amount of business activity. Examples of fixed costs might be rent, property taxes, fire insurance, or light and heat costs. Variable costs might be material costs, labor costs, excise taxes, or transportation expenses. In many cases, costs vary linearly with the amount of goods sold and we will assume so in this problem.
   (a) If the cost of producing 10,000 radios in a certain factory is \$170,000 and the cost of producing 12,000 radios is \$200,000, express the costs in terms of N, the number of radios produced.
   (b) What is the cost of producing 15,000 radios?
   (c) What are the fixed costs of the business; that is, what are the costs if zero radios are produced?
9. Suppose the profits in a business vary linearly with sales and that profits are \$4000 a month on sales of \$100,000, and \$5000 on sales of \$110,000.
   (a) Let y denote profits and x sales, and express the relationship between profits and sales.
   (b) What would be the monthly profits on sales of \$130,000?
   (c) What sales are necessary in order to break even, that is, produce a zero profit?
10. The cost of a bottle of a particular soft drink should vary linearly with the amount of drink in the bottle.
    (a) If a 12-oz bottle of the drink sells for 14¢ and a 16-oz bottle sells for 17¢, what should be the price of a 26-oz bottle?
    (b) What should be the price of an 8-oz bottle?

5. Circles and Ellipses

If a plane meets a right circular cone as illustrated in Figure 3–19, the intersection is called a *conic section*. These curves are quite common and their properties can be studied using only the geometric definition. In this section we will give alternate definitions for two of these curves using the concept of distance and will derive the general forms of their equations by using the distance formula.

The most familiar conic section is a circle. In geometry, a circle with center P and radius r is defined to be the set of all points in the plane that are a distance of r units from P.

Figure 3-19

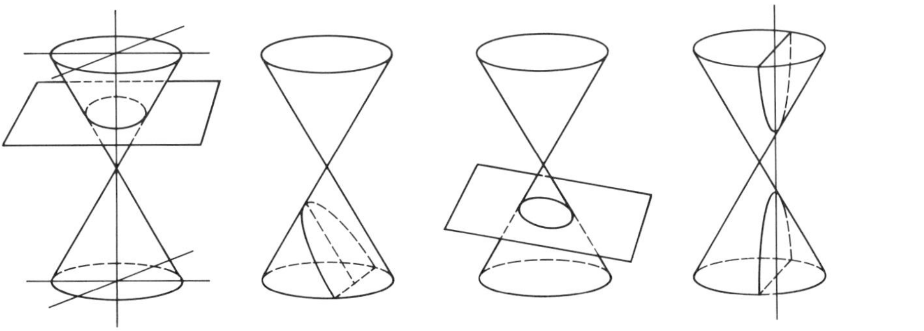

Figure 3-20

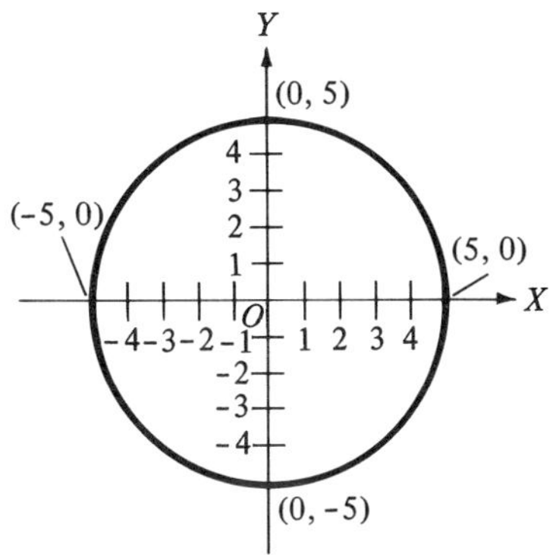

Example 1

Find an equation whose graph is a circle with center at (0, 0) and radius 5.
This circle is

$\{(x, y)$: the distance from (x, y) to $(0, 0)$ is $5\}$

By the distance formula this is

$$\{(x, y): \sqrt{(x - 0)^2 + (y - 0)^2} = 5\} = \{(x, y): \sqrt{x^2 + y^2} = 5\}$$
$$= \{(x, y): x^2 + y^2 = 25\}$$

Thus $x^2 + y^2 = 25$ is an equation whose graph is the circle with radius 5 and center at (0, 0). See Figure 3–20. □

In the more general case, the circle with center (0, 0) and radius r is the set of points whose distance from (0, 0) is r.

$$\begin{aligned}\{(x, y)\text{: the distance from } (x, y) \text{ to } (0, 0) \text{ is } r\} &= \{(x, y)\colon \sqrt{(x-0)^2 + (y-0)^2} = r\} \\ &= \{(x, y)\colon \sqrt{x^2 + y^2} = r\} \\ &= \{(x, y)\colon x^2 + y^2 = r^2\}\end{aligned}$$

Thus

$$x^2 + y^2 = r^2$$

is an equation whose graph is a circle with center at (0, 0) and radius r.

Example 2

$x^2 + y^2 = 9$ is an equation for the circle with center at the origin and radius 3. □

Example 3

$x^2 + y^2 = \frac{49}{4}$ is an equation for the circle with center at the origin and radius $\frac{7}{2}$. □

Example 4

What is an equation for the circle with center (4, 1) and radius 5? The circle is the set

$$\{(x, y)\colon \sqrt{(x-4)^2 + (y-1)^2} = 5\} = \{(x, y)\colon (x-4)^2 + (y-1)^2 = 25\}$$

Thus

$$(x-4)^2 + (y-1)^2 = 25$$

is the desired equation. See Figure 3–21. □

Figure 3–21

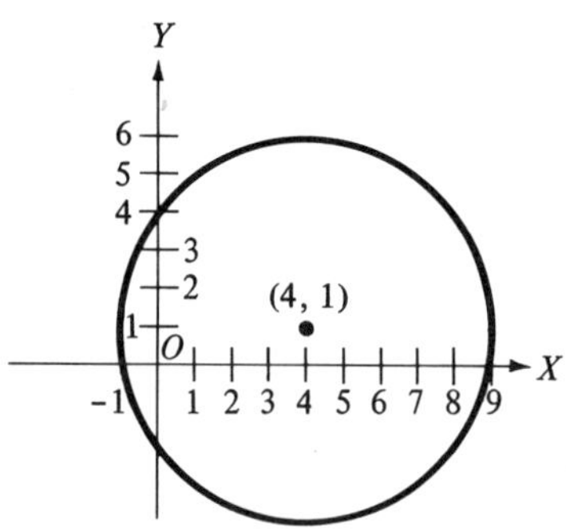

We can generalize this result to a circle with radius r and center at the point (h, k). This circle is

$$\{(x, y)\text{: the distance from } (x, y) \text{ to } (h, k) \text{ is } r\}$$
$$= \{(x, y)\colon \sqrt{(x-h)^2 + (y-k)^2} = r\}$$
$$= \{(x, y)\colon (x-h)^2 + (y-k)^2 = r^2\}$$

Thus

$$(x-h)^2 + (y-k)^2 = r^2$$

is an equation for the circle with center (h, k) and radius r. To graph such an equation, we need only to locate the center, determine the radius, and use a compass.

Example 5

$$(x-2)^2 + (y-3)^2 = 16$$

See Figure 3–22. □

Figure 3-22

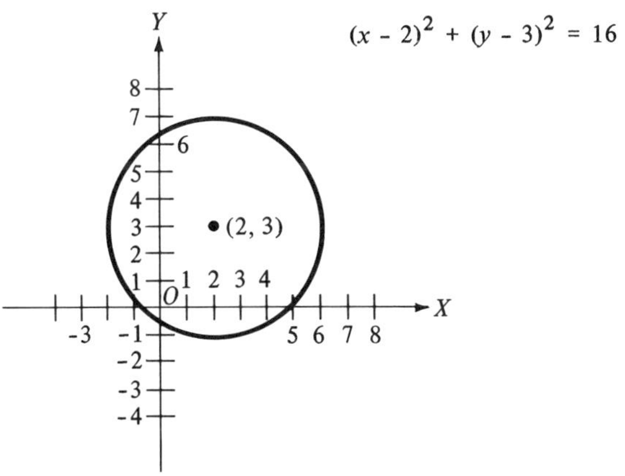

Example 6

$$(x+3)^2 + (y-\tfrac{1}{2})^2 = 4$$

See Figure 3–23. □

Sometimes an equation does not fall immediately into this pattern but can be transformed into one of the desired form.

Figure 3-23

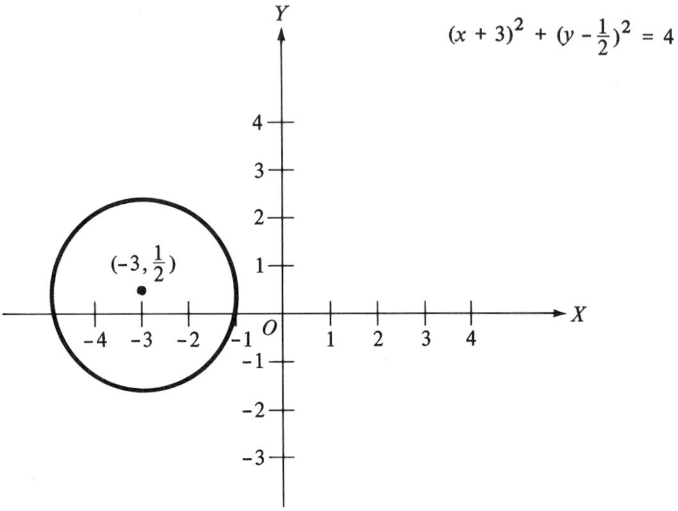

Example 7

$$x^2 - 2x + y^2 - 6y = 6$$
$$x^2 - 2x + 1 + y^2 - 6y + 9 = 6 + 1 + 9$$
$$(x - 1)^2 + (y - 3)^2 = 16 \ \square$$

The graph of this equation is a circle with center (1, 3) and radius 4. In transforming the equation, we used the "completing the square technique" twice in that we added suitable numbers to both sides in order to make the left side the sum of two perfect squares of the desired form.

Example 8

$$x^2 - 4x + y^2 - 10y = 7$$
$$x^2 - 4x + 4 + y^2 - 10y + 25 = 7 + 4 + 25$$
$$(x - 2)^2 + (y - 5)^2 = 36$$

The graph is a circle with center (2, 5) and radius 6. □

Example 9

$$x^2 + 5x + y^2 - 3y = 4$$
$$x^2 + 5x + \frac{25}{4} + y^2 - 3y + \frac{9}{4} = 4 + \frac{25}{4} + \frac{9}{4}$$
$$\left(x + \frac{5}{2}\right)^2 + \left(y - \frac{3}{2}\right)^2 = \frac{50}{4}$$

Figure 3–24

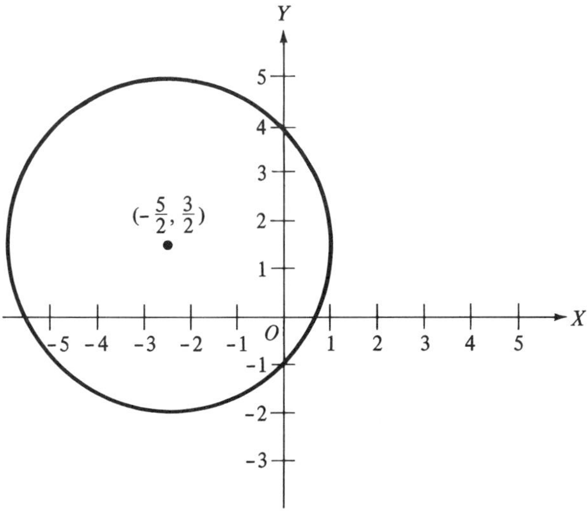

The graph is a circle with center $(-\frac{5}{2}, \frac{3}{2})$ and radius $\frac{5}{2}\sqrt{2}$ as in Figure 3–24. □

An *ellipse* is a set of points having the property that the sum of the distances from any point in the set to two fixed points, called the *foci*, is a constant. See Figure 3–25.

Figure 3–25

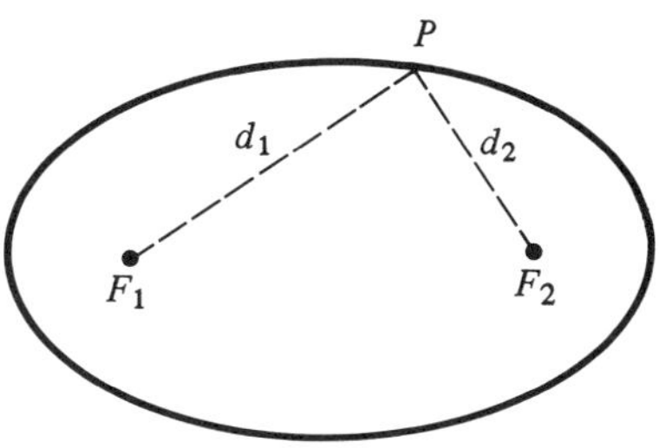

The orbits of the planets take the form of an ellipse with the sun as a focus. See Figure 3–26.

Before deriving the general form of the equation, let us consider a special case.

Figure 3-26

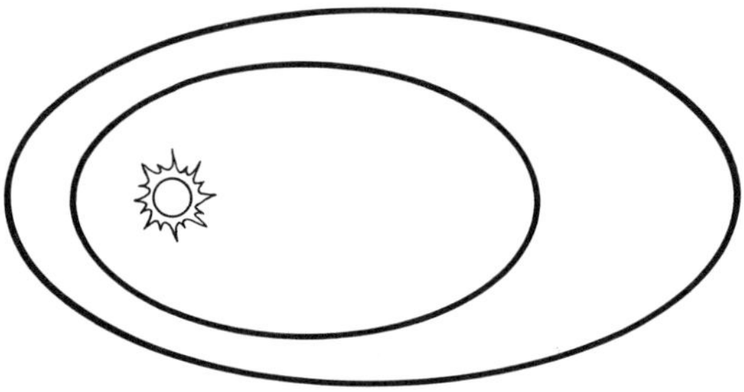

Example 10

Suppose the foci of an ellipse are $(-12, 0)$ and $(12, 0)$ and the sum of the distances from any point on the ellipse to these two points is 26. Then a point (x, y) is on the ellipse if and only if the distance from (x, y) to $(-12, 0)$ plus the distance from (x, y) to $(12, 0)$ is 26. Using the distance formula we have

$$\sqrt{(x+12)^2 + (y-0)^2} + \sqrt{(x-12)^2 + (y-0)^2} = 26$$

See Figure 3–27.

Notice that $(13, 0)$ and $(-13, 0)$ are on the graph since

$$\sqrt{25^2 + 0} + \sqrt{1^2 + 0} = 26$$

and

$$\sqrt{(-1)^2 + 0} + \sqrt{(-25)^2 + 0} = 26$$

Figure 3-27

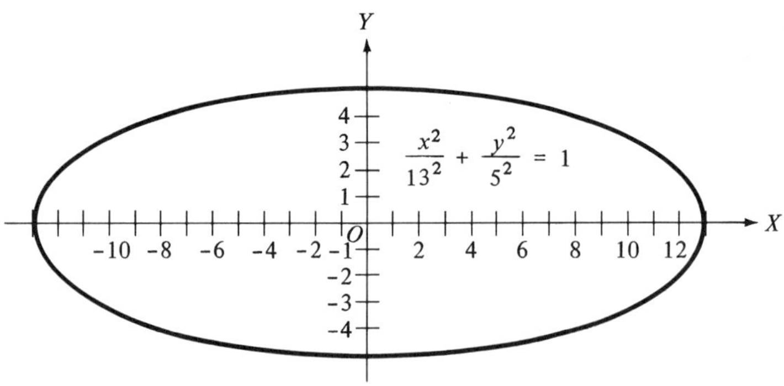

Furthermore, (0, 5) and (0, −5) are on the graph. (To verify this, just check in the equation.) Since (0, 5) and (0, −5) are each equidistant from (−12, 0) and (12, 0) and the distance from (0, 5) to (12, 0) plus the distance from (0, 5) to (−12, 0) is 26, the distance from (0, 5) to (12, 0) is 13 and the distance from (0, 5) to (−12, 0) is 13. By similar reasoning the distance from (0, −5) to (12, 0) is 13 and the distance from (0, −5) to (−12, 0) is also 13. Of course, all of these distances can be verified by the distance formula without the geometric arguments above.

The equation

$$\sqrt{(x + 12)^2 + y^2} + \sqrt{(x - 12)^2 + y^2} = 26$$

can be simplified to yield

$$\frac{x^2}{13^2} + \frac{y^2}{5^2} = 1 \quad \square$$

Instead of completing this simplification, we will consider the general case and perform the simplification in the more general situation.

The *center* of an ellipse is the midpoint of the segment between the foci. We will limit our discussion to cases where the two foci are on the same horizontal line or on the same vertical line and first will consider the case where both foci are on the X-axis with the center at the origin. Let $a > 0$, $c > 0$; let the foci be $(c, 0)$ and $(-c, 0)$, and let the sum of the distances from a point on the ellipse to the foci be $2a$. (See Figure 3–28.)

Figure 3–28

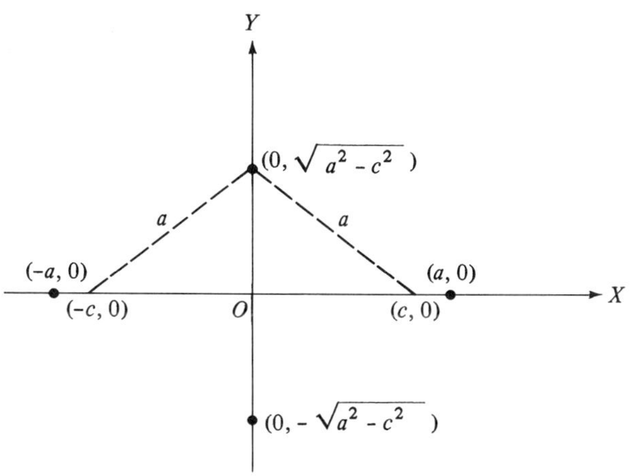

Since the distance from $(-c, 0)$ to $(c, 0)$ is $2c$, the distance from any point (p, q) in the plane to $(-c, 0)$ plus the distance from (p, q) to $(c, 0)$ must be at least $2c$ and, in fact, the only points where the sum of the distances is exactly $2c$ are points between the foci. Thus for the ellipse to have points not on the segment between the foci, we require $a > c$.

Notice that $(a, 0)$ and $(-a, 0)$ are on the ellipse since

$$|a - c| + |a - (-c)| = (a - c) + (a + c) = 2a$$

and

$$\begin{aligned} |-a - c| + |-a - (-c)| &= |-(a + c)| + |-(a - c)| \\ &= (a + c) + (a - c) = 2a \end{aligned}$$

This is rather apparent from Figure 3–28.

Furthermore, by the Pythagorean theorem, the points $(0, \sqrt{a^2 - c^2})$ and $(0, -\sqrt{a^2 - c^2})$ are points on the ellipse. We will denote $\sqrt{a^2 - c^2}$ by b. Thus we have four points

$$(a, 0), \quad (-a, 0), \quad (0, b), \quad \text{and} \quad (0, -b)$$

Figure 3-29

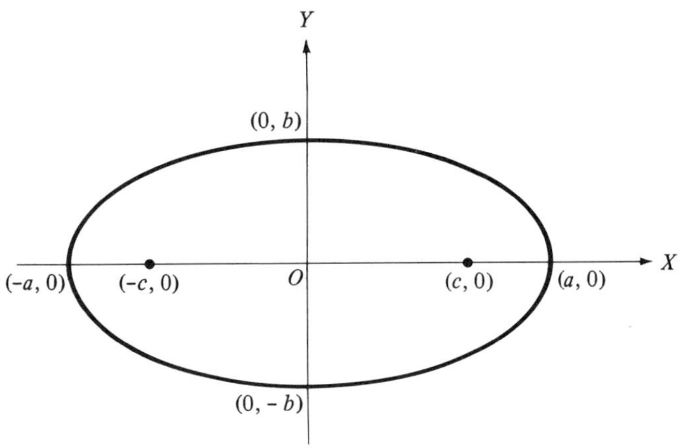

on the ellipse. These points are called the *vertices* of the ellipse. The *major axis* is the segment between $(a, 0)$ and $(-a, 0)$ and the *minor axis* is the segment between $(0, b)$ and $(0, -b)$. The major axis contains the foci. The minor axis is perpendicular to the major axis and contains the center of the ellipse.

In general, (x, y) is on the ellipse if and only if

$$\sqrt{(x-c)^2+y^2}+\sqrt{(x+c)^2+y^2}=2a$$

This equation is equivalent to

$$\sqrt{(x-c)^2+y^2}=2a-\sqrt{(x+c)^2+y^2}$$

Squaring both sides yields

$$(x-c)^2+y^2=4a^2-4a\sqrt{(x+c)^2+y^2}+(x+c)^2+y^2$$

Thus

$$\begin{aligned} x^2-2xc+c^2+y^2 &= 4a^2-4a\sqrt{(x+c)^2+y^2}+x^2+2xc+c^2+y^2 \\ -4xc &= 4a^2-4a\sqrt{(x+c)^2+y^2} \end{aligned}$$

and

$$\begin{aligned} 4a\sqrt{(x+c)^2+y^2} &= 4a^2+4xc \\ a\sqrt{(x+c)^2+y^2} &= a^2+xc \end{aligned}$$

Squaring both sides yields

$$\begin{aligned} a^2(x^2+2xc+c^2+y^2) &= a^4+2a^2xc+x^2c^2 \\ a^2x^2+2a^2xc+a^2c^2+a^2y^2 &= a^4+2a^2xc+x^2c^2 \\ a^2x^2-c^2x^2+a^2y^2 &= a^4-a^2c^2 \\ (a^2-c^2)x^2+a^2y^2 &= a^2(a^2-c^2) \end{aligned}$$

Since $b^2=a^2-c^2$ we have

$$b^2x^2+a^2y^2=a^2b^2$$

Dividing both sides by a^2b^2 yields

$$\frac{x^2}{a^2}+\frac{y^2}{b^2}=1$$

Thus the ellipse with foci $(-c, 0)$ and $(c, 0)$ and sum of distances equaling $2a$ is the graph of the equation

$$\frac{x^2}{a^2}+\frac{y^2}{b^2}=1$$

where $b^2=a^2-c^2$.

If we choose our foci on the Y-axis with coordinates $(0, c)$ and $(0, -c)$, a similar process yields the equation

$$\frac{x^2}{b^2}+\frac{y^2}{a^2}=1 \qquad \text{with } b^2=a^2-c^2$$

and the vertices will be $(b, 0)$, $(-b, 0)$, $(0, a)$, and $(0, -a)$. In this case the major axis, the one containing the foci, is the segment between $(0, a)$ and $(0, -a)$ and the minor axis is the segment between $(b, 0)$ and $(-b, 0)$. See Figure 3–30.

Figure 3–30

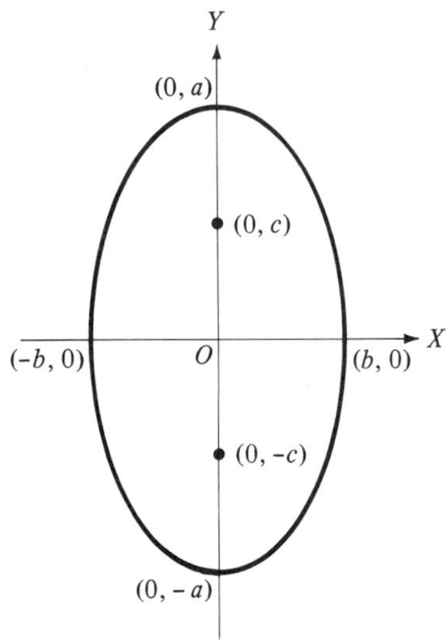

Example 11

Draw the graph of

$$\frac{x^2}{9} + \frac{y^2}{25} = 1$$

In this case $b^2 = 9$ and $a^2 = 25$. Thus

$$c^2 = 25 - 9 = 16$$
$$c = 4$$

The vertices are $(3, 0)$, $(-3, 0)$, $(0, 5)$, and $(0, -5)$. The foci are $(0, 4)$ and $(0, -4)$. (See Figure 3–31.) □

Notice that when the foci are on the X-axis, the width of the ellipse is greater than its height. When the foci are on the Y-axis, the height exceeds the width.

For the ellipse in Example 10 and in the general cases described above, the center is $(0, 0)$. What is the equation of an ellipse with some other point

Figure 3-31

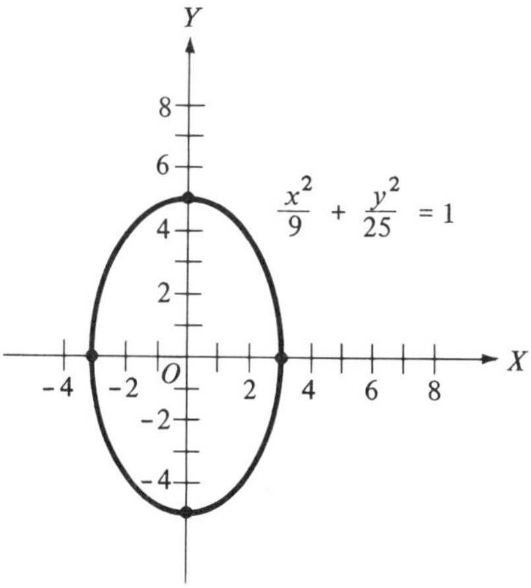

as center but with the same shape, size, and orientation as the ellipse in Example 10?

The shape and size of the ellipse depend on the distance between the foci and the sum of the distances from a point to the foci; it does not depend on the location of the foci. The orientation depends on the direction of the major axis. In order for the ellipse to have the same shape, size, and orientation as that in Example 10, the sum of the distances must be 26, the foci must be 24 units apart, and the major axis must be horizontal.

Example 12

What is the equation for the ellipse with the same shape and size as the ellipse in Example 10 but with center at (2, 3)?

In this case the distance between the foci is 24 and they are on the line $y = 3$ with (2, 3) halfway between them. Hence the foci are

$$(-10, 3) \quad \text{and} \quad (14, 3)$$

Thus the equation must be

$$\sqrt{(x + 10)^2 + (y - 3)^2} + \sqrt{(x - 14)^2 + (y - 3)^2} = 26$$

This equation can be simplified to

$$\frac{(x - 2)^2}{13^2} + \frac{(y - 3)^2}{5^2} = 1$$

We leave the simplification as an exercise. (See Figure 3–32.) □

Figure 3-32

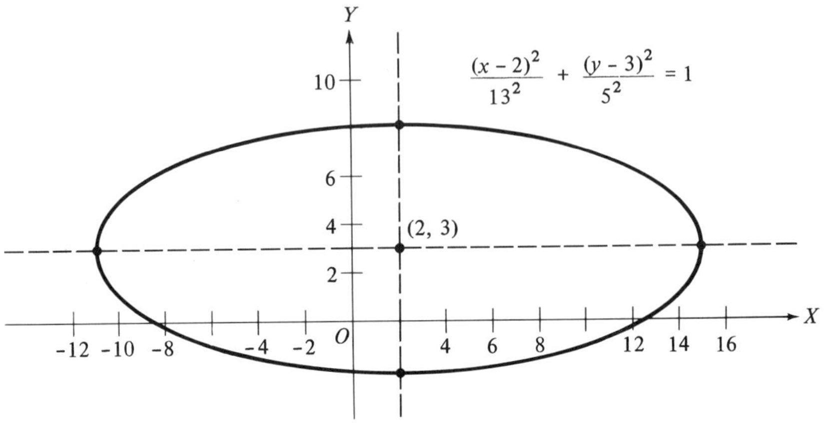

In general, an ellipse with center (h, k), foci $(h - c, k)$ and $(h + c, k)$, and the distance sum $2a$ will have the same shape, size, and orientation as the ellipse with equation

$$\frac{x^2}{a^2} + \frac{y^2}{b^2} = 1$$

where $b^2 = a^2 - c^2$. By the distance formula, the equation is

$$\sqrt{[x - (h - c)]^2 + (y - k)^2} + \sqrt{[x - (h + c)]^2 + (y - k)^2} = 2a$$

which can be simplified to

$$\frac{(x - h)^2}{a^2} + \frac{(y - k)^2}{b^2} = 1$$

We leave the simplification as an exercise. The process is similar to that done before for the case where the center is at the origin.

In this case the vertices are $(h + a, k)$, $(h - a, k)$, $(h, k + b)$, and $(h, k - b)$. The major axis is the segment joining $(h + a, k)$ and $(h - a, k)$; the minor axis is the segment joining $(h, k + b)$ and $(h, k - b)$.

By an analogous argument, the ellipse with center (h, k), foci $(h, k - c)$, $(h, k + c)$, and distance sum $2a$ has the equation

$$\frac{(x - h)^2}{b^2} + \frac{(y - k)^2}{a^2} = 1$$

where $b^2 = a^2 - c^2$.

In this case, the vertices are $(h, k + a)$, $(h, k - a)$, $(h + b, k)$, and $(h - b, k)$. The major axis is the segment from $(h, k + a)$ to $(h, k - a)$ and the minor axis is the segment from $(h + b, k)$ to $(h - b, k)$.

To draw the graph of the equation

$$\frac{x^2}{p^2} + \frac{y^2}{q^2} = 1$$

we plot the points $(p, 0)$, $(-p, 0)$, $(0, q)$, $(0, -q)$ and some others, if necessary, and sketch the ellipse as in Examples 13 and 14.

Example 13

$$\frac{x^2}{16} + \frac{y^2}{4} = 1$$

See Figure 3–33. □

Figure 3–33

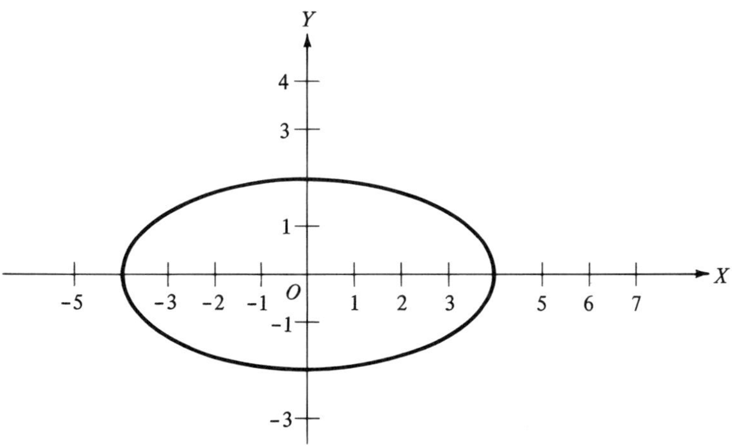

Example 14

$$\frac{x^2}{2} + \frac{y^2}{9} = 1$$

See Figure 3–34. □

Figure 3–34

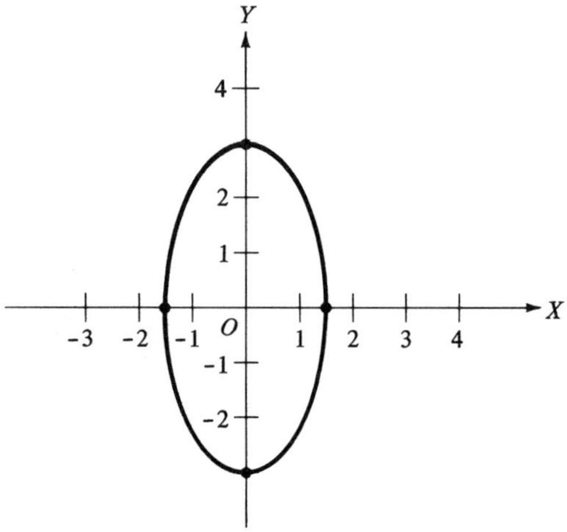

If the center of the ellipse is at (h, k), we can draw the dotted lines $x = h$ and $y = k$, and then draw the graph of

$$\frac{(x-h)^2}{p^2} + \frac{(y-k)^2}{q^2} = 1$$

just like we draw the graph of

$$\frac{x^2}{p^2} + \frac{y^2}{q^2} = 1$$

We just treat the lines $x = h$ and $y = k$ as if they were the X- and Y-axes.

Figure 3–35

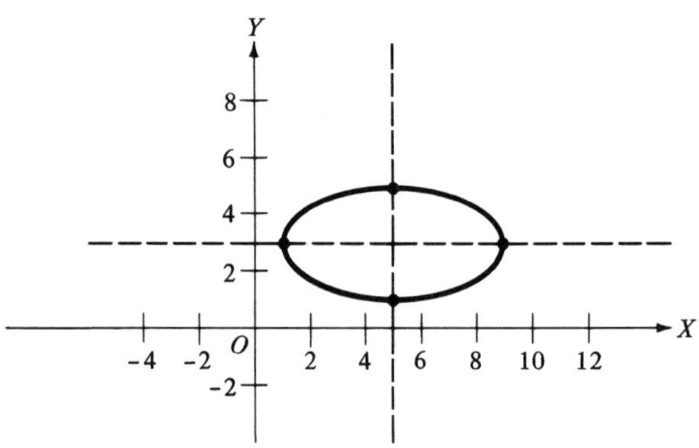

Example 15

$$\frac{(x-5)^2}{16}+\frac{(y-3)^2}{4}=1$$

See Figure 3–35. □

Example 16

$$\frac{(x-4)^2}{2}+\frac{(y+1)^2}{9}=1$$

See Figure 3–36. □

Figure 3–36

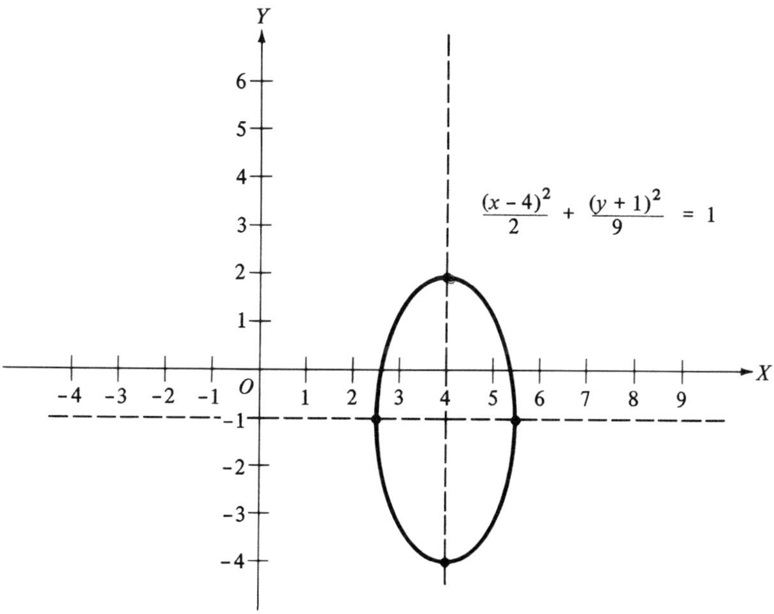

In summary, the equation

$$\frac{(x-h)^2}{p^2}+\frac{(y-k)^2}{q^2}=1$$

is the graph of an ellipse with:

center at (h, k)

vertices $(h+p, k)$, $(h-p, k)$, $(h, k+q)$, $(h, k-q)$

This information is sufficient to enable us to draw the graph. The axes of the ellipse are the line segments joining opposite vertices. The major axis is the longer and the minor axis is the shorter.

The "completing the square" technique can be used to transform an equation of the form

$$Ax^2 + Bx + Cy^2 + Dy = E \qquad \text{with } A > 0 \quad \text{and} \quad C > 0$$

into one of the form

$$\frac{(x - h)^2}{p^2} + \frac{(y - k)^2}{q^2} = 1$$

providing E is sufficiently large. Furthermore, if E is not sufficiently large, the process will show that there are no points or just one point satisfying the equation. The technique transforms the equation into one in which the left side is the sum of two squares, which cannot be negative. If the right side is zero, then just one point satisfies the equation; if the right side is negative, no points satisfy it.

Figure 3-37

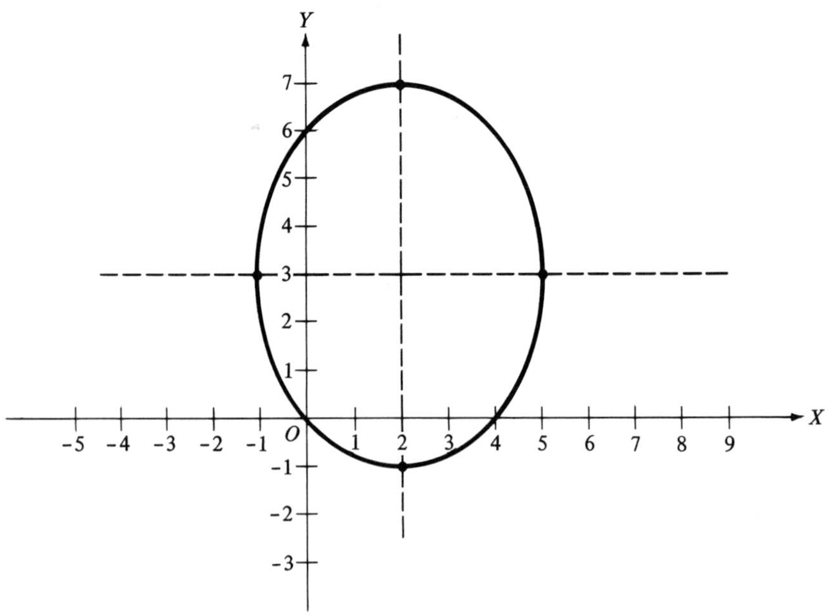

Example 17

$$\begin{aligned}
16x^2 - 64x + 9y^2 - 54y &= -1 \\
16x^2 - 64x + 64 + 9y^2 - 54y + 81 &= -1 + 64 + 81 \\
16(x^2 - 4x + 4) + 9(y^2 - 6y + 9) &= 144 \\
16(x - 2)^2 + 9(x - 3)^2 &= 144 \\
\frac{16(x - 2)^2}{144} + \frac{9(x - 3)^2}{144} &= 1 \\
\frac{(x - 2)^2}{9} + \frac{(y - 3)^2}{16} &= 1
\end{aligned}$$

We now can draw the graph as before. See Figure 3–37. □

Example 18

$$\begin{aligned}
4x^2 + 24x + 9y^2 + 18y &= 4 \\
4(x^2 + 6x) + 9(y^2 + 2y) &= 4 \\
4(x^2 + 6x + 9) + 9(y^2 + 2y + 1) &= 36 + 9 + 4 \\
\frac{4(x + 3)^2}{49} + \frac{9(y + 1)^2}{49} &= 1 \\
\frac{(x + 3)^2}{\frac{49}{4}} + \frac{(y + 1)^2}{\frac{49}{9}} &= 1 \\
\frac{(x + 3)^2}{(\frac{7}{2})^2} + \frac{(y + 1)^2}{(\frac{7}{3})^2} &= 1 \; \square
\end{aligned}$$

Example 19

$$\begin{aligned}
2x^2 + 3x + 6y^2 + 12y &= 2 \\
2(x^2 + \tfrac{3}{2}x) + 6(y^2 + 2y) &= 2 \\
2(x^2 + \tfrac{3}{2}x + \tfrac{9}{16}) + 6(y^2 + 2y + 1) &= 2 + \tfrac{9}{8} + 6 \\
2(x + \tfrac{3}{4})^2 + 6(y + 1)^2 &= \tfrac{71}{8} \\
\frac{2(x + \frac{3}{4})^2}{\frac{71}{8}} + \frac{6(y + 1)^2}{\frac{71}{8}} &= 1 \\
\frac{(x + \frac{3}{4})^2}{\frac{71}{16}} + \frac{(y + 1)^2}{\frac{71}{48}} &= 1 \\
\frac{(x + \frac{3}{4})^2}{\left(\frac{\sqrt{71}}{4}\right)^2} + \frac{(y + 1)^2}{\left(\frac{\sqrt{71}}{4\sqrt{3}}\right)^2} &= 1 \; \square
\end{aligned}$$

Example 20

$$\begin{aligned}
2x^2 + 8x + 6y^2 - 12y &= -20 \\
2(x^2 + 4x) + 6(y^2 - 2y) &= -20 \\
2(x^2 + 4x + 4) + 6(y^2 - 2y + 1) &= (-20) + 8 + 6 \\
2(x + 2)^2 + 6(y - 1)^2 &= -6
\end{aligned}$$

This equation has no solutions because the left side must be positive for all possible values of x and y. □

EXERCISES

1. Draw graphs of the following equations.
   (a) $x^2 + y^2 = 16$ (b) $x^2 + y^2 = 12$
   (c) $x^2 + y^2 = 15$ (d) $(x - 1)^2 + (y - 2)^2 = 9$
   (e) $(x + 2)^2 + (y - 3)^2 = 16$ (f) $(x - 2)^2 + (y - 5)^2 = 81$
   (g) $x^2 + 2x + y^2 + 4y = 7$ (h) $x^2 - x + y^2 - 2y = 4$
   (i) $x^2 + 5x + y^2 + 3y = 2$ (j) $x^2 - 10x + y^2 - 12y = 0$
   (k) $x^2 - 5x + y^2 = 6$ (l) $x^2 + 11x + y^2 - 20y = 15$
2. Draw graphs of the following equations.
   (a) $\dfrac{x^2}{4} + \dfrac{y^2}{25} = 1$ (b) $\dfrac{x^2}{16} + \dfrac{y^2}{49} = 1$ (c) $\dfrac{x^2}{1} + \dfrac{y^2}{16} = 1$
   (d) $\dfrac{x^2}{25} + \dfrac{y^2}{9} = 1$ (e) $\dfrac{x^2}{49} + \dfrac{y^2}{4} = 1$ (f) $\dfrac{x^2}{10} + \dfrac{y^2}{16} = 1$
   (g) $\dfrac{x^2}{9} + \dfrac{y^2}{5} = 1$ (h) $\dfrac{x^2}{11} + \dfrac{y^2}{6} = 1$
3. Draw graphs of the following equations.
   (a) $\dfrac{(x - 1)^2}{4} + \dfrac{(y - 3)^2}{25} = 1$ (b) $\dfrac{(x + 2)^2}{16} + \dfrac{(y - 4)^2}{49} = 1$
   (c) $\dfrac{x^2}{16} + \dfrac{(y - 7)^2}{4} = 1$ (d) $\dfrac{(x + 5)^2}{1} + \dfrac{(y - 2)^2}{16} = 1$
   (e) $\dfrac{(x + 2)^2}{10} + \dfrac{(y - \frac{3}{2})^2}{16} = 1$ (f) $\dfrac{(x + 1.3)^2}{11} + \dfrac{(y - 4.7)^2}{6} = 1$
   (g) $\dfrac{(x - \pi)^2}{6} + \dfrac{(y - \sqrt{7})^2}{10} = 1$ (h) $\dfrac{(x + \sqrt{19})^2}{32} + \dfrac{(y - 5.3)^2}{27} = 1$
4. Draw graphs of the following equations (if such graphs exist).
   (a) $x^2 - 6x + 4y^2 + 8y = 3$ (b) $5x^2 - 10x + 3y^2 - 12y = 4$
   (c) $2x^2 - 2x + y^2 + y = 9$ (d) $4x^2 + 3x + 5y^2 + 8y = 3$
   (e) $4x^2 + 12x + 3y^2 + 9y + 30 = 0$
5. Complete the simplification to show that the equation

$$\sqrt{[x - (h - c)]^2 + (y + k)^2} + \sqrt{[x - (h + c)]^2 + (y - k)^2} = 2a$$

yields

$$\frac{(x-h)^2}{a^2} + \frac{(y-k)^2}{b^2} = 1$$

where $b^2 = a^2 - c^2$.

6. If the two foci of an ellipse coincide, what is the shape of the ellipse?
7. The following questions refer to the general ellipse

$$\frac{(x-h)^2}{a^2} + \frac{(y-k)^2}{b^2} = 1$$

where $b^2 = a^2 - c^2$.

(a) If we fix a, what happens to the shape of the ellipse if we increase c? If we decrease c?
(b) If we fix c, what happens to the shape of the ellipse if we increase a? If we decrease a?
*(c) What happens to $a - c$ if we fix b and increase a?

6. Parabolas

A third type of conic is a *parabola* as illustrated in Figure 3–38. As we did with circles and ellipses, we wish to define such curves in terms of distances so that we may derive equations for them.

Figure 3-38

A *parabola* is a set of points all of which are equidistant from a fixed line, called the *directrix*, and a fixed point, called the *focus*.

The curve in Figure 3–39 is an example of a parabola with the point F as the focus and the line d as the directrix. Notice that the point halfway between the focus and directrix is one of the points on the parabola. This special point is called the *vertex* of the parabola. The shape of the parabola depends on the distance from the focus to the directrix; the larger this distance, the "flatter" the curve. The *axis* of a parabola is the line through the focus perpendicular to the directrix.

Figure 3–39

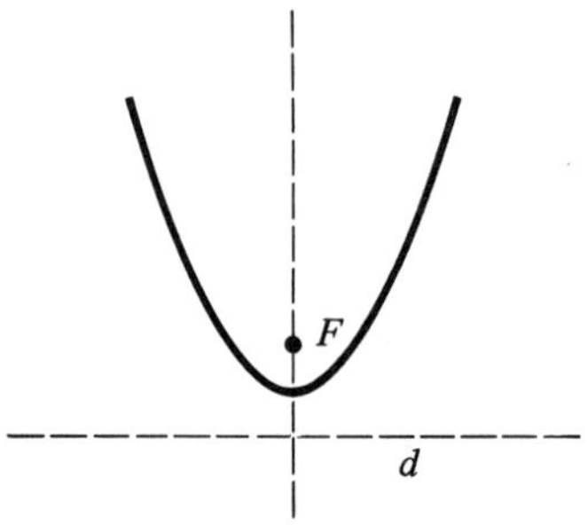

In order to determine equations for such curves, let us first consider a particular case.

Example 1

Determine an equation for the parabola with vertex at the origin (0, 0), focus at (1, 0), and directrix the line $y = -1$. (See Figure 3–40.)

Figure 3–40

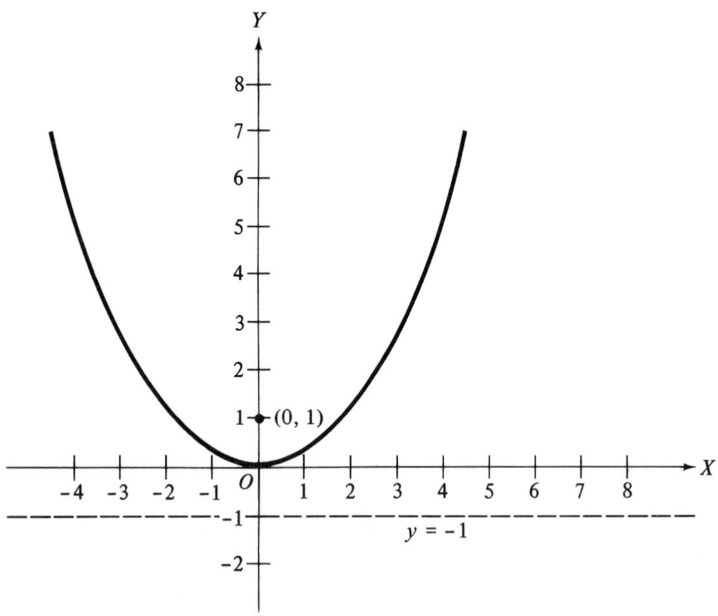

In this case, the distance from the focus to the directrix is 2 units and the axis is the Y-axis. A point (x, y) is on the parabola if and only if the distance from (x, y) to $(0, 1)$ is equal to the distance from (x, y) to the line $y = -1$. By the distance formula, the distance from (x, y) to $(0, 1)$ is

$$\sqrt{(x-0)^2 + (y-1)^2} = \sqrt{x^2 + (y-1)^2}$$

and the distance from (x, y) to the line $y = -1$ is just

$$|y - (-1)| = |y + 1|$$

Thus we have

$$\sqrt{x^2 + (y-1)^2} = |y + 1|$$

Squaring both sides yields

$$x^2 + (y-1)^2 = (y+1)^2$$

Then

$$\begin{aligned} x^2 + y^2 - 2y + 1 &= y^2 + 2y + 1 \\ x^2 - 2y &= 2y \end{aligned}$$

and

$$4y = x^2$$

or

$$y = \tfrac{1}{4}x^2 \quad \square$$

Now to generalize this example, suppose we have a parabola with vertex at $(0, 0)$, focus $(0, p)$, and directrix the line $y = -p$. A point (x, y) will be on the parabola if the distance from (x, y) to $(0, p)$ is the same as the distance from (x, y) to the line $y = -p$. The distance from (x, y) to $(0, p)$ is

$$\sqrt{x^2 + (y-p)^2}$$

and the distance from (x, y) to the line $y = -p$ is

$$|y - (-p)| = |y + p|$$

Thus we have

$$\sqrt{x^2 + (y-p)^2} = |y + p|$$

Squaring both sides yields

$$x^2 + (y-p)^2 = (y+p)^2$$

Then

$$\begin{aligned} x^2 + y^2 - 2yp + p^2 &= y^2 + 2yp + p^2 \\ x^2 - 2yp &= 2yp \\ 4py &= x^2 \end{aligned}$$

or

$$y = \frac{1}{4p}x^2$$

As in Figures 3–41 and 3–42, if $p > 0$, the curve opens upward and if $p < 0$ the curve opens downward. The parabola always opens away from the directrix on the side of the focus.

Figure 3–41

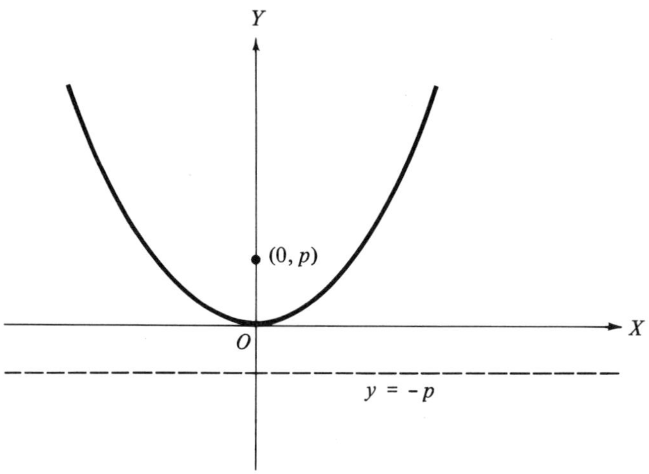

Figure 3–42

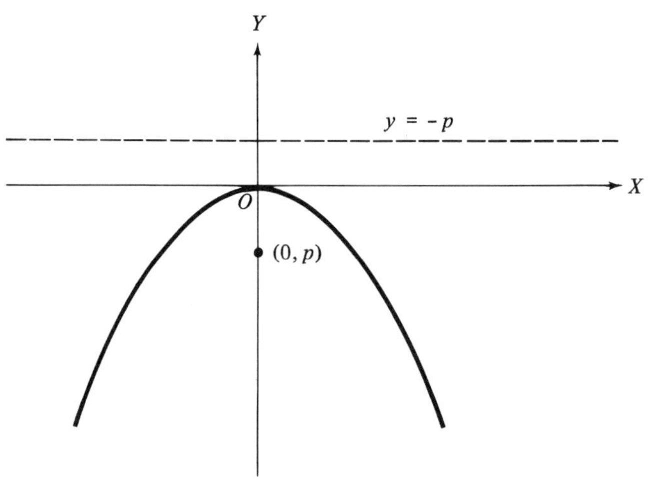

If the vertex is at (0, 0), the focus at $(p, 0)$, and the directrix the vertical line $x = -p$, the resulting equation will be

$$4px = y^2$$

or

$$x = \frac{1}{4p} y^2$$

The graph will be as in Figure 3–43 if $p > 0$ and as in Figure 3–44 if $p < 0$.

Figure 3–43

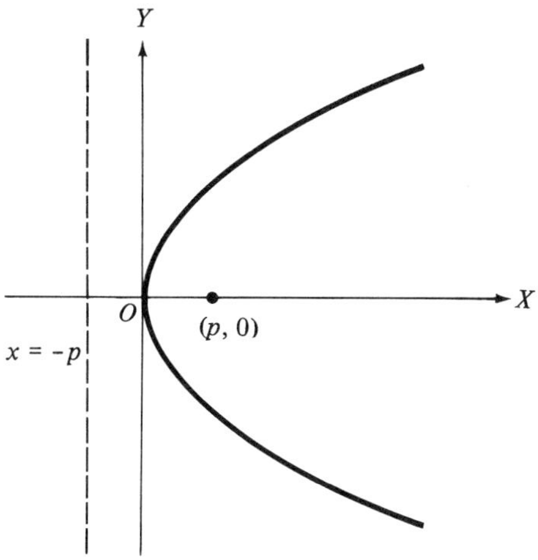

Figure 3–44

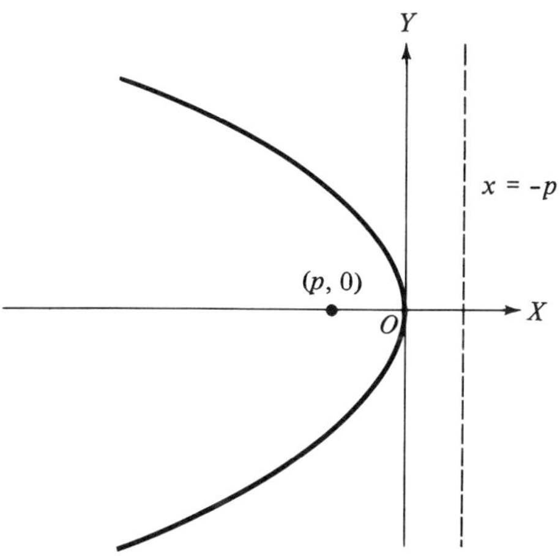

Example 2

$y = 2x^2$

Since the general form is $y = \dfrac{1}{4p}x^2$, we have

$$\frac{1}{4p} = 2$$

and

$p = \frac{1}{8}$

Thus the focus is $(0, \frac{1}{8})$ and the directrix is $y = -\frac{1}{8}$. The graph is as in Figure 3–45. □

Figure 3–45

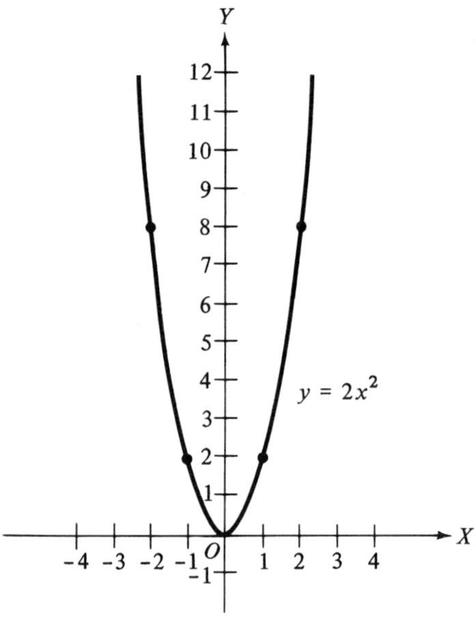

Example 3

$x = \frac{1}{2}y^2$

Since the general form is $x = \dfrac{1}{4p}y^2$, we have

$$\frac{1}{4p} = \frac{1}{2}$$

and

$$p = \frac{1}{2}$$

Thus the focus is $(\frac{1}{2}, 0)$ and the directrix is $x = -\frac{1}{2}$. The graph is as in Figure 3–46. □

Figure 3-46

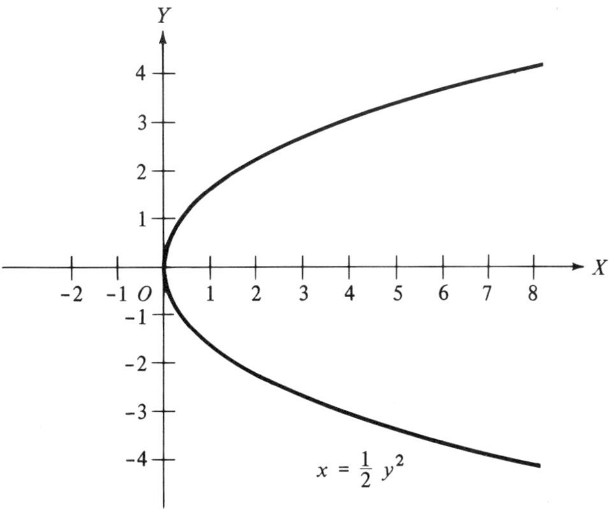

Example 4

$$y = -6x^2$$

From the general form $y = \frac{1}{4p} x^2$, we have

$$\frac{1}{4p} = -6$$

$$p = -\frac{1}{24}$$

The focus is $(0, -\frac{1}{24})$ and the directrix is $y = \frac{1}{24}$. The graph is given by Figure 3–47. □

Thus far we have limited our discussion to parabolas with vertex at the origin and with a horizontal or vertical directrix. We will continue to limit our discussion to the cases where the directrix is horizontal or vertical but will generalize to vertices located at any point.

Figure 3–47

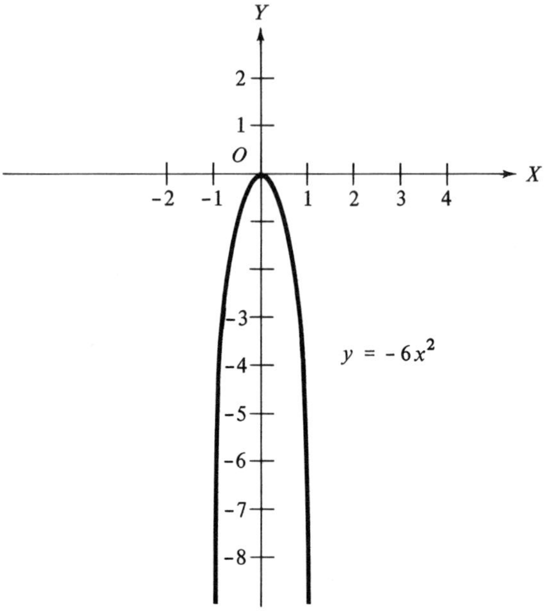

Example 5

What is the equation of a parabola with the same shape as the parabola in Example 1 and which has vertex at (3, 5), opens upward, and has a horizontal directrix?

Since the shape of the parabola depends only on the distance of the focus from the directrix, and in Example 1 this distance was 2, the focus must be 1 unit above the vertex and the directrix must be 1 unit below the vertex. Thus the focus must be (3, 6) and the directrix must be the line $y = 4$. (See Figure 3–48.)

A point (x, y) is on the parabola if its distance from (3, 5), which is $\sqrt{(x-3)^2 + (y-6)^2}$, is equal to its distance from the directrix, that is, $|y - 4|$. Thus (x, y) is on the parabola if and only if

$$\sqrt{(x-3)^2 + (y-6)^2} = |y - 4|$$

Squaring both sides yields

$$\begin{aligned} (x-3)^2 + (y-6)^2 &= (y-4)^2 \\ (x-3)^2 + y^2 - 12y + 36 &= y^2 - 8y + 16 \\ (x-3)^2 &= 4y - 20 \\ 4(y-5) &= (x-3)^2 \end{aligned}$$

Figure 3-48

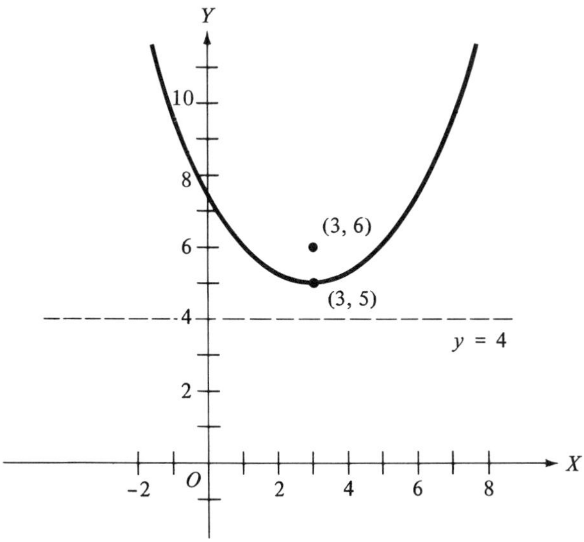

or

$$y - 5 = \tfrac{1}{4}(x - 3)^2 \; \square$$

In general, suppose we have a parabola with vertex at (h, k) and a horizontal directrix. Let the equation of the directrix be $y = k - p$. Then the focus is at $(h, k + p)$. The focus is above the directrix if $p > 0$ as in Figure 3–49 and below the directrix if $p < 0$ as in Figure 3–50.

The point (x, y) is on the parabola if and only if the distance from (x, y) to $(h, k + p)$ is the same as the distance from (x, y) to the line $y = k - p$. The distance from (x, y) to $(h, k + p)$ is

$$\sqrt{(x - h)^2 + [y - (k + p)]^2}$$

and the distance from (x, y) to the line $y = k - p$ is

$$|y - (k - p)|$$

Thus the equation is

$$\sqrt{(x - h)^2 + [y - (k + p)]^2} = |y - (k - p)|$$

Squaring both sides yields

$$(x - h)^2 + [y - (k + p)]^2 = [y - (k - p)]^2$$

Then

$$\begin{aligned} (x - h)^2 + [(y - k) - p]^2 &= [(y - k) + p]^2 \\ (x - h)^2 + (y - k)^2 - 2p(y - k) + p^2 &= (y - k)^2 + 2p(y - k) + p^2 \\ (x - h)^2 &= 4p(y - k) \end{aligned}$$

Figure 3–49

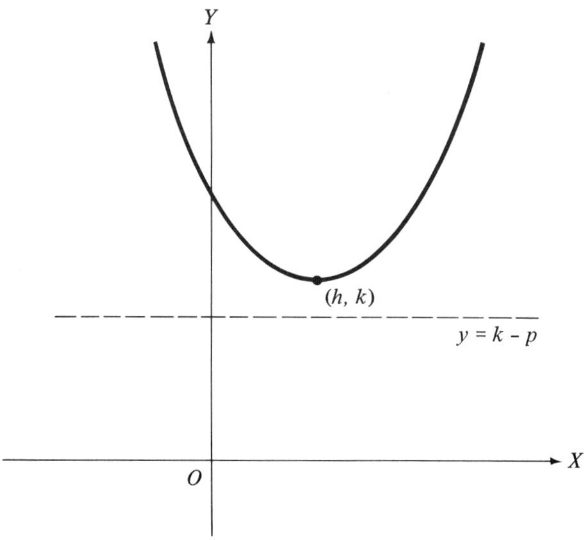

Figure 3–50

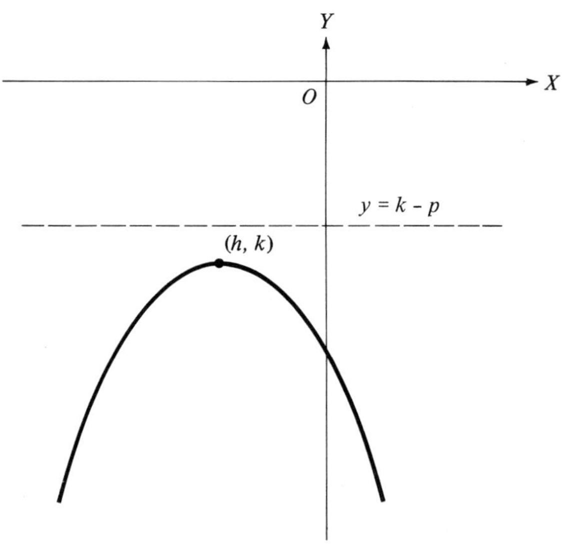

or

$$y - k = \frac{1}{4p}(x - h)^2$$

Thus the general form for the equation of a parabola with a horizontal directrix is

$$y - k = \frac{1}{4p}(x - h)^2$$

where the vertex is at (h, k) and the distance from the focus to the vertex is $|p|$. The curve opens upward if $p > 0$ and opens downward if $p < 0$.

Example 6

The graph of

$$y - 7 = \tfrac{1}{5}(x - 4)^2$$

is a parabola with vertex (4, 7). The distance from the focus to the vertex is $\frac{5}{4}$ and the curve opens upward. (See Figure 3–51.) □

Figure 3-51

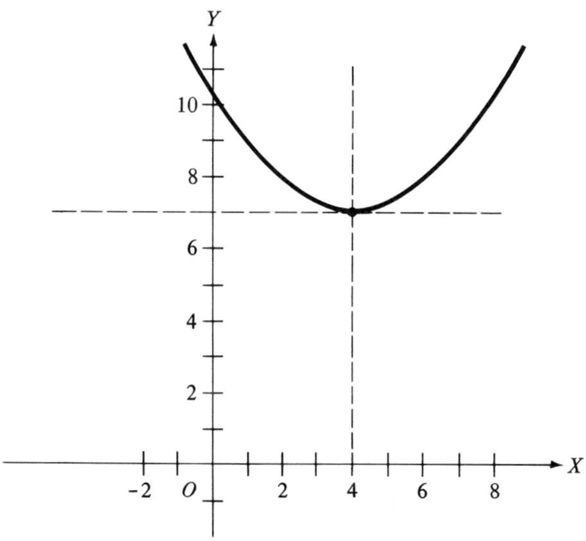

Example 7

The graph of

$$y + 2 = -3(x + 1)^2$$

is a parabola with vertex $(-1, -2)$. The distance from the vertex to the focus is $\frac{1}{12}$, $p = -\frac{1}{12}$, and the curve opens downward. (See Figure 3–52). □

Figure 3–52

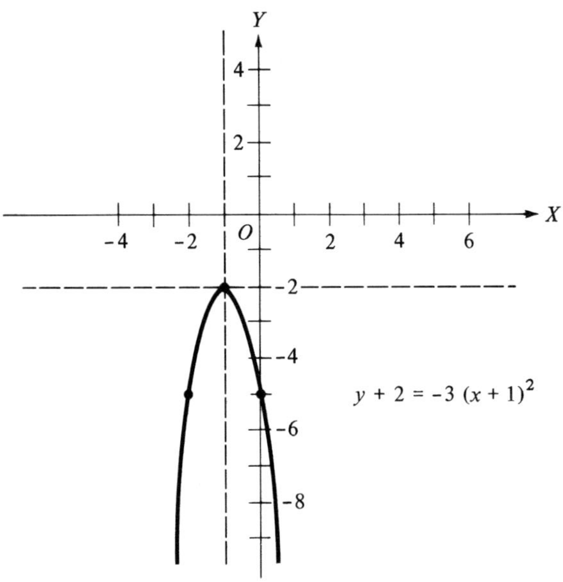

By an analogous argument, the equation of a parabola with vertex (h, k) and a vertical directrix $x = h - p$ is

$$x - h = \frac{1}{4p}(y - k)^2$$

We leave the verification as an exercise. If $p > 0$, the curve opens to the right; if $p < 0$, the curve opens to the left.

Example 8

$$x - 3 = 2(y - 5)^2$$

See Figure 3–53. □

Figure 3-53

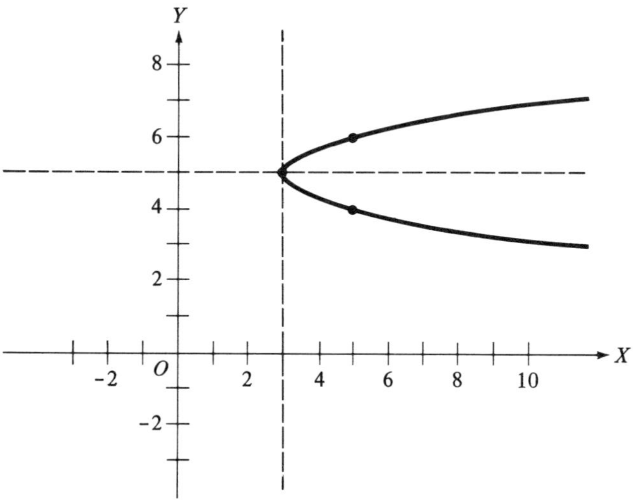

Consider now the equation

$$y = x^2 - 4x + 7$$

If we apply the "completing the square" technique to this equation in order to obtain an equation in which the right side is a perfect square, we obtain

$$y - 3 = x^2 - 4x + 4$$
$$y - 3 = (x - 2)^2$$

This equation is an equation for a parabola with vertex at (2, 3) with $p = \frac{1}{4}$, since $1/4p = 1$.

In general, any equation of the form

$$y = Ax^2 + Bx + C \qquad \text{with } A \neq 0$$

can be rewritten in the form

$$y - k = \frac{1}{4p}(x - h)^2$$

Thus any equation of the form

$$y = Ax^2 + Bx + C \qquad \text{with } A \neq 0$$

has a graph which is a parabola. We could apply the completing the square technique for each such equation. However, since we have a general technique, we can develop a general formula.

Dividing both sides of

$$y = Ax^2 + Bx + C$$

by A yields

$$\frac{1}{A}y = x^2 + \frac{B}{A}x + \frac{C}{A}$$

Then

$$\frac{1}{A}y - \frac{C}{A} = x^2 + \frac{B}{A}x$$

$$\frac{1}{A}y - \frac{C}{A} + \frac{B^2}{4A^2} = x^2 + \frac{B}{A}x + \frac{B^2}{4A^2}$$

$$\frac{1}{A}y - \frac{4AC}{4A^2} + \frac{B^2}{4A^2} = \left(x + \frac{B}{2A}\right)^2$$

$$\frac{1}{A}\left(y - \frac{4AC - B^2}{4A}\right) = \left(x + \frac{B}{2A}\right)^2$$

$$y - \frac{4AC - B^2}{4A} = A\left(x + \frac{B}{2A}\right)^2$$

The vertex is $(-B/2A, (4AC - B^2)/4A)$ and $p = 1/4A$.

(You may find it easier to remember the coordinates of the vertex in the form $(-B/2A, C - B^2/4A)$.

Thus for the graph of $y = Ax^2 + Bx + C$:

(i) Since $p = 1/4A$, the shape of the graph depends only on A. The smaller $|A|$, the "flatter" the curve. If $A < 0$, the parabola opens downward; if $A > 0$, it opens upward.

(ii) The X-coordinate of the vertex is $-B/2A$ and, therefore, the axis depends only on A and B.

(iii) The Y-coordinate of the vertex is $C - B^2/4A$ and thus depends on A, B, and C. However, once A and B are fixed, the vertex varies up and down with C. The larger C is, the higher the vertex.

Example 9

Determine the vertex and the shape of the graph of

$$y = x^2 + 3x + 5$$

In this case, $A = 1$, $B = 3$, and $C = 5$. Thus the vertex is at the point $[-3/2(1), 5 - (3)^2/4(1)]$ which is $(-3/2, 11/4)$. The curve is shaped like the graph of $y = x^2$ and opens upward. See Figure 3–54. □

Example 10

The graph of

$$y = 3x^2 - 2x + 7$$

has vertex $(1/3, 20/3)$ and is shaped like the graph of $y = 3x^2$. (See Figure 3–54.) □

Figure 3–54

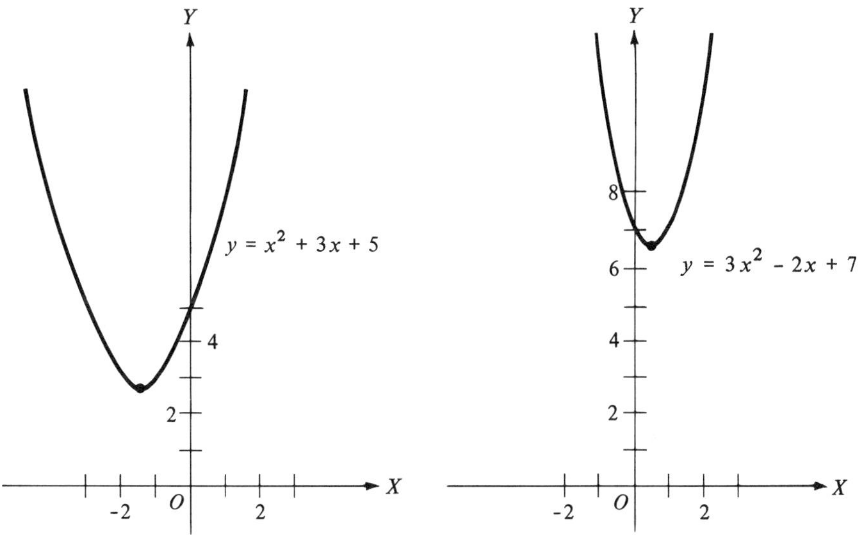

If you do not recall all of the formula, the Y-coordinate of the vertex is obtainable by substituting $-B/2A$ for x in the equation.

Since the graph of

$$y - k = \frac{1}{4p}(x - h)^2$$

has the same shape as the graph of

$$y = \frac{1}{4p}x^2$$

we can draw all parabolas once we can draw the graphs of equations of the form

$$y = \frac{1}{4p}x^2$$

All we need to do is draw the dotted lines $y = k$ and $x = h$ and draw the graph of

$$y = \frac{1}{4p}x^2$$

as if these lines were the axes.

Example 11

$y - 2 = \frac{1}{2}(x - 3)^2$

See Figure 3–55. □

Figure 3–55

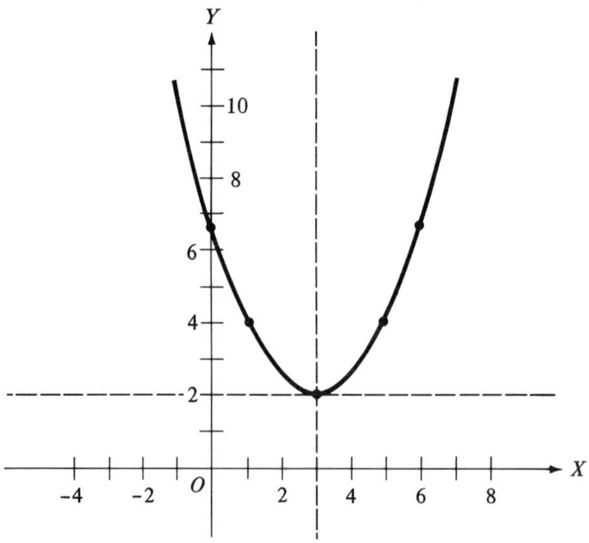

Example 12

$y + 3 = 2(x - 1)^2$

See Figure 3–56. □

Figure 3-56

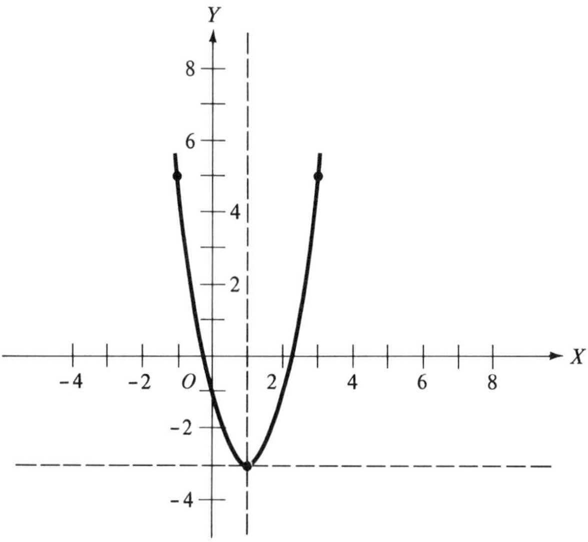

The graph of $y = Ax^2$ is obtained easily from the graph of $y = x^2$ just by multiplying each Y-coordinate by A. This is illustrated by the following tables and by Figure 3–57. The multiplication can be done mentally once you become familiar with the Y-coordinates of $y = x^2$.

In summary:

(a) The graph of an equation of the form $y = Ax^2 + Bx + C$ with $A \neq 0$ is a parabola.
(b) The vertex of the parabola is at $(-B/2A, C - B^2/4A)$.
(c) The parabola is shaped like the graph of $y = Ax^2$.

To draw the graph:

(i) Determine the coordinates of the vertex, $(-B/2A, C - B^2/4A)$.
(ii) Draw the graph of $y = Ax^2$ as if the lines $x = -B/2A$ and $y = C - B^2/4A$ were the axes.
(iii) The graph of $y = Ax^2$ is drawn by multiplying each Y-coordinate of $y = x^2$ by A.

Figure 3–57

$y = x^2$

x	−4	−3	−2	−1	0	1	2	3	4
y	16	9	4	1	0	1	4	9	16

$y = 2x^2$

x	−4	−3	−2	−1	0	1	2	3	4
y	32	18	8	2	0	2	8	18	32

$y = \frac{1}{3}x^2$

x	−4	−3	−2	−1	0	1	2	3	4
y	$\frac{16}{3}$	3	$\frac{4}{3}$	$\frac{1}{3}$	0	$\frac{1}{3}$	$\frac{4}{3}$	3	$\frac{16}{3}$

$y = -5x^2$

x	−4	−3	−2	−1	0	1	2	3	4
y	−80	−45	−20	−5	0	−5	−20	−45	−80

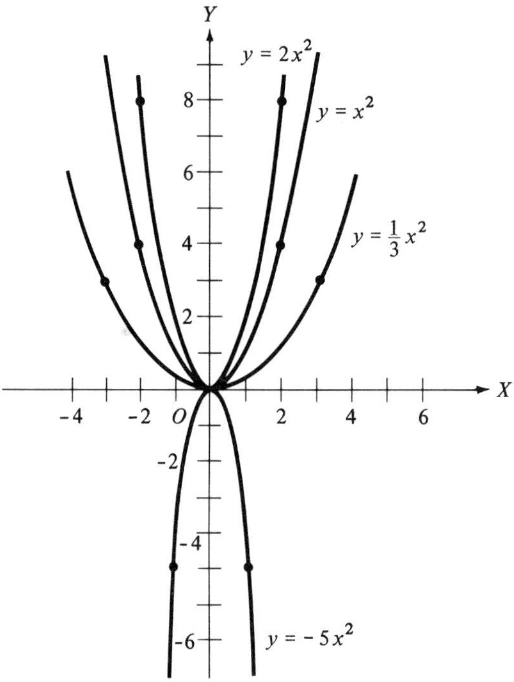

Example 13

Draw the graph of

$y = x^2 + 4x + 9$

The vertex is at $(-2, 5)$. The graph has the same shape as the graph of $y = x^2$. (See Figure 3–58.) □

Figure 3–58

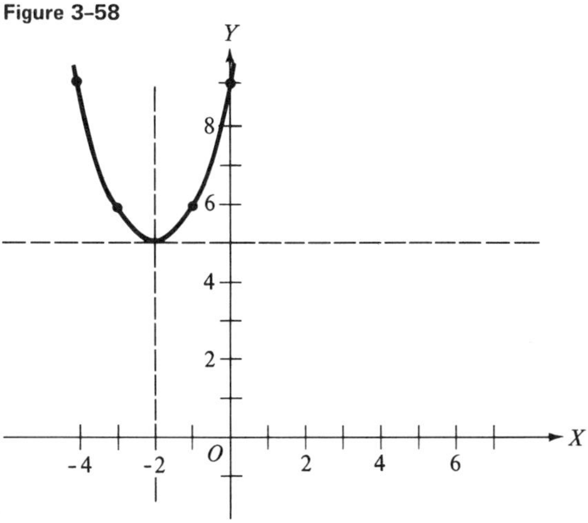

Example 14

Draw the graph of

$y = -2x^2 + 8x - 7$

The vertex is at (2, 1). The graph has the same shape as the graph of $y = -2x^2$. (See Figure 3–59.) □

Figure 3–59

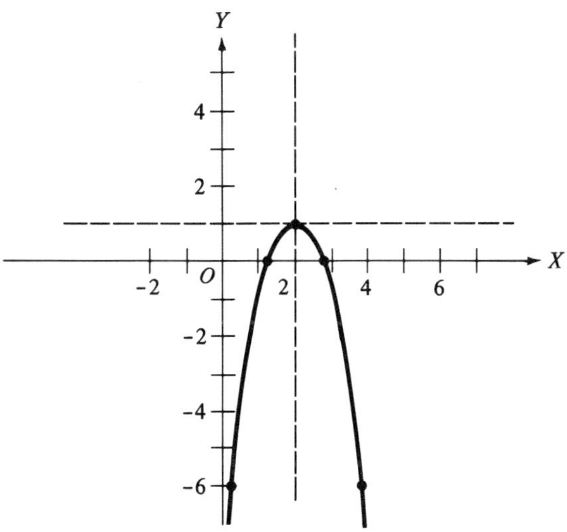

Analogous arguments, formulas, and methods can be developed for equations of the form

$$x = Ay^2 + By + C \qquad \text{with } A \neq 0.$$

We only need to interchange x and y in the previous work.

From the examples given, we see that a parabola is symmetric about the axis and for this reason the axis is also called the *axis of symmetry* of the parabola.

EXERCISES

1. Draw graphs of the following equations.
(a) $y = 4x^2$ (b) $y = -2x^2$ (c) $y = \frac{1}{2}x^2$ (d) $y = \frac{2}{3}x^2$
(e) $y = -\frac{4}{3}x^2$ (f) $y = 6x^2$ (g) $x = 3y^2$ (h) $x = -\frac{1}{4}y^2$
2. Draw graphs of the following equations.
(a) $y - 2 = 4(x - 3)^2$ (b) $y + 3 = -2(x + 4)^2$
(c) $(y - \frac{2}{3}) = \frac{1}{2}(x + 1)^2$ (d) $y - \frac{2}{3} = \frac{2}{3}(x - 1)^2$
(e) $y + 5 = -\frac{4}{3}(x + 3)^2$ (f) $y + 3 = 6(x - 7)^2$
3. Draw graphs of the following equations.
(a) $y = 2x^2 + 8x - 3$ (b) $y = x^2 + 6x - 7$
(c) $y = 4x^2 - 12x + 3$ (d) $y = -x^2 - 2x + 5$
(e) $y = -3x^2 - 6x + 5$ (f) $y = 3x^2 + 4x - 1$
(g) $y = -3x^2 + 4x - 2$ (h) $y = -5x^2 + 30x - 4$
(i) $y = \frac{1}{3}x^2 + \frac{1}{2}x - 2$ (j) $y = \frac{4}{5}x^2 + 3x$
4. Draw graphs of the following equations.
(a) $x - 2 = 3(y - 1)^2$ (b) $x + 1 = 2(y + 5)^2$ (c) $x = y^2 + 4y + 9$
(d) $x = 2y^2 + 6y - 7$ (e) $x = -y^2 - 4y + 5$ (f) $x = 5y^2 - 2y + 1$
5. What is the relationship between the axis of symmetry of the graph of $y = Ax^2 + Bx + C$ and the average of the roots of $Ax^2 + Bx + C = 0$?

*6. Verify directly from the definition that the equation of the parabola with vertex (h, k) and directrix $x = h - p$ is $x - h = (1/4p)(y - k)^2$.

7. If the equation $Ax^2 + Bx + C = 0$ has no solutions, what does this say about the graph of $y = Ax^2 + Bx + C$? What can be said if $A > 0$? If $A < 0$?
8. Find all pairs of numbers (x, y) such that $y = x^2 + 2x + 3$ and $y = x^2 - 7x + 1$.
9. Find all pairs of numbers (x, y) such that $y = x^2 + 4x + 5$ and $y = 3x^2 - 7x + 2$.
10. Under what conditions will there exist at least two pairs (x, y) satisfying both $y = Ax^2 + Bx + C$ and $y = \alpha x^2 + \beta x + \gamma$?
11. Find all pairs (x, y) which are solutions of $y = x^2 + 3$ and $x^2 + y^2 = 4$.

*12. If a line is drawn through the focus, parallel to the directrix, it will meet the parabola in two points. The segment joining these two points is called the *latus rectum* of the parabola. Determine the length of the latus rectum of the graph of $y = (1/4p)x^2$.

7. Hyperbolas

The fourth type of conic section is the *hyperbola* as illustrated in Figure 3–60. As with circles, ellipses, and parabolas, we wish to define these curves in

Figure 3–60

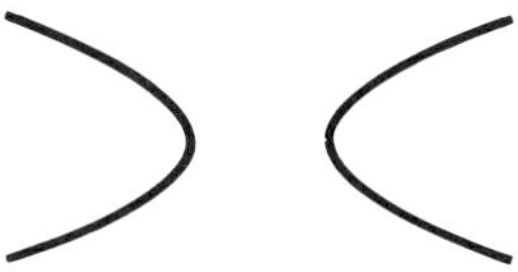

terms of distances so that we can derive the equation using the distance formula.

A *hyperbola* is a set of points such that the difference of the distances from any point in the set to two fixed points, called the *foci*, is a constant. The shape of a hyperbola is dependent upon this constant and the distance between the foci; the orientation of the foci has no effect on the shape of the curve, only on the orientation of the curve. We will limit our discussion to the cases where both foci are on the same horizontal line or on the same vertical line. The line that contains the foci is called the *axis* of the hyperbola. The *center* of the hyperbola is the midpoint of the segment between the foci. (See Figure 3–61.)

Figure 3–61

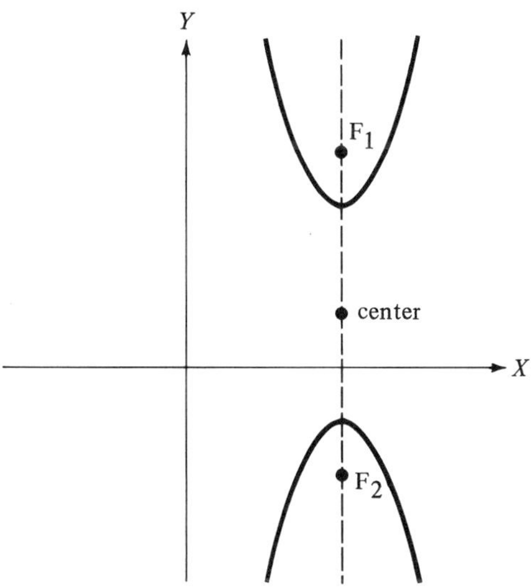

Example 1

Consider the hyperbola having foci $(-5, 0)$ and $(5, 0)$ with the difference of the distances equal to 8. Notice that $(4, 0)$ and $(-4, 0)$ are both on the curve since the distance from (4, 0) to (5, 0) is 1 unit and the distance from (4, 0) to $(-5, 0)$ is 9 units. A similar computation holds for $(-4, 0)$. Notice also that the center is the origin.

A point (x, y) is on the hyperbola if and only if the difference between the distance from (x, y) to $(-5, 0)$ and the distance from (x, y) to $(5, 0)$ is 8, that is,

$$\left|\sqrt{(x + 5)^2 + y^2} - \sqrt{(x - 5)^2 + y^2}\right| = 8$$

This equation can be transformed to yield

$$\frac{x^2}{4^2} - \frac{y^2}{3^2} = 1$$

(See Figure 3–62.)

We leave the simplification as an exercise. □

Figure 3–62

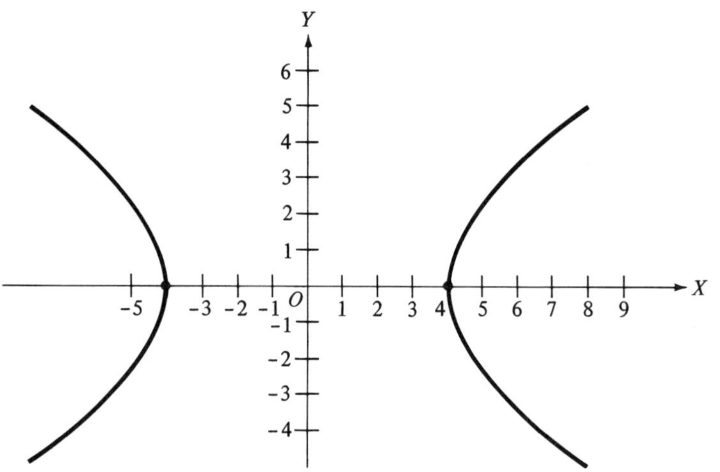

To generalize this example, consider the general case of a hyperbola with foci on the X-axis with center at the origin. Let the difference of the distances be $2a$ and let the foci be $(-c, 0)$ and $(c, 0)$ with $c > 0$. (See Figure 3–63.)

If P is any point in the plane, then the difference in the distance from P to $(-c, 0)$ and the distance from P to $(c, 0)$ is not greater than $2c$ for

Figure 3–63

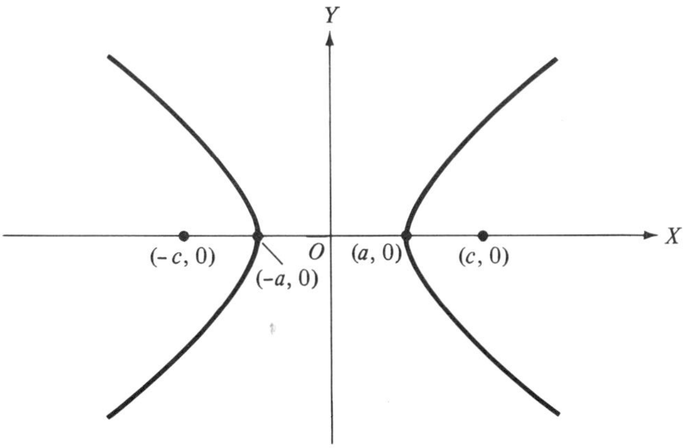

otherwise the sum of the lengths of two sides of a triangle would be less than the length of the third side. If the difference is exactly $2c$, then P is on the X-axis but not between the foci. Thus for a hyperbola to contain points not on the X-axis, we must have $2a < 2c$ and therefore we limit ourselves to the case where $a < c$.

From Figure 3–63 we see that the distance from $(a, 0)$ to $(-c, 0)$ is $a + c$ and the distance from $(a, 0)$ to $(c, 0)$ is $c - a$. The difference of these distances is $(a + c) - (c - a) = 2a$, and therefore, $(a, 0)$ is a point in the hyperbola. A similar computation shows that $(-a, 0)$ is also in the set. The points $(a, 0)$ and $(-a, 0)$ are called the *vertices* of the hyperbola. In general, the *vertices* of a hyperbola are the points of the hyperbola that are on the axis of the hyperbola.

In general, the point (x, y) is on the hyperbola if and only if the difference in the distance from (x, y) to $(c, 0)$ and the distance from (x, y) to $(-c, 0)$ is $2a$, that is,

$$|\sqrt{(x - c)^2 + y^2} - \sqrt{(x + c)^2 + y^2}| = 2a$$

This can be rewritten

$$\sqrt{(x - c)^2 + y^2} - \sqrt{(x + c)^2 + y^2} = \pm 2a$$

where the $\pm$ sign means that we are indicating two equations. We deal with both of them at the same time.

To simplify this equation, first add $\sqrt{(x + c)^2 + y^2}$ to both sides:

$$\sqrt{(x - c)^2 + y^2} = \pm 2a + \sqrt{(x + c)^2 + y^2}$$

Squaring both sides yields

$$\begin{aligned}
(x-c)^2+y^2 &= 4a^2 \pm 4a\sqrt{(x+c)^2+y^2}+(x+c)^2+y^2\\
c^2+x^2-2xc+y^2 &= 4a^2 \pm 4a\sqrt{(x+c)^2+y^2}+x^2+2xc+c^2+y^2\\
-2xc &= 4a^2 \pm 4a\sqrt{(x+c)^2+y^2}+2xc\\
-4xc &= 4a^2 \pm 4a\sqrt{(x+c)^2+y^2}\\
-4xc-4a^2 &= \pm 4a\sqrt{(x+c)^2+y^2}\\
(-1)(xc+a^2) &= \pm a\sqrt{(x+c)^2+y^2}
\end{aligned}$$

Squaring both sides yields

$$\begin{aligned}
x^2c^2+2xca^2+a^4 &= a^2[(x+c)^2+y^2]\\
x^2c^2+2xca^2+a^4 &= a^2(x^2+2xc+c^2+y^2)\\
x^2c^2+2xca^2+a^4 &= a^2x^2+2xca^2+a^2c^2+a^2y^2\\
x^2c^2-x^2a^2-a^2y^2 &= a^2c^2-a^4\\
(c^2-a^2)x^2-a^2y^2 &= a^2(c^2-a^2)
\end{aligned}$$

Letting $b=\sqrt{c^2-a^2}$, that is, $b^2=c^2-a^2$, we have

$$b^2x^2-a^2y^2=a^2b^2$$

Dividing both sides by a^2b^2 yields

$$\frac{x^2}{a^2}-\frac{y^2}{b^2}=1$$

If the foci are on the Y-axis with the center at the origin, say at $(0, c)$ and $(0, -c)$, then a similar process yields

$$\frac{y^2}{a^2}-\frac{x^2}{b^2}=1$$

with $b^2=c^2-a^2$ as before. In this case, the vertices are $(0, a)$ and $(0, -a)$.

Example 2

Draw the graph of

$$\frac{y^2}{9}-\frac{x^2}{16}=1$$

The graph is a hyperbola with a vertical axis. (See Figure 3–64.) Since $c^2=a^2+b^2=9+16=25$, $c=5$ and the foci must be $(0, 5)$ and $(0, -5)$. The difference in distances is $2(3)=6$ and the vertices are $(0, 3)$ and $(0, -3)$. Thus the center must be $(0, 0)$. See Figure 3–64. □

Figure 3-64

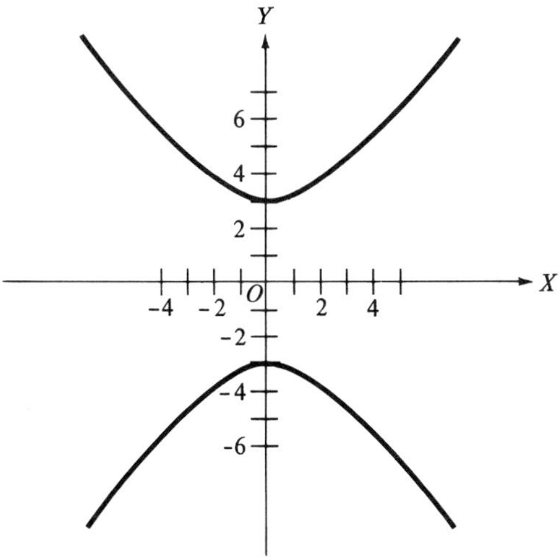

The hyperbolas discussed thus far are symmetric about the origin and the center of each is, in fact, the origin.

Example 3

What is the equation of the hyperbola with center (2, 7) and having the same shape and orientation as the hyperbola in Example 1?

First of all, since the axis was horizontal in Example 1, the axis of this hyperbola must be horizontal and pass through the center (2, 7). Thus the axis is the line $y = 7$.

Since the shape depends on the distance between the foci and the difference in the distances, the distance between the foci must be 10 as in Example 1. Since (2, 7) is the midpoint of the segment joining the foci, the foci must be the points $(-3, 7)$ and $(7, 7)$.

A point (x, y) is on the hyperbola if and only if the difference between the distance from (x, y) to $(-3, 7)$ and the distance from (x, y) to $(7, 7)$ is 6, that is, if and only if

$$\left|\sqrt{(x+3)^2 + (y-7)^2} - \sqrt{(x-7)^2 + (y-7)^2}\right| = 6$$

This equation can be simplified to yield

$$\frac{(x-2)^2}{16} - \frac{(y-7)^2}{9} = 1$$

The vertices are $(-2, 7)$ and $(6, 7)$. (See Figure 3–65.) □

Figure 3-65

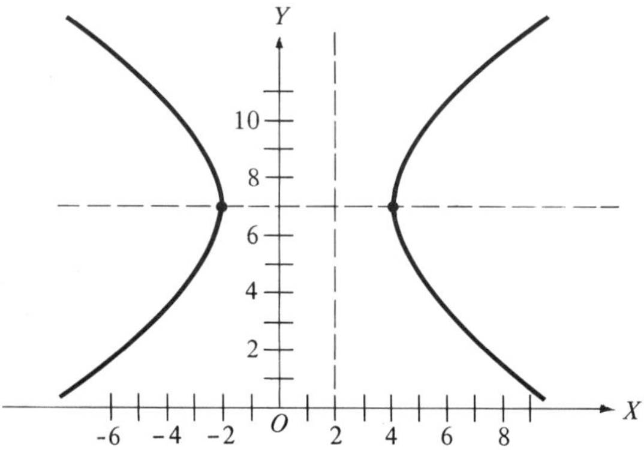

In general, if we know that the center of a hyperbola is (h, k), the axis is *horizontal*, the distance between the foci is $2c$, and the difference in the distances is $2a$, then we know that the axis is the line $y = k$ and that the foci are $(h - c, k)$ and $(h + c, k)$. A point (x, y) is on the hyperbola if and only if

$$\left|\sqrt{[x - (h - c)]^2 + (y - k)^2} - \sqrt{[x - (h + c)]^2 + (y - k)^2}\right| = 2a$$

This equation can be simplified to yield

$$\frac{(x - h)^2}{a^2} - \frac{(y - k)^2}{b^2} = 1$$

where $b^2 = c^2 - a^2$ as before. We leave the simplification as an exercise.

By analogous reasoning, if the center of the hyperbola is at (h, k), the axis is *vertical*, the distance between foci is $2c$, and the difference in distances is $2a$, then an equation for the hyperbola is given by

$$\frac{(y - k)^2}{a^2} - \frac{(x - h)^2}{b^2} = 1$$

where $b^2 = c^2 - a^2$ as before.

Example 4

The graph of

$$\frac{(y - 4)^2}{36} - \frac{(x - 1)^2}{64} = 1$$

is a hyperbola with vertical axis $x = 1$ and center $(1, 4)$. Since $c^2 = a^2 + b^2$, $c = 10$ and the foci are at $(1, 14)$ and $(1, -6)$. The vertices are $(1, 10)$ and $(1, -2)$. (See Figure 3–66.) □

Figure 3-66

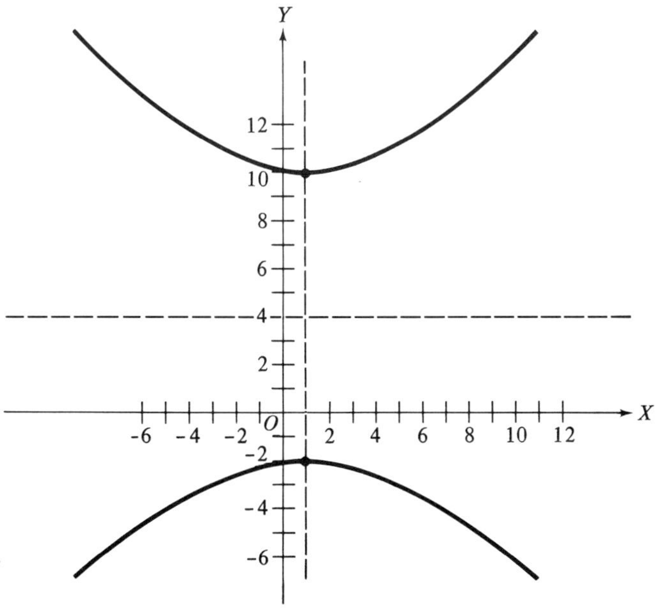

Since graphs of equations of the forms

$$\frac{(x-h)^2}{a^2} - \frac{(y-k)^2}{b^2} = 1 \quad \text{and} \quad \frac{(y-k)^2}{a^2} - \frac{(x-h)^2}{b^2} = 1$$

have the same shape and orientation as graphs of

$$\frac{x^2}{a^2} - \frac{y^2}{b^2} = 1 \quad \text{and} \quad \frac{y^2}{a^2} - \frac{x^2}{b^2} = 1$$

we can draw graphs of hyperbolas with center (h, k) by drawing the lines $x = h$ and $y = k$ and drawing the graph as if we were drawing the graph of the simpler equation with these lines as axes. Thus, in order to be able to draw the graphs of hyperbolas, we only need to develop techniques for drawing those with center at the origin.

Example 5

Draw the graph of

$$\frac{x^2}{9} - \frac{y^2}{16} = 1$$

The graph is a hyperbola with center (0, 0) and the X-axis as its axis. The vertices are $(-3, 0)$ and $(3, 0)$. The problem now is to determine the shape of the curve.

The curve does not intersect the lines $y = \frac{4}{3}x$ and $y = -\frac{4}{3}x$ because for any point (x, y) on either of these lines,

$$y^2 = \frac{16}{9}x^2$$

$$9y^2 = 16x^2$$

or

$$\frac{y^2}{16} = \frac{x^2}{9}$$

whereas if (x, y) is on the hyperbola,

$$\frac{x^2}{9} - \frac{y^2}{16} = 1$$

However, the further away from the origin we choose the point (x, y), the closer it comes to one of these lines. To see this, notice first that if we solve the equation of the hyperbola for y in terms of x, we have

$$y = \pm\tfrac{4}{3}\sqrt{x^2 - 9}$$

In quadrant I, the vertical distance between the line $y = \frac{4}{3}x$ and the hyperbola is just

$$\left|\tfrac{4}{3}x - \tfrac{4}{3}\sqrt{x^2 - 9}\right| = \left|\tfrac{4}{3}(x - \sqrt{x^2 - 9})\right|$$

Since

$$x - \sqrt{x^2 - 9} = \frac{9}{x + \sqrt{x^2 - 9}}$$

this distance is

$$\left|\frac{4}{3}\left(\frac{9}{x + \sqrt{x^2 - 9}}\right)\right| = \frac{12}{x + \sqrt{x^2 - 9}}$$

Now as we choose values of x larger and larger, the denominator gets larger and larger and the value of the fraction gets arbitrarily small but is never 0. Thus the curve gets closer to the line. A similar result occurs in the other quadrants with the line $y = \frac{4}{3}x$ or with $y = -\frac{4}{3}x$.

To sketch the curve, we now draw the dotted lines $y = \frac{4}{3}x$ and $y = -\frac{4}{3}x$. (See Figure 3–67.) Then starting from the vertices (3, 0) and (−3, 0), we draw smooth curves which approach the dotted lines as values for x are chosen further and further from the origin. □

In general, the graph of

$$\frac{x^2}{a^2} - \frac{y^2}{b^2} = 1$$

Figure 3–67

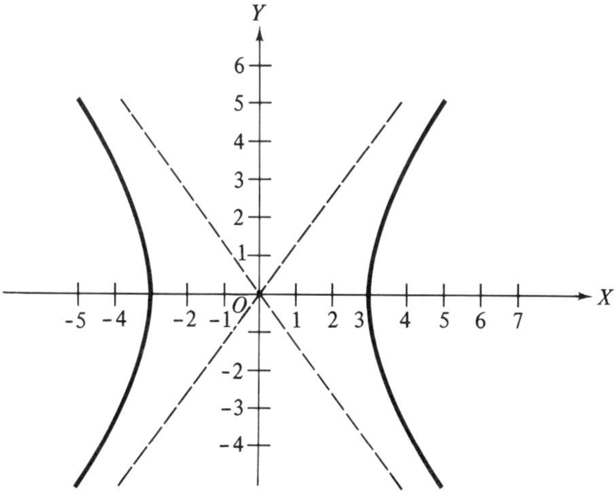

gets close to the lines $y = (b/a)x$ and $y = (-b/a)x$ as values of x are chosen further from the origin. The argument given in Example 1 generalizes as follows:

Solving the equation

$$\frac{x^2}{a^2} - \frac{y^2}{b^2} = 1$$

for y in terms of x yields

$$y = \pm\frac{b}{a}\sqrt{x^2 - a^2}$$

In quadrants I and II, the sign is positive and in quadrants III and IV, it is negative. In quadrant I, the horizontal distance between the line $y = (b/a)x$ and the curve is

$$\left|\frac{b}{a}x - \frac{b}{a}\sqrt{x^2 - a^2}\right| = \left|\frac{b}{a}(x - \sqrt{x^2 - a^2})\right|$$

$$= \left|\frac{b}{a}\frac{a^2}{x + \sqrt{x^2 - a^2}}\right|$$

$$= \frac{ba}{x + \sqrt{x^2 - a^2}}$$

See Figure 3–68.

Figure 3–68

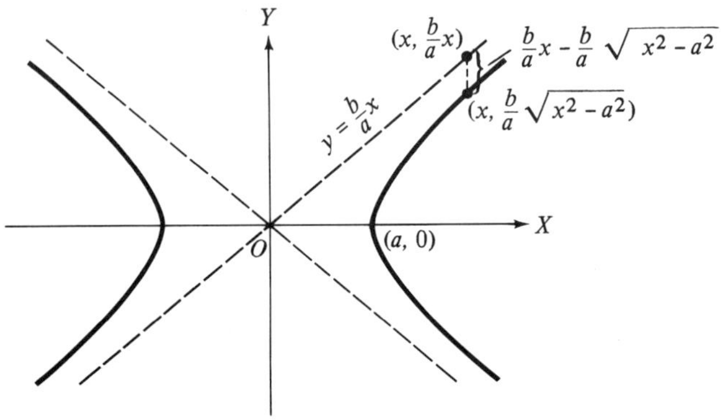

As values of x are chosen larger, the denominator gets larger, the fraction gets smaller, and the distance arbitrarily small but is never 0. Similar justifications can be made for other quadrants and we leave them as exercises. The lines $y = (b/a)x$ and $y = (-b/a)x$ are called the *asymptotes* of the curve.

The graph of

$$\frac{y^2}{a^2} - \frac{x^2}{b^2} = 1$$

gets arbitrarily close to the lines $y = (a/b)x$ and $y = (-a/b)x$. The justification is similar to that given above and we leave it to the reader. As before, we plot the vertices, draw the dotted lines $y = (a/b)x$ and $y = (-a/b)x$, and sketch smooth curves that approach these lines. Again, these lines are called *asymptotes*.

Example 6

The graph of

$$\frac{y^2}{4} - \frac{x^2}{16} = 1$$

is a hyperbola with center at the origin, vertical axis, vertices at $(0, -2)$ and $(0, 2)$. The asymptotes are $y = \frac{1}{2}x$ and $y = -\frac{1}{2}x$. (See Figure 3–69.) □

In summary:

(a) The graph of an equation of the form

$$\frac{(x-h)^2}{a^2} - \frac{(y-k)^2}{b^2} = 1$$

Figure 3–69

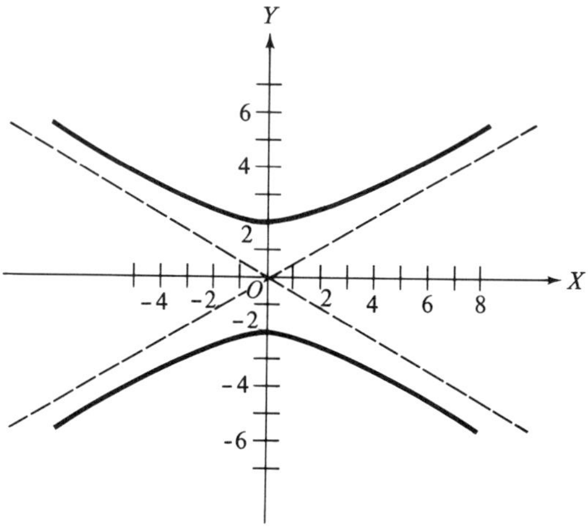

is a hyperbola with center (h, k), axis $y = k$, and vertices $(h - a, k)$ and $(h + a, k)$.

(b) The graph of an equation of the form

$$\frac{(y - k)^2}{a^2} - \frac{(x - h)^2}{b^2} = 1$$

is a hyperbola with center (h, k), axis $x = h$, and vertices $(h, k + a)$ and $(h, k - a)$.

(c) The graph of

$$\frac{x^2}{a^2} - \frac{y^2}{b^2} = 1 \quad \text{or} \quad \frac{y^2}{a^2} - \frac{x^2}{b^2} = 1$$

is drawn by drawing the asymptotes, plotting the vertices, and sketching smooth curves through the vertices approaching the asymptotes.

(d) The graph of

$$\frac{(x - h)^2}{a^2} - \frac{(y - k)^2}{b^2} = 1 \quad \text{or} \quad \frac{(y - k)^2}{a^2} - \frac{(x - h)^2}{b^2} = 1$$

is drawn by drawing the lines $x = h$ and $y = k$ and drawing the graph of

$$\frac{x^2}{a^2} - \frac{y^2}{b^2} = 1 \quad \text{or} \quad \frac{y^2}{a^2} - \frac{x^2}{b^2} = 1$$

as if these lines were the X- and Y-axes.

Example 7

Draw the graph of

$$\frac{(x-4)^2}{9} - \frac{(y-1)^2}{25} = 1$$

The center is at (4, 1) and the axis is horizontal. Draw the dotted lines $x = 4$ and $y = 1$. Now draw the graph of

$$\frac{x^2}{9} - \frac{y^2}{25} = 1$$

on this "new system of axes." The asymptotes are $y = \frac{5}{3}x$ and $y = -\frac{5}{3}x$ (on the "new" system). (See Figure 3–70.) □

Figure 3–70

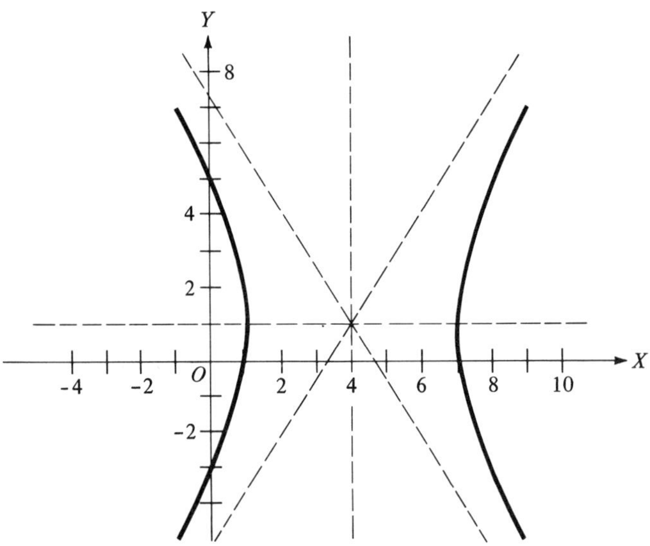

Example 8

Draw the graph of

$$\frac{(y+1)^2}{4} - \frac{(x-2)^2}{1} = 1$$

The center is at $(2, -1)$ and the axis is vertical. Draw the dotted lines $x = 2$ and $y = -1$. Now draw the graph of

$$\frac{y^2}{4} - \frac{x^2}{1} = 1$$

on this "new coordinate system." The asymptotes are $y = 2x$ and $y = -2x$ (on the "new" system). (See Figure 3–71.) □

Figure 3–71

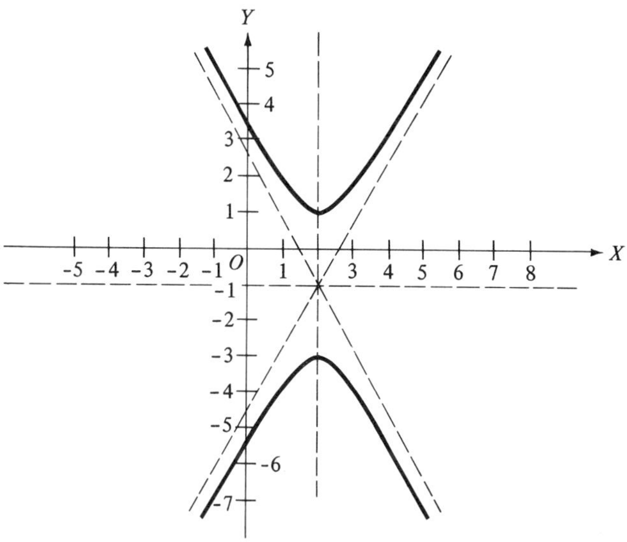

By using the completing the square technique, an equation of the form

$$Ax^2 + Bx + Cy^2 + Dy = E$$

can be transformed into one of the above forms or into an equation of the form

$$\frac{(x-h)^2}{a^2} - \frac{(y-k)^2}{b^2} = 0$$

In this case, the graph is a pair of lines. We leave the verification as an exercise.

Example 9

$$4x^2 - 8x - 9y^2 - 18y = 41$$
$$4(x^2 - 2x + 1) - 9(y^2 + 2y + 1) = 4 - 9 + 41$$
$$4(x - 1)^2 - 9(y + 1)^2 = 36$$
$$\frac{4(x - 1)^2}{36} - \frac{9(y + 1)^2}{36} = 1$$
$$\frac{(x - 1)^2}{9} - \frac{(y - 1)^2}{4} = 1$$

The graph is a hyperbola with horizontal axis $y = -1$, center at $(1, -1)$, and vertices $(-2, -1)$ and $(4, -1)$. We draw the dotted lines $x = 1$ and $y = -1$ as the "new" axes and draw the asymptotes $y = \frac{2}{3}x$ and $y = -\frac{2}{3}x$ on the "new" coordinate system and sketch curves through the vertices and approaching the asymptotes. (See Figure 3–72.) □

Figure 3–72

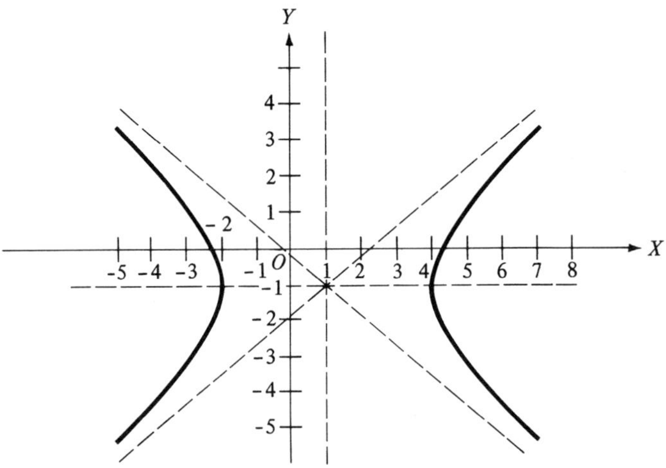

EXERCISES

In Exercises 1–10 draw graphs of the equations.

1. $\frac{x^2}{4} - \frac{y^2}{16} = 1$

2. $\frac{y^2}{4} - \frac{x^2}{16} = 1$

3. $\frac{(x - 1)^2}{4} - \frac{(y - 2)^2}{9} = 1$

4. $\frac{(x + 3)^2}{9} - \frac{(y - 1)^2}{25} = 1$

5. $\dfrac{(y-4)^2}{3} - \dfrac{(x-1)^2}{5} = 1$

6. $9x^2 - 4y^2 = 36$

7. $x^2 - 6x - 4y^2 - 8y = 11$

8. $4x^2 + 16x - y^2 + 2y = 10$

9. $2y^2 - 3x^2 + 6x = 4$

10. $5y^2 - 8y - 6x^2 - 7x = 3$

11. Verify directly that the equation of the hyperbola with center (h, k), foci $(h, k + c)$ and $(h, k - c)$, and difference of distances $2a$ is

$$\frac{(y-k)^2}{a^2} - \frac{(x-h)^2}{b^2} = 1, \qquad \text{with } b^2 = c^2 - a^2$$

12. (a) Show that the graph of

$$\frac{x^2}{4} - \frac{y^2}{9} = 0$$

is a pair of lines.

(b) Generalize part (a) for the equation

$$\frac{(x-h)^2}{a^2} - \frac{(y-k)^2}{b^2} = 0$$

13. Draw the graph of $x^2 - y^2 = 1$ and the graph of $y = 1/x$. Do the curves have the same shape?

14. Show that in quadrant II, the graph of

$$\frac{x^2}{a^2} - \frac{y^2}{b^2} = 1$$

gets close to the line $y = (-b/a)x$ as $|x|$ is chosen larger.

15. Show that in quadrant I, the graph of

$$\frac{y^2}{a^2} - \frac{x^2}{b^2} = 1$$

gets close to the line $y = (a/b)x$ as x is chosen larger.

16. Explain why there are no points on the graph of

$$\frac{x^2}{a^2} - \frac{y^2}{b^2} = 1$$

for values of x such that $-a < x < a$.

17. What relationship is there between the graph of

$$\frac{x^2}{a^2} - \frac{y^2}{b^2} = 0$$

and the graph of

$$\frac{x^2}{a^2} - \frac{y^2}{b^2} = 1?$$

18. Verify the fact that

$$x - \sqrt{x^2 - a^2} = \frac{a}{x + \sqrt{x^2 - a^2}}$$

8. Functions

The words "function" and "mapping" are used interchangeably in modern mathematics and, although we will use the word "function" most of the time in this text, the word "mapping" may give a better intuitive feeling for the concept. Suppose we have two sets of numbers X and Y, and a rule or process by which each element of X is associated with exactly one element of Y. This rule or process can be said to "map" X into Y. If we denote the process by a letter, such as f, a diagram such as Figure 3–73 gives us an intuitive grasp of the notion.

Figure 3–73

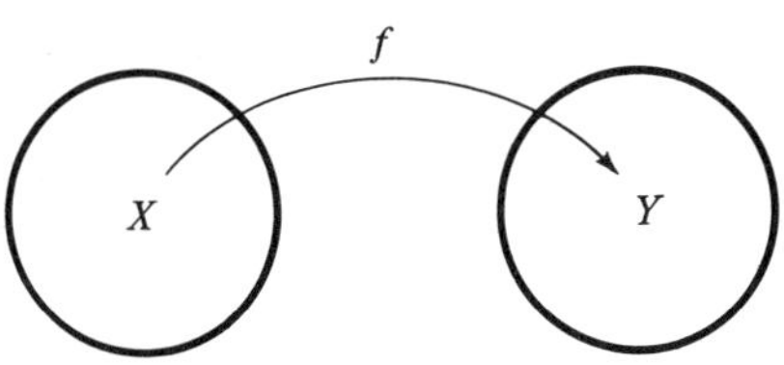

If X and Y are both R, the set of all real numbers, we could use number lines and draw a diagram as in Figure 3–74.

Figure 3–74

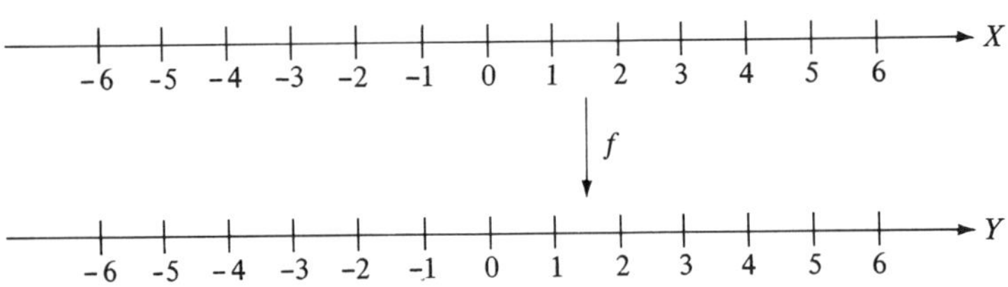

Example 1

Let f be the process that associates each real number with its square. This is illustrated by Figure 3–75. □

Figure 3–75

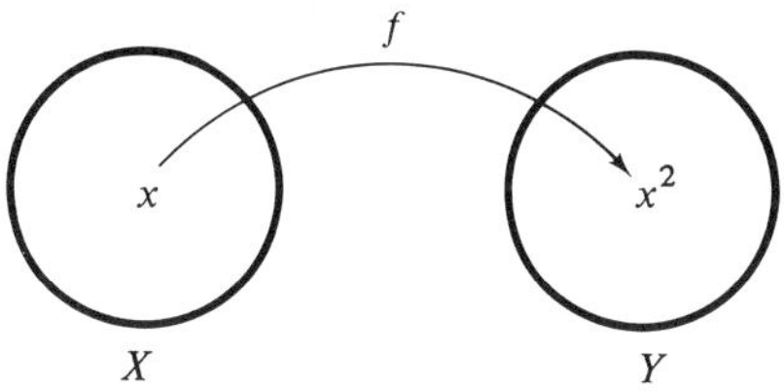

Example 2

Let g be the process that associates each number with its cube root. This is illustrated by Figure 3–76. □

Figure 3–76

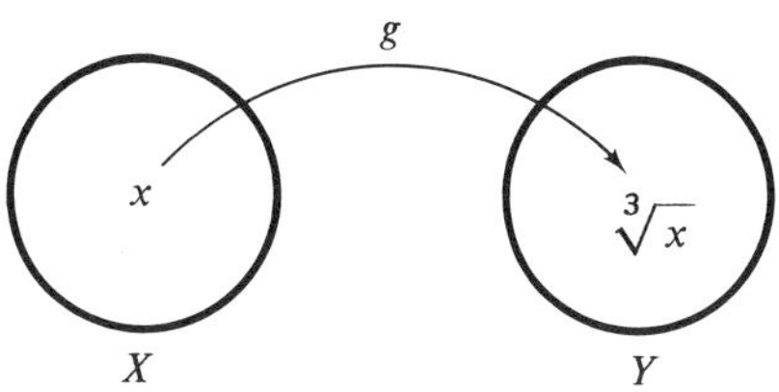

Example 3

Let h be the process that associates each number with its double. This is illustrated by Figure 3–77. □

Figure 3–77

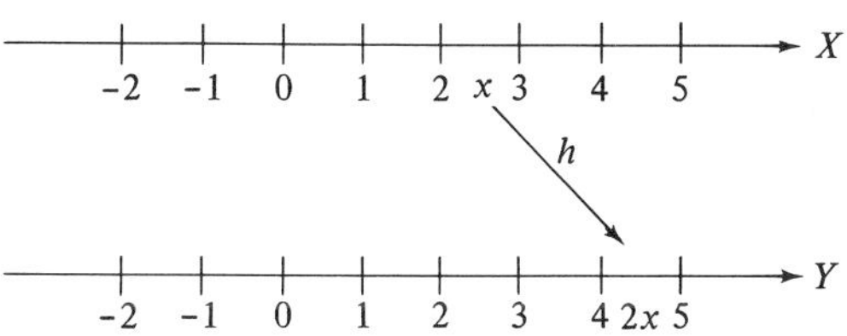

If f is a function and a is a number, we let $f(a)$ denote the unique number that f associates with a or that f "maps a to." We can use algebraic expressions to describe many mappings and the usual method is to use a variable, such as x, and an algebraic expression involving the variable to indicate the number that each value of x is associated with. [As with polynomials, "$f(x)$" is read "f of x" and "$f(2)$" is read "f of 2."]

Thus for the function in Example 1 we could write

$$f(x) = x^2$$

For Example 2 we could write

$$g(x) = \sqrt[3]{x}$$

and for Example 3 we could write

$$h(x) = 2x$$

To determine the number that such a function, say k, associates with a number, we just substitute that number for x in the algebraic expression.

Example 4

Let

$$k(x) = 2x^2 - 3$$

Then

$$k(0) = 0 - 3 = -3$$
$$k(1) = 2 \cdot 1 - 3 = -1$$
$$k(5) = 2 \cdot 25 - 3 = 47 \quad \square$$

Example 5

Let

$$p(x) = \frac{x^2 - 7x}{x^2 + 1}$$

Then

$$p(0) = 0$$
$$p(1) = -3$$
$$p(4) = -\frac{12}{17} \quad \square$$

Example 6

Let

$$q(x) = \sqrt{x} \qquad \text{for all } x \geq 0$$

Then

$$q(0) = 0$$
$$q(4) = 2$$
$$q(19) = \sqrt{19} \ \square$$

The set of all numbers that a function f "acts upon" or "maps to some number" is called the *domain* of f and the set of all numbers that f "maps something to" is the *range* of f. It is customary to assume, unless otherwise stated, that when a function is given by means of an algebraic or geometric rule or expression involving a variable, say x, the domain of the function is the set of all numbers which can meaningfully be substituted for x.

Example 7

Let

$$s(x) = \sqrt{x - 3}$$

The domain of s is the set of all real numbers greater than or equal to 3. The range is the set of all nonnegative numbers. $\square$

Example 8

Let

$$w(x) = \frac{4}{x^2 - 1}$$

The domain of w is the set of all real numbers except 1 and -1. The range is $\{y: y \leq -4 \quad \text{or} \quad y > 0\}$. $\square$

Example 9

Let

$$z(x) = \frac{x^2 - 2x + 5}{x^2 + 4}$$

The domain of $z(x)$ is the set of all real numbers. $\square$

If we wish to use algebraic expressions to describe functions that have a certain set as the domain, all we need do is specify the domain in our description.

Example 10

Let

$$v(x) = x^2 + 3x - 2 \qquad \text{for } x \geq 5$$

In this case we have, by our description, limited the domain to numbers greater than or equal to 5. □

Example 11

Let

$u(x) = x^2 + 1$ for x an integer

In this case, the domain is the set of integers. □

The range of a function described by an algebraic expression is more difficult to determine. We need to treat each function individually but some of the simpler ones are rather easy to analyze.

Example 12

Let

$m(x) = x^2 + 5$

Since $x^2 \geq 0$ for all x, $x^2 + 5 \geq 5$ for all x. Furthermore, $x^2 + 5$ will assume all possible values greater than or equal to 5 and therefore the range of m is the set of numbers greater than or equal to 5. □

Example 13

Let

$n(x) = \sqrt[3]{x}$

Since each real number has exactly one cube, each real number is the cube root of some number and the range of $n(x)$ is the set of all real numbers.

Thus far we have discussed only the intuitive idea of a function and all our examples were functions that can be described by algebraic expressions. In Chapter 5 we will be concerned with six very important functions that are described by geometric rules.

The intuitive notion of a function can be made precise by the following formal set-theoretic definition:

A *function* is a subset of $R \times R$ in which no two distinct pairs have the same first element.

This meets our intuitive notion if we consider each first number in an ordered pair to be associated with the second number in the pair. The domain is the set of all numbers which appear as first elements and the range is the set of all numbers which appear as second elements. We will not use this definition to any great extent but it does put the concept of a function on a more firm foundation.

The calculus and, more generally, real analysis are primarily concerned with the study of functions and one of the more useful tools for studying a function is its graph. The graph of a function, like the graph of an equation, is a picture of a set of ordered pairs. Since a function is a set of ordered pairs, the graph of a function is, in fact, a picture of the function. In order to draw the graph of a function, we draw the graph of the equation

$$y = f(x)$$

Example 14

To draw the graph of

$$f(x) = x^2 + 2x + 3$$

we draw the graph of

$$y = x^2 + 2x + 3$$

which we know is a parabola with vertex $(-1, 2)$, opening upward, and shaped like $y = x^2$. (See Figure 3–78.) □

Figure 3–78

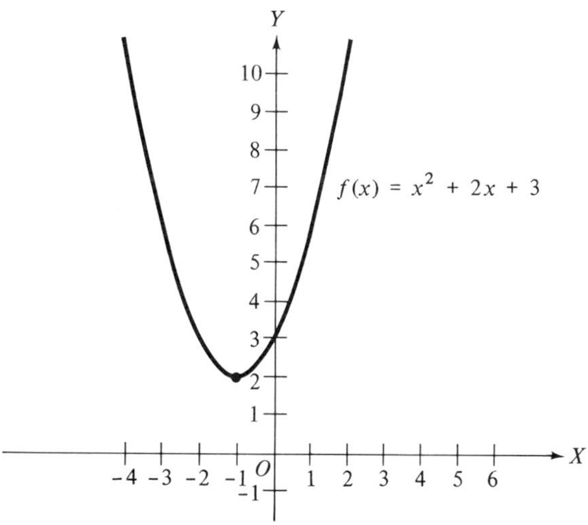

Example 15

To draw the graph of

$$g(x) = \frac{1}{x}$$

Figure 3–79

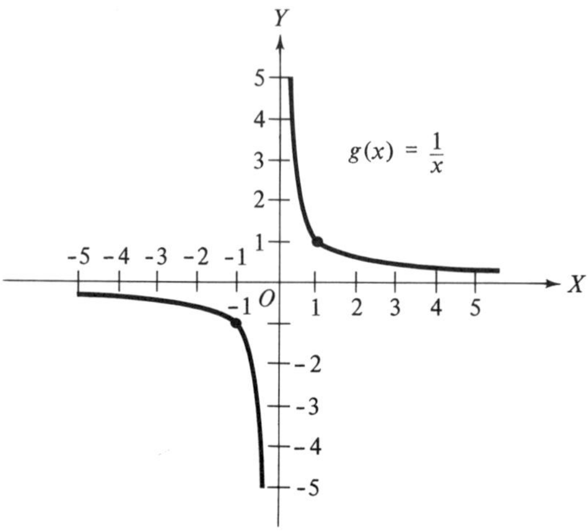

we draw the graph of

$$y = \frac{1}{x}$$

as in Figure 3–79. Notice that the domain of g is the set of all nonzero real numbers. □

Our techniques for graphing functions are, at this point, limited to our techniques for graphing equations. If a function can be described using an algebraic expression which is related to one of the special forms we have previously discussed, then we can use our knowledge to draw the graph as we did in Examples 14 and 15. Otherwise, for now, we must use the "plot and sketch" technique.

Example 16

If $p(x) = 3x - 2$, the graph of p is a line with slope 3 and Y-intercept -2. □

Example 17

If $h(x) = \sqrt{16 - x^2}$, the graph of h will be the graph of the equation $y = \sqrt{16 - x^2}$. Squaring both sides yields $y^2 = 16 - x^2$ which is equivalent to $x^2 + y^2 = 16$. This is the equation of a circle with center at (0, 0) and radius 4. Since the original equation is $y = \sqrt{16 - x^2}$, the only admissible values for y are nonnegative and the graph consists of the upper semicircle. The graph is given in Figure 3–80. □

Figure 3–80

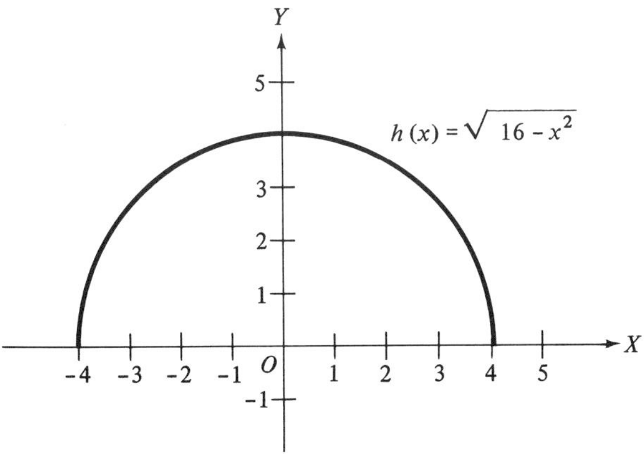

Example 18

If

$$k(x) = \frac{4}{x^2 + 1}$$

the graph has the same shape to the left of 0 as to the right because $k(a) = k(-a)$ for each real number a. Since $x^2 \geq 0$ for all x, 1 is the smallest value of the denominator and, therefore, is the largest value of $k(x)$. As we choose x larger, $k(x)$ gets smaller. The graph is given by Figure 3–81. □

Figure 3–81

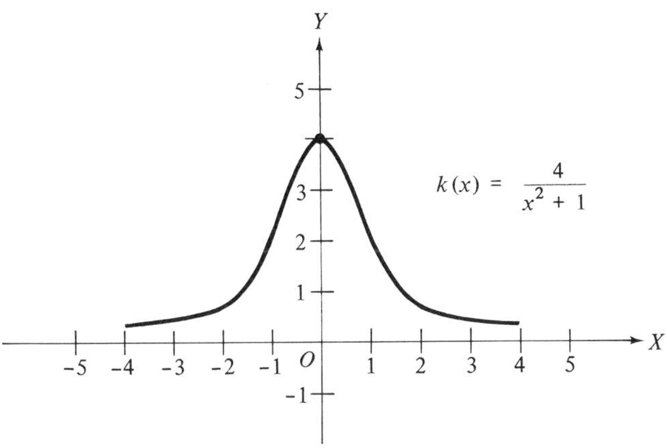

With the limited tools available, most graphs we draw must be fairly rough approximations but there are a few general procedures we can follow:

(i) Examine the function f to see if its equation $y = f(x)$ has some familiar form. If so, use your previous knowledge to draw it.
(ii) Otherwise try to determine sufficiently many points to give an outline of the general picture and then sketch from the outline. Among the more useful points are those where $f(x) = 0$ if such points exist. Such points are called the *zeros* of the function. It is also helpful to determine just where $f(x) > 0$ and where $f(x) < 0$.

Example 19

$f(x) = x^3 - 5x^2 + 6x$

Factoring yields

$f(x) = x(x - 3)(x - 2)$

Thus

$f(x) = 0 \qquad \text{if } x = 0, 2, \text{ or } 3$

and the graph crosses the X-axis at (0, 0), (2, 0), and (3, 0). Also

$f(x) > 0 \qquad \text{if } x > 3 \quad \text{or} \quad 0 < x < 2$

and

$f(x) < 0 \qquad \text{if } x < 0 \quad \text{or} \quad 2 < x < 3$

Using this information we can sketch a graph as in Figure 3–82. □

Figure 3–82

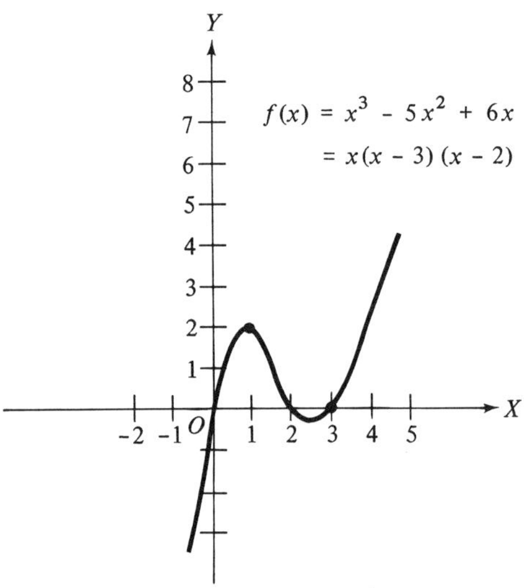

EXERCISES

1. Determine the domain and range of each of the following functions.

   (a) $f(x) = \sqrt{25 - x^2}$ (b) $g(x) = \dfrac{4}{x - 1}$

   (c) $h(x) = \dfrac{18}{x^2 + 2}$ (d) $k(x) = x^3 + 3x^2 - 2x + 1$

   (e) $p(x) = \dfrac{x^3 + 5x^2 + 2x + 7}{x^2 - 3x - 4}$

2. Draw graphs of the following functions.
   (a) $f(x) = x^2 - 5x - 6$ (b) $g(x) = \sqrt{9 - x^2}$
   (c) $h(x) = \sqrt{9 + x^2}$ (d) $k(x) = x^3 + 7x^2 + 12x$
   (e) $p(x) = -\sqrt{16 - 4x^2}$ (f) $q(x) = (x + 4)/\sqrt{3}$
   (g) $r(x) = 6/(x^2 + 3)$ (h) $t(x) = \sqrt{5 - x^2 - 4x}$
3. Draw graphs of the following functions.
   (a) $f(x) = 1/x$ (b) $g(x) = \sqrt{x^2 - 6x}$
   (c) $h(x) = x/(x^2 + 1)$ (d) $k(x) = x^3 + 3x^2 - 4x - 12$
   (e) $p(x) = x^4 - 13x^2 + 36$ (f) $q(x) = x^4 - 2x^3 - 9x^2 + 18x$
4. Draw the graphs of the following functions. What can be said about their shapes?

   $f(x) = x^3,\quad g(x) = (x - 1)^3,\quad h(x) = (x + 2)^3,\quad k(x) = (x - 1)^3 + 4$

5. Draw the graphs of the following functions. What can be said about their shapes?

   $p(x) = 1/x^2,\quad q(x) = 1/(x + 1)^2,\quad w(x) = 1/(x - 2)^2,\quad v(x) = 1/(x - 1)^2 + 3$

6. Draw the graphs of the following functions.
   (a) $f(x) = x$ (b) $f(x) = x^2$ (c) $g(x) = x^3$ (d) $h(x) = x^4$
   (e) $k(x) = x^5$ (f) $m(x) = x^6$ (g) $n(x) = x^7$ (h) $p(x) = x^8$
7. Using Exercise 6 above, what can be said about the graph of $F(x) = x^n$ for x^n a positive integer?
8. A function is said to be *piecewise linear* if its graph is made up of pieces of lines. An example of such a graph is given in Figure 3–83.

Figure 3–83

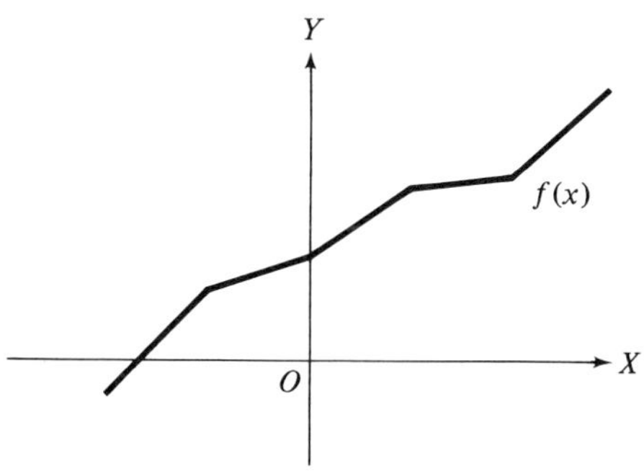

The income tax function $T(x)$ is an example of such a function as the table below indicates. From 0 to \$1000 it has slope .14 and passes through the points (0, 0) and (1000, 140). Use the accompanying table to draw the graph of $T(x)$.

Married Taxpayers Filing Joint Returns and Certain Widows and Widowers

If the amount on Form 1040, line 48, is: — Enter on Form 1040, line 16:

Not over \$1,000 — 14% of the amount on line 48.

Over—	*But not over—*		*of excess over—*
\$ 1,000	\$ 2,000	\$ 140 + 15%	\$ 1,000
\$ 2,000	\$ 3,000	\$ 290 + 16%	\$ 2,000
\$ 3,000	\$ 4,000	\$ 450 + 17%	\$ 3,000
\$ 4,000	\$ 8,000	\$ 620 + 19%	\$ 4,000
\$ 8,000	\$ 12,000	\$ 1,380 + 22%	\$ 8,000
\$ 12,000	\$ 16,000	\$ 2,260 + 25%	\$ 12,000
\$ 16,000	\$ 20,000	\$ 3,260 + 28%	\$ 16,000
\$ 20,000	\$ 24,000	\$ 4,380 + 32%	\$ 20,000
\$ 24,000	\$ 28,000	\$ 5,660 + 36%	\$ 24,000
\$ 28,000	\$ 32,000	\$ 7,100 + 39%	\$ 28,000
\$ 32,000	\$ 36,000	\$ 8,660 + 42%	\$ 32,000
\$ 36,000	\$ 40,000	\$ 10,340 + 45%	\$ 36,000
\$ 40,000	\$ 44,000	\$ 12,140 + 48%	\$ 40,000
\$ 44,000	\$ 52,000	\$ 14,060 + 50%	\$ 44,000
\$ 52,000	\$ 64,000	\$ 18,060 + 53%	\$ 52,000
\$ 64,000	\$ 76,000	\$ 24,420 + 55%	\$ 64,000
\$ 76,000	\$ 88,000	\$ 31,020 + 58%	\$ 76,000
\$ 88,000	\$100,000	\$ 37,980 + 60%	\$ 88,000
\$100,000	\$120,000	\$ 45,180 + 62%	\$100,000
\$120,000	\$140,000	\$ 57,580 + 64%	\$120,000
\$140,000	\$160,000	\$ 70,380 + 66%	\$140,000
\$160,000	\$180,000	\$ 83,580 + 68%	\$160,000
\$180,000	\$200,000	\$ 97,180 + 69%	\$180,000
\$200,000	....	\$110,980 + 70%	\$200,000

9. Draw the graphs of the following functions.
   (a) $f(x) = 1/x$ (b) $p(x) = 1/x^2$ (c) $q(x) = 1/x^3$
   (d) $v(x) = 1/x^4$ (e) $w(x) = 1/x^5$ (f) $z(x) = 1/x^6$
10. Using Exercise 9 above, what can be said about the graph of $G(x) = 1/x^n$ for n a positive integer?

9. Applications of Functions

In scientific, social, or economic problems, we frequently deal with one quantity that depends on other quantities.

Example 1

In physics, the intensity of light depends (among other things) on the distance from the source. From an exercise in Chapter 2, we know that the relationship is of the form $I = k/d^2$ where I is the intensity, d is the distance from the source, and k is a constant (which we determined once we knew one pair of associated values for d and I). It may be more convenient to write this as a function $I(d) = k/d^2$ to indicate that values of I are determined by values of d once k is fixed. Figure 3–84 provides a graph of this function for the case when $k = 4$. Notice that the domain of this function consists only of positive values since the distance must be positive. □

Figure 3–84

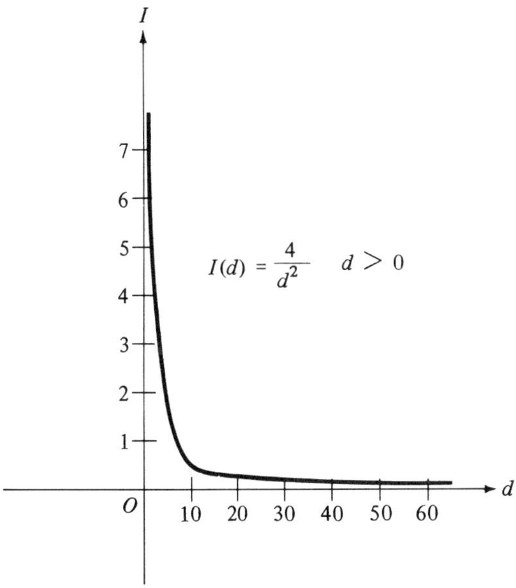

Example 2

In many cases, the volume of business can be expected to depend (among other things) on the price charged. In Example 2 in Section 4, we considered the simple case of a donut shop in which the sales and price were related by the equation $S = -600P + 800$ for values of P between \$.50 and \$1.00. We could also write this relationship as a function: $S(P) = -600P + 800$. The graph is given in Figure 3–85. Notice that the domain is the set of numbers from \$.50 to \$1.00. □

Figure 3-85

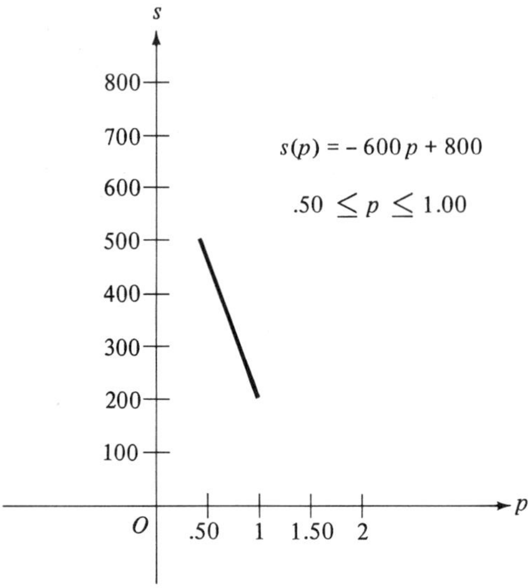

If we can express one quantity as a function of another quantity, then we can derive information by analyzing this function. Such analysis is one of the main topics of calculus but we will explore a few examples that do not need the power of calculus.

In analyzing functions, it frequently is useful to know the maximum and minimum values attained and also to know the "local" maximum values and minimum values or "peaks and valleys." If we have a good graph we can approximate such values directly from the graph.

Example 3

Consider the function $f(x) = x^3 - 12x$ with its graph given in Figure 3–86. If the graph is carefully drawn, we can use a ruler to determine the values of x that correspond to the high points and low points. From the drawing it appears that the graph has a "peak" or local maximum at $x = -2$ and a "valley" or local minimum at $x = 2$. Computing shows that $f(-2) = 16$ and $f(2) = -16$. One way to help decide whether or not we did pick just the right value of x would be to compute $f(-1.99)$ and $f(-2.01)$ to see if they are both less than 16 and to compute $f(1.99)$ and $f(2.01)$ to see if they are greater than -16. If you do so, you will

Figure 3-86

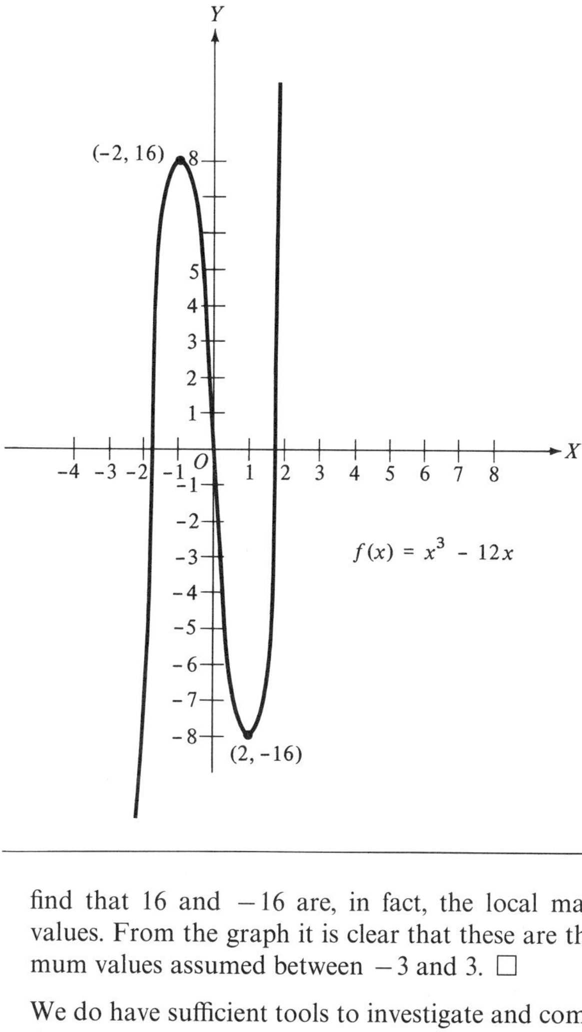

find that 16 and -16 are, in fact, the local maximum and minimum values. From the graph it is clear that these are the maximum and minimum values assumed between -3 and 3. □

We do have sufficient tools to investigate and completely determine such peaks and valleys for some kinds of functions. In particular, if the graph is a parabola opening upward or downward, then it has just one "valley" or one "peak" and this minimum or maximum value occurs at the vertex.

Example 4

Suppose $g(x) = 2x^2 - 4x + 7$ and we wish to determine the minimal value assumed by g. We know the graph of g is a parabola opening upward and the minimum value of g will be the y-coordinate of the vertex. Recall that the vertex of the graph of $y = Ax^2 + Bx + C$ is at the point where $x = -B/2A$. Thus the minimum value of g is

$$g\left(\frac{4}{2 \cdot 2}\right) = g(1) = 5 \square$$

Example 5

Suppose $p(x)$ denotes the profit function for a manufacturing company where x is the number of units manufactured per day. Let us suppose further that, through previous knowledge, we have determined that the profit function is given by

$$p(x) = 360x - .1x^2 \qquad \text{where } 0 \leq x \leq 2500$$

Thus

$$p(0) = 0 \qquad p(3600) = 0$$

and for $x > 3600$ the profit is negative, that is, there is a loss. (This could easily happen if 3600 was the maximum per day with our equipment and one shift. Employing another shift could cost more than it could make.) If we wish to maximize our profit, how many units do we manufacture per day?

The graph of $p(x)$ is a parabola and the maximum value of $p(x)$ is at the vertex of the graph. Thus, the x-coordinate is given by the formula $-B/2A$ and $x = 1800$. The y-coordinate of the vertex is $p(1800) = 32{,}400$. Thus producing 1800 units per day yields \$32,400 in profits. $\square$

If we have access to an electronic computer, we can have the graph carefully drawn by the machine (see Figure 3–87) and then use drawing instruments to help us approximate the maximum and minimum values. The machine can also be used to check to see how good the approximations are and to compute other estimates until we are satisfied with the degree of accuracy. Remember that in real applications, the original data arises from measurements and, by its very nature, is approximate. We can, therefore, only expect approximation in our answers.

Example 6

Let

$$g(x) = x^4 - 8x^3 + 6x^2 + 40x - 10$$

Figure 3-87

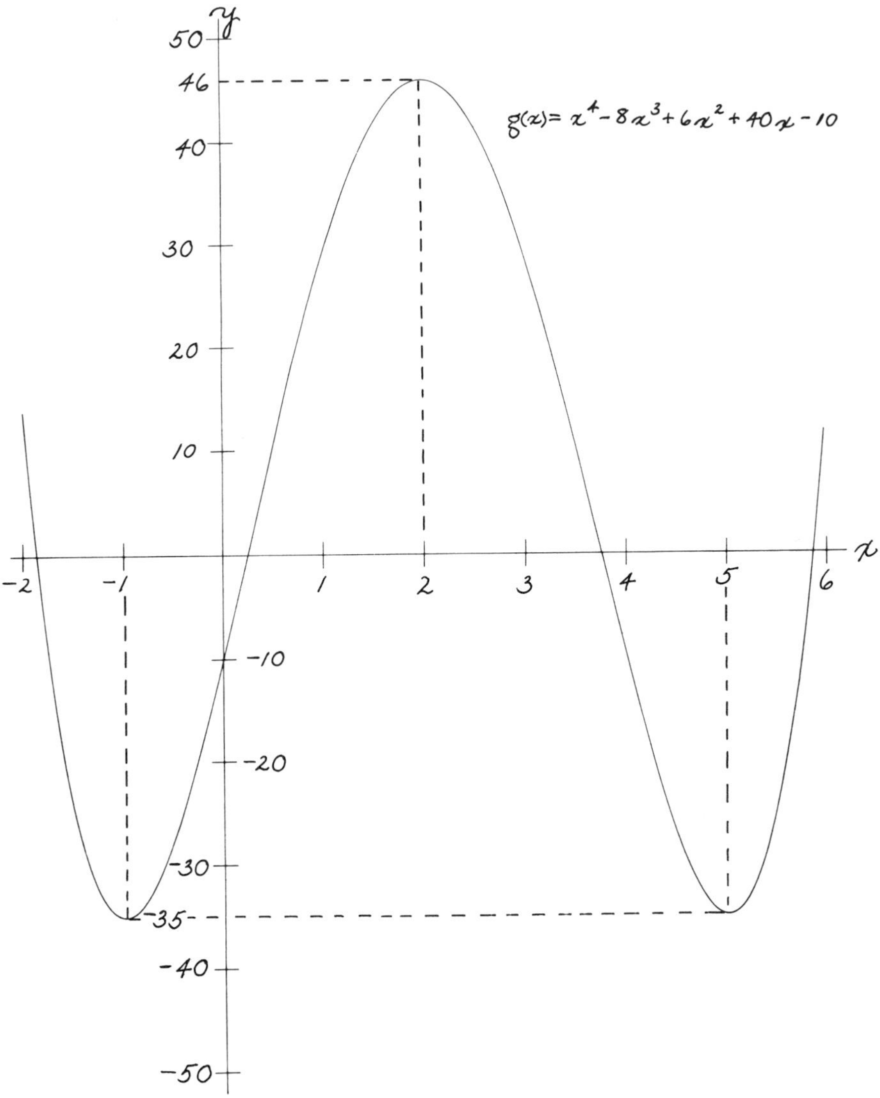

Figure 3–87 provides a graph of g drawn by an electronic computer. From the graph it appears that the local minima or "valleys" are at points where $x = -1$ or $x = 5$. Computing shows that $g(-1) = -35$ and $g(5) = -35$. The only local maximum or "peak" is the point where $x = 2$ and $g(2) = 46$. □

EXERCISES

1. Draw graphs of the following functions and determine the "peaks" and "valleys," that is, the local maxima and minima.
   (a) $f(x) = x^2 - 4x + 19$
   (b) $g(x) = 2x^2 + 5x - 7$
   (c) $h(x) = 13 + 6x - 3x^2$
   (d) $k(x) = 24 - 2x - 4x^2$
   (e) $p(x) = x^3 - 27x, \quad \text{for } -4 \leq x \leq 4$
   (f) $q(x) = 2x^3 - 6x + 1, \quad \text{for } -2 \leq x \leq 2$
   (g) $r(x) = x^4 - 8x^2 + 16$
   *(h) $t(x) = x^4 - 20x^2 + 64$
2. If an object is thrown upward from a point d_0 ft above the ground at a velocity of v_0 at time 0, then its distance above the ground at time t is given by

   $$d(t) = -16t^2 + v_0t + d_0$$

   (This relationship was discussed in Chapter 2 under the assumption that $d_0 = 0$.)
   (a) If a bullet is shot upward with an initial velocity of 1000 ft/sec from a point 8 ft above the ground, how high will it go?
   (b) How long will it take it to reach the highest point?
   (c) How long will it take it to reach the highest point if the bullet is shot from 20 ft above the ground?
3. In throwing a ball upward, assume that the average man releases the ball about 8 ft above the ground. What must be the initial velocity, that is, how fast must he throw it, in order for the ball to reach a height of at least 136 ft?
4. Suppose in a particular business the profits are related to the sales by the relationship

   $$P(s) = 2400s - s^2$$

   What volume of sales yields the maximum profit?
5. Suppose that in your business costs per unit are related to the number of units sold by the relationship

   $$C(u) = 83 - .1u + .0001u^2$$

   (a) How many units per day yields the minimum per unit cost?
   (b) What is the minimum per unit cost per day?
   (c) How does the problem change if we replace 83 by 93?
6. Consider again the example of the donut shop where sales are a function of price, $S = -600P + 800$. Assume that the fixed costs are \$70 per day and the variable costs are .32 per dozen produced.
   (a) Express the costs C as a function of S, the sales. (Assume that you sell all that you produce.)
   (b) Express the earnings E in terms of the sales S and price P. Recall that

   $$\text{Earnings} = \text{income} - \text{costs}$$

   (c) Using substitution, express earnings as a function of price.
   (d) What price yields maximal earnings?

10. The Arithmetic of Functions

If f and g are functions and a is a number in the domains of both f and g, then $f(a)$ and $g(a)$ are numbers and we can perform the usual arithmetic operations with them, that is, we can compute

$$f(a) + g(a), \qquad f(a) - g(a), \qquad g(a) - f(a),$$

$$[f(a)][g(a)], \qquad \frac{f(a)}{g(a)} \quad [\text{if } g(a) \neq 0], \quad \text{and} \quad \frac{g(a)}{f(a)} \quad [\text{if } f(a) \neq 0]$$

Example 1

If $f(x) = x^2$ and $g(x) = x + 1$, then the number 4 is in the domain of both functions and

$$\begin{aligned} f(4) + g(4) &= 16 + 5 = 21 \\ f(4) - g(4) &= 16 - 5 = 11 \\ g(4) - f(4) &= 5 - 16 = -11 \\ [f(4)][g(4)] &= (16)(5) = 80 \\ \frac{f(4)}{g(4)} &= \frac{16}{5} \\ \frac{g(4)}{f(4)} &= \frac{5}{16} \end{aligned}$$ □

Example 2

If $p(x) = x - 7$ and $q(x) = \sqrt{x - 1}$, then $p(-2) + q(-2)$ has no meaning because $q(-2)$ does not exist, that is, -2 is not in the domain of q. Furthermore, $p(1)/q(1)$ has no meaning because $q(1) = 0$. □

If f and g are functions, we can use these arithmetic processes to define new functions as follows:

(i) The *sum* of f and g is denoted by $f + g$ and defined by

$$(f + g)(x) = f(x) + g(x)$$

for each number x that is in both the domain of f and the domain of g.

(ii) The *difference* of f and g is denoted by $f - g$ and defined by

$$(f - g)(x) = f(x) - g(x)$$

for each number x that is in both domains.

(iii) The *product* of f and g is denoted by fg and defined by

$$(fg)(x) = f(x)\,g(x)$$

for each number x in both domains.

(iv) The *quotient* of f and g is denoted by f/g or $f \div g$ and defined by

$$\frac{f}{g}(x) = \frac{f(x)}{g(x)}$$

for each number x in both domains under the further restriction that $g(x) \neq 0$.

These operations are called, as expected, addition, subtraction, multiplication, and division of functions. They are not the same as the corresponding arithmetic operations for numbers but, because they are defined using these operations, they do inherit the associative and commutative properties for addition and multiplication and the distributive law. In fact, most of the basic arithmetic properties of the real number system extend to the arithmetic of functions. The constant function $z(x) = 0$ takes the place of the number 0 and the constant function $I(x) = 1$ behaves like the number 1.

If we have functions that are given by algebraic expressions, the simplest way to express the sum, product, difference, or quotient may be by indicating the sum, product, difference, or quotient of the original expressions and perhaps simplifying them. Care must be used in doing so to make sure that the new domain is specified since the resulting expression may have meaning for substitutions that were meaningless for one of the original functions.

Example 3

If $h(x) = 1/(x - 1)$ and $k(x) = \sqrt{x}$, then the domain of $h + k$ is the set of all nonnegative numbers except 1 and

$$(h + k)(x) = h(x) + k(x) = \frac{1}{x - 1} + \sqrt{x} = \frac{x\sqrt{x} - \sqrt{x} + 1}{x - 1}$$

The domain of the product is the same as that of the sum and

$$(hk)(x) = h(x)k(x) = \frac{1}{x - 1}\sqrt{x} = \frac{\sqrt{x}}{x - 1}$$

The difference also has this domain and

$$(h - k)(x) = \frac{1}{x - 1} - \sqrt{x} = \frac{1 - x\sqrt{x} + \sqrt{x}}{x - 1}$$

Also

$$\frac{h}{k}(x) = \frac{h(x)}{k(x)} = \frac{1}{x-1}\sqrt{x} = \frac{1}{(x-1)\sqrt{x}}$$

with the domain being the set of all positive numbers except 1. □

Example 4

If $f(x) = x^2 - 4$ and $g(x) = x - 2$, then

$$(f + g)(x) = f(x) + g(x) = (x^2 - 4) + (x - 2) = x^2 + x - 6$$
$$(fg)(x) = (x^2 - 4)(x - 2) = x^3 - 2x^2 - 4x + 8$$
$$\frac{f}{g}(x) = \frac{f(x)}{g(x)} = \frac{x^2 - 4}{x - 2} = x + 2 \qquad \text{for } x \neq 2$$

Although $x + 2$ has meaning for $x = 2$, $(x^2 - 4)/(x - 2)$ does not and the domain of f/g is the set of all numbers except 2. □

Example 5

If $p(x) = x^2 + \sqrt{x}$ and $q(x) = \sqrt{x}$, then

$$(p - q)(x) = (x^2 + \sqrt{x}) - \sqrt{x} = x^2$$

The domain of $p - q$, however, is the set of all nonnegative numbers because these are the only numbers for which both p and q have meaning. □

These arithmetic operations for functions can be used to simplify the graphing of functions. If we know the graph of f and the graph of g, we can draw the graph of $f + g$ just by adding Y-coordinates of points with the same X-coordinate. Similarly, the graphs of $f - g$, fg, and f/g can be drawn by subtracting, multiplying, or dividing these Y-coordinates.

The simplest cases are those in which one of the functions is a constant.

Example 6

To draw the graph of $f(x) = x^2 + 3$, first notice that f is the sum of $s(x) = x^2$ and the constant function $c(x) = 3$.

In order to draw the graph of this function, we need only add 3 to each Y-coordinate on the graph of $s(x) = x^2$. (See Figure 3–88.) □

Figure 3–88

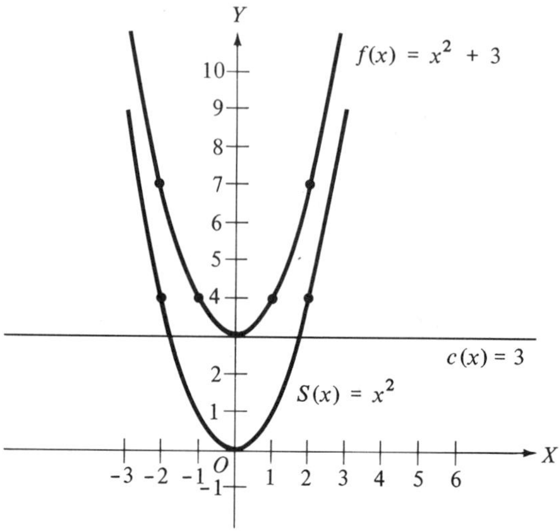

Example 7

Draw the graph of $g(x) = x^3 - 1$.

We need only to subtract 1 from each Y-coordinate of the graph of $p(x) = x^3$. (See Figure 3–89.) □

Figure 3–89

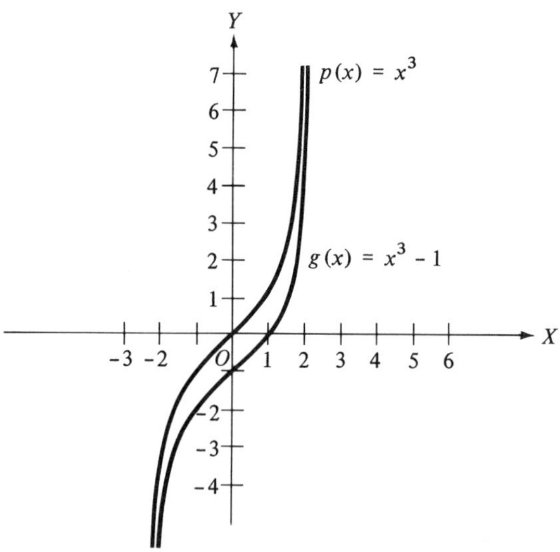

Figure 3–90

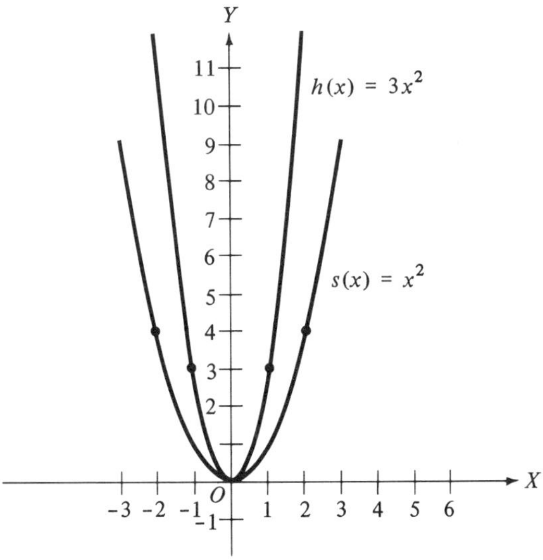

Example 8

The graph of $h(x) = 3x^2$ can be drawn by multiplying each Y-coordinate of $s(x) = x^2$ by 3. (See Figure 3–90.) □

Example 9

The graph of $w(x) = x^3 + x^2$ can be drawn by adding the Y-coordinates of the graphs of $p(x) = x^3$ and $q(x) = x^2$. (See Figure 3–91.) □

Figure 3–91

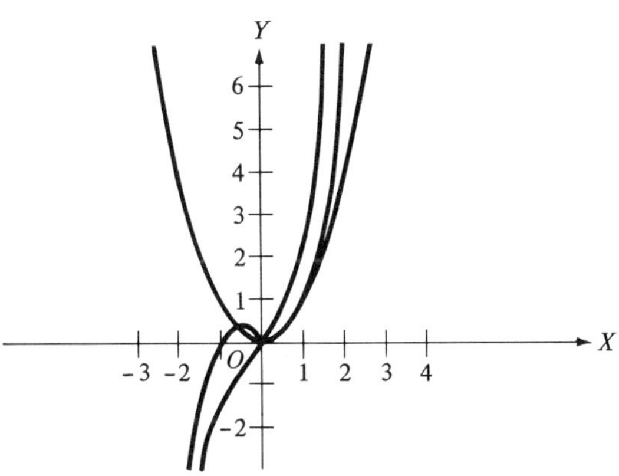

Example 10

The graph of

$$z(x) = \frac{x^3}{x^2 + 1}$$

can be drawn by dividing the Y-coordinates of $p(x) = x^3$ by the Y-coordinates of $v(x) = x^2 + 1$. (See Figure 3–92.) □

Figure 3–92

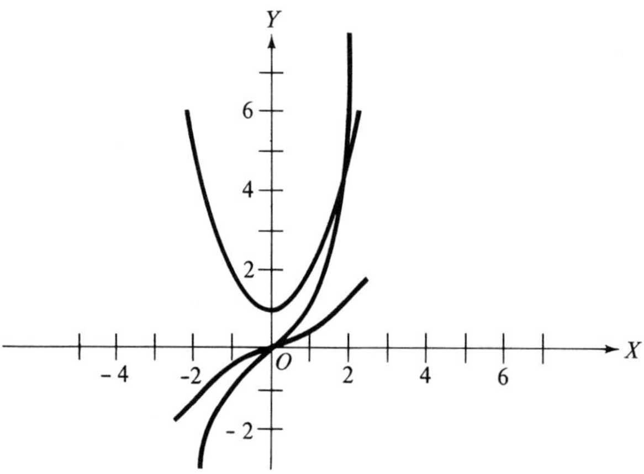

These methods are easiest to use when one of the functions is a constant or when adding or subtracting. Remember:

(i) Adding or subtracting a constant function just shifts the curve upward or downward by the absolute value of the constant.
(ii) Multiplying or dividing by a constant function just distorts the curve vertically.

Composition is another very useful way to combine two functions to produce a new function. If f and g are functions, the *composite of f and g* or *f composed with g* is denoted by $f \circ g$ and defined by

$$(f \circ g)(x) = f(g(x))$$

for each number x in the domain of g such that the number $g(x)$ is in the domain of f.

You have seen many functions which are, in fact, the result of composing functions.

Example 11

Let $s(x) = x^2$ and $g(x) = 2x - 1$. Then

$$(s \circ g)(x) = s(g(x)) = s(2x - 1) = (2x - 1)^2$$

Since both domains are R, the domain of $s \circ g$ is also R. Notice that

$$(g \circ s)(x) = g(s(x)) = g(x^2) = 2x^2 - 1 \; \square$$

Example 12

Let $s(x) = x^2$ and $p(x) = \sqrt{x - 2}$. Then

$$(s \circ p)(x) = s(p(x)) = s(\sqrt{x - 2}) = (\sqrt{x - 2})^2 = x - 2$$

Although the expression $x - 2$ has meaning for all substitutions, the domain of p is all numbers greater than or equal to 2. Since $p(x)$ is in the domain of $s(x)$ for all x, the domain of $s \circ p$ is just the domain of p. Notice that

$$(p \circ s)(x) = p(s(x)) = p(x^2) = \sqrt{x^2 - 2}$$

Since $\sqrt{x^2 - 2}$ has meaning only for values of x such that $|x| \geq \sqrt{2}$, the domain of $p \circ s$ is the set of all numbers with absolute value at least $\sqrt{2}$. $\square$

From the examples, we see that, in general, $f \circ g \neq g \circ f$ and composition of functions does not satisfy the commutative law. Suppose now that f, g, and h are functions and a is a number such that $h(a)$ is in the domain of g and $g(h(a))$ is in the domain of f. Then

$$[(f \circ g) \circ h](a) = (f \circ g)(h(a)) = f(g(h(a)))$$

and

$$[(f \circ (g \circ h)](a) = f((g \circ h)(a)) = f(g(h(a)))$$

Thus for each number a where $f \circ (g \circ h)$ or $(f \circ g) \circ h$ has meaning,

$$[f \circ (g \circ h)](a) = [(f \circ g) \circ h](a)$$

Thus

$$[f \circ (g \circ h)] = [(f \circ g) \circ h]$$

and composition of functions satisfies the associative law.

Example 13

Let $f(x) = x^2, g(x) = 2x + 1, h(x) = 1/(x - 1)$.

$$\begin{aligned}
[f \circ (g \circ h)](x) &= f((g \circ h)(x)) \\
&= [(g \circ h)(x)]^2 \\
&= [g(h(x))]^2 \\
&= [2(h(x)) + 1]^2 \\
&= \left[2\left(\frac{1}{x - 1}\right) + 1\right]^2 \\
&= \left[\frac{2}{x - 1} + 1\right]^2 \\
&= \left(\frac{x + 1}{x - 1}\right)^2 \\
[(f \circ g) \circ h](x) &= (f \circ g)(h(x)) \\
&= (f \circ g)\left(\frac{1}{x - 1}\right) \\
&= f\left[g\left(\frac{1}{x - 1}\right)\right] \\
&= f\left[2\left(\frac{1}{x - 1}\right) + 1\right] \\
&= f\left[\frac{x + 1}{x - 1}\right] \\
&= \left(\frac{x + 1}{x - 1}\right)^2 \quad \square
\end{aligned}$$

In Exercise 4 of Section 9 you were asked to draw the graphs of $f(x) = x^3$, $g(x) = (x - 1)^3$, and $h(x) = (x + 2)^3$. All three of these graphs have the same shape, they are just located differently on the coordinate system. (See Figure 3–93.)

In Section 6 you noted also that the graphs of the equations $y = (x + 4)^2$, $y = (x - 3)^2$, and $y = (x - \pi)^2$ all have the same shape.

In general, the graphs of $f(x)$ and $f(x - a)$ have exactly the same shape. The function $f(x - a)$, after all, is just $(f \circ g)(x)$ where $g(x) = x - a$. (See Figure 3–94.)

The result is to move the graph of f a distance of a units to the right if $a > 0$ and $|a|$ units to the left if $a < 0$. This can be seen by noticing that if b is any number, then

$$f(b) = f[(b + a) - a]$$

Thus $f(x - a)$ takes the same value at $b + a$ as $f(x)$ does at b and the graph of $f(x - a)$ is just the graph of $f(x)$ shifted $|a|$ units to the right or left, right if $a > 0$ and left if $a < 0$.

Figure 3–93

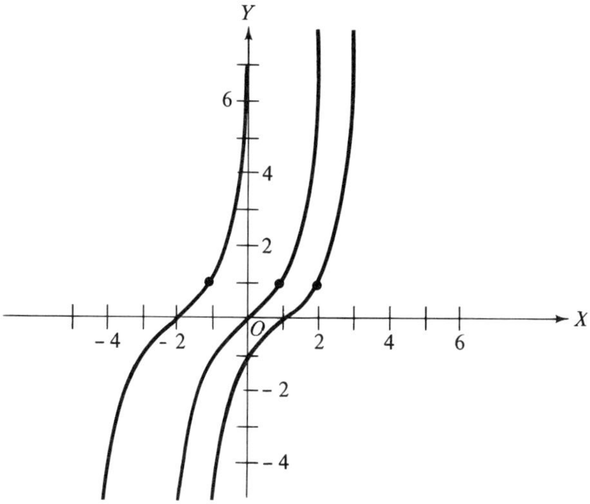

Figure 3–94

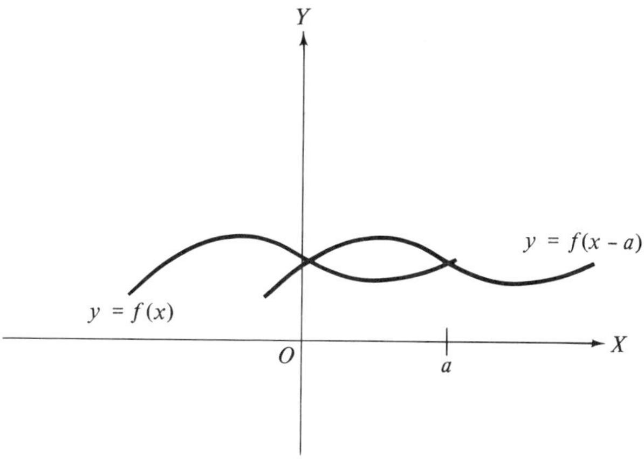

Example 14

Draw the graph of $f(x) = (x - 3)^5 + 1$.

The graph of $g(x) = x^5 + 1$ is drawn by adding 1 to each Y-coordinate of the graph of $y = x^5$. The graph of $f(x)$ is just the graph of $g(x)$ shifted 3 units to the right. (See Figure 3–95.) □

Figure 3–95

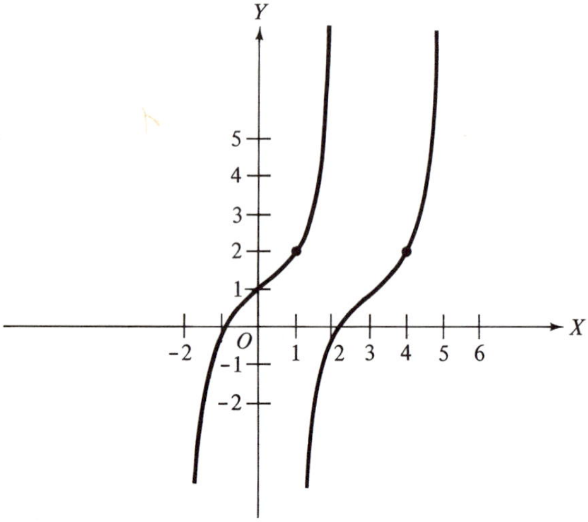

In summary:

(i) The graphs of $f + g$, $f - g$, fg, and f/g can be drawn using the graphs of f and g simply by adding, subtracting, multiplying, or dividing the two Y-coordinates associated with each number x. In practice, we compute enough of these values to enable us to sketch the graph. When one of the functions is a constant, the computation is quite simple.

(ii) The graph of $f(x - a)$ has the same shape as the graph of $f(x)$, it is just shifted right or left by $|a|$ units depending on whether a is positive or negative.

EXERCISES

1. Let $f(x) = x^2 + 1$, $g(x) = x^3$, $h(x) = \sqrt{x - 2}$, and $k(x) = 1/(x + 1)$. Determine simplified expressions for the following functions and specify the domain of each.
   (a) $f + g$ (b) fg (c) g/h (d) hk
   (e) fk (f) $g - f$ (g) g/k (h) k/f
2. Draw the graphs of the functions f, g, h, and k given in Exercise 1 and the graphs of those you were asked to express.
3. Let f, g, h, and k be as in Exercise 1. Determine simplified expressions for the following functions and specify the domain of each.
   (a) $f \circ g$ (b) $g \circ f$ (c) $f \circ h$ (d) $h \circ k$
   (e) $f \circ k$ (f) $g \circ h$ (g) $(h \circ f) \circ k$ (h) $k \circ (g \circ k)$
4. Use the method illustrated in Example 14 to draw the graphs of the following functions.
   (a) $w(x) = (x^2 - 3)^3$ (b) $v(x) = (x^3 + 2)^2$

(c) $p(x) = \sqrt{2x - 7}$ (d) $q(x) = (3x + 2)^3$

5. Draw the graphs of the following functions.
   (a) $f(x) = x^2 - 5$ (b) $g(x) = (x - 4)^3 + 3$
   (c) $h(x) = (2x - 5)^4 - 1$ (d) $k(x) = \sqrt{(x - 7)^3 + 2}$
   (e) $p(x) = [\sqrt{x + 1} - 3]^4$ (f) $q(x) = 1/x^2$
   (g) $w(x) = 1/(x - 2)^2$ (h) $v(x) = x + 2/x$
   (i) $u(x) = x + 2/(x - 3)$ (j) $z(x) = x^2 + 1/x$
6. Which of the axioms for R from Chapter 1 are satisfied by the arithmetic of functions?

11. Graphs of Inequalities

Just as the graph of an equation involving x and y as the only variables is a picture of the set of all pairs (x, y) satisfying the equation, the graph of an inequality involving only the variables x and y is just a picture of the set of all pairs (x, y) satisfying the inequality.

Example 1

To draw the graph of $y > x^2$, we first draw the graph of $y = x^2$ as a dotted curve. Then since our set is the set of all points (x, y) where $y > x^2$, it must be all points above the curve. We indicate this by shading this region. (See Figure 3–96.) □

Figure 3–96

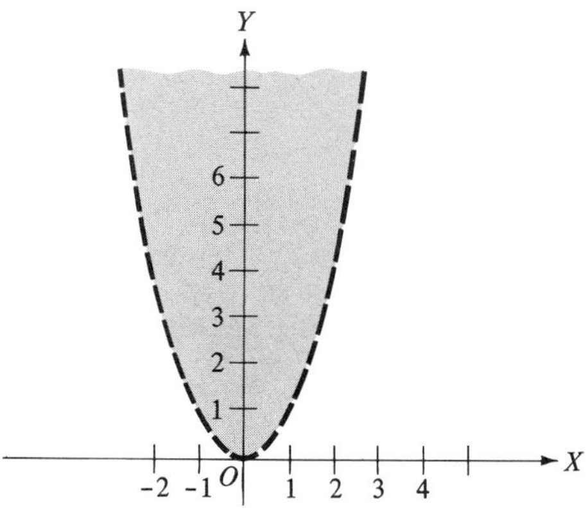

Example 2

The graph of $y < 2x + 2$ is drawn by drawing the dotted line $y = 2x + 2$ and then shading all points below it. (See Figure 3–97.) □

Figure 3–97

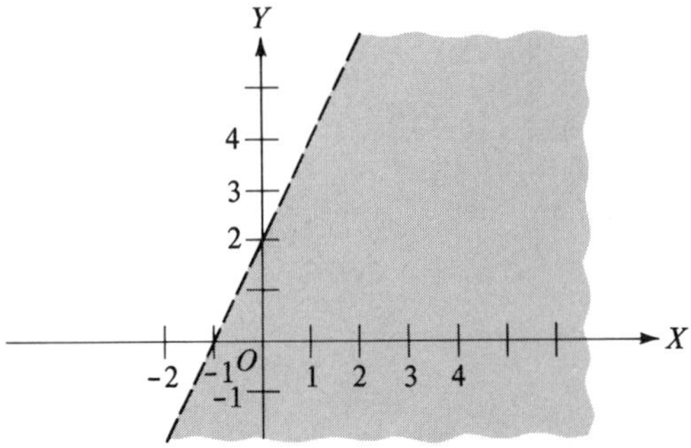

Example 3

The graph of $x^2 + y^2 < 4$ is the set of all points whose distance from (0, 0) is less than 2. This is just the interior of the circle with equation $x^2 + y^2 = 4$. Again, we shade the solution set. (See Figure 3–98.) □

Figure 3–98

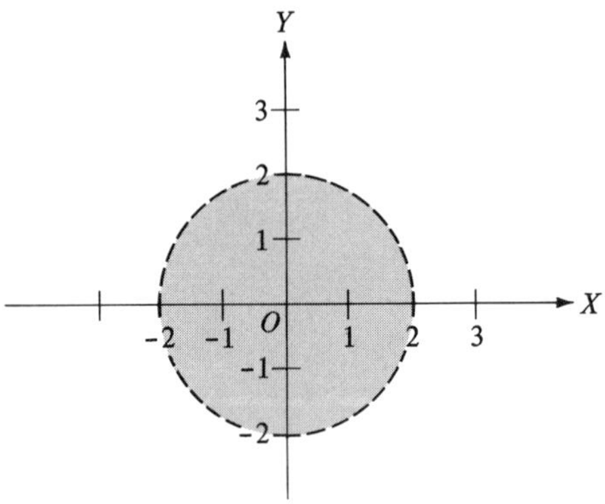

Example 4

To draw the graph of $y \leq -4x + 1$, notice that this is just the line $y = -4x + 1$ and all points below it. Draw the line and shade all points below. (See Figure 3–99.) □

To graph a linear equality it is generally simpler if we transform it into a form in which the corresponding linear equation is in slope-intercept form.

Figure 3-99

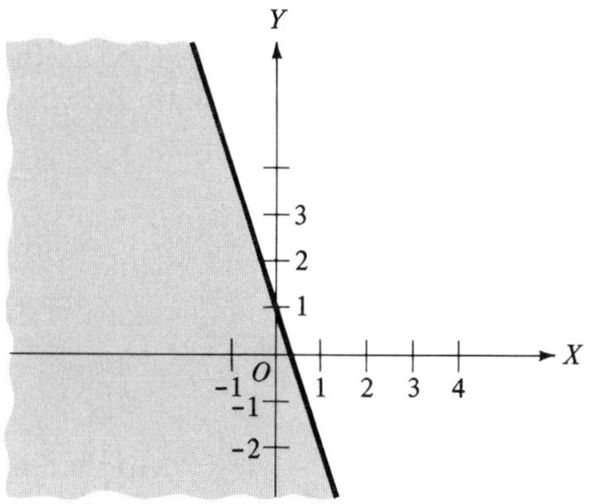

Example 5

$$3x + 2y > 10$$
$$2y > -3x + 10$$
$$y > -\tfrac{3}{2}x + 5$$

See Figure 3–100. □

Figure 3-100

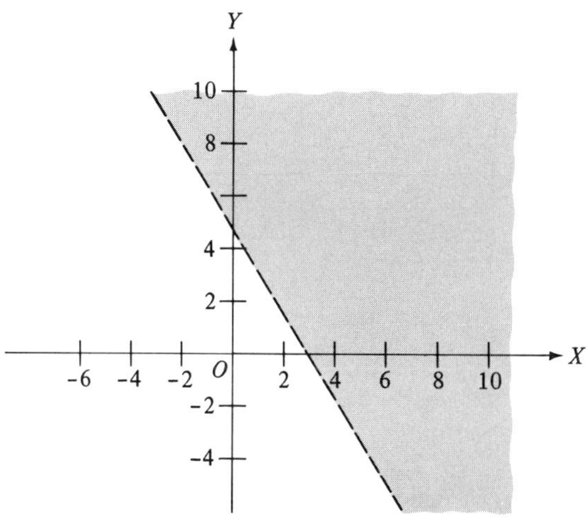

In Chapter 2 we discussed systems of linear inequalities in one variable and systems of linear equations involving two or more variables. Suppose now we have two or more inequalities involving two variables, x and y. To *solve the system* means to find all pairs (x, y) that are solutions for all the inequalities in the system. Frequently the set of all such solutions is best indicated by means of a graph.

Example 6

Solve the system

$y < 2x + 1$

$y < -3x + 2$

Drawing the dotted lines $y = 2x + 1$ and $y = -3x + 2$ and shading all points under both lines yields a graph of the solution set. (See Figure 3–101.) □

Figure 3–101

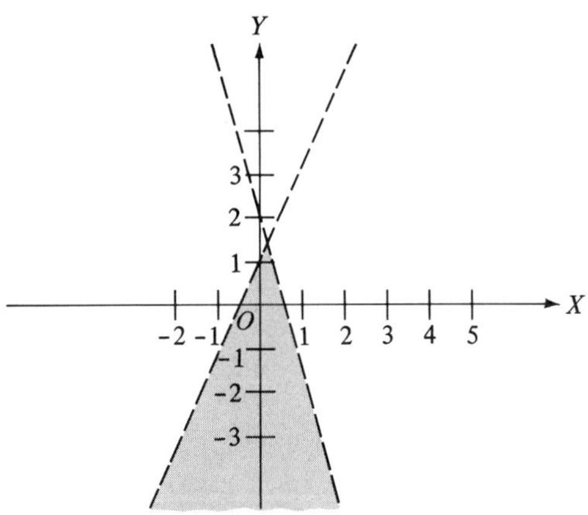

Example 7

Solve the system

$y \leq x + 2$

$y \leq -2x + 1$

$y \geq \frac{1}{2}x - 3$

We draw the graphs of $y = x + 2$, $y = -2x + 1$, and $y = \frac{1}{2}x - 3$. The solution set is any point on or inside the shaded triangle. (See Figure 3–102.) □

Figure 3-102

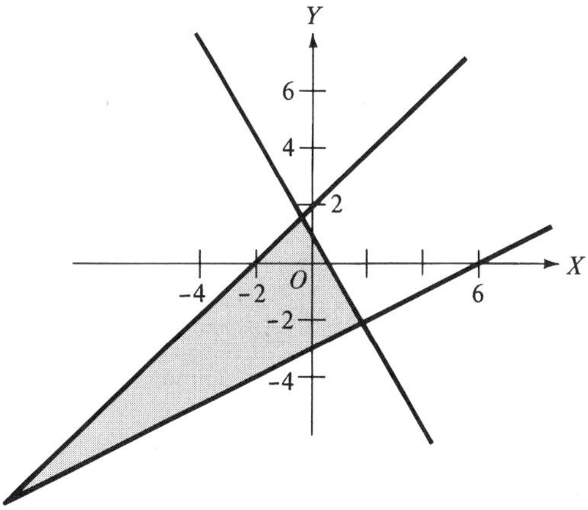

Example 8

Solve the system

$$y > 2x + 1$$
$$x^2 + y^2 < 4$$

The solution set is the set of all points inside the circle $x^2 + y^2 = 4$ that are above the line $y = 2x + 1$. (See Figure 3–103.) □

Figure 3-103

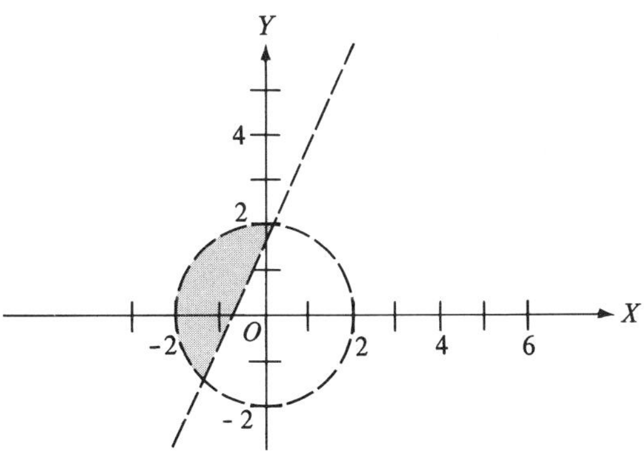

EXERCISES

1. Draw the graphs of the following inequalities.
   (a) $y < 3x - 1$ (b) $y \geq -2x - 3$ (c) $2x + y < 7$
   (d) $3x + 4y \geq 5$ (e) $5x - 7y < 35$ (f) $\sqrt{2}x + \sqrt{5}y < \sqrt{7}$
   (g) $\frac{2}{3}x - \frac{4}{5}y + \frac{1}{9} > 0$ (h) $.7x + .9y - .43 < 0$
2. Draw the graphs of the following inequalities.
   (a) $x^2 + y^2 \leq 9$ (b) $x^2 + y^2 > 1$ (c) $\dfrac{x^2}{4} + \dfrac{y^2}{9} < 1$
   (d) $\dfrac{x^2}{16} + \dfrac{y^2}{4} \geq 1$ (e) $\dfrac{x^2}{4} - \dfrac{y^2}{9} < 1$ (f) $\dfrac{x^2}{16} - \dfrac{y^2}{9} > 1$
   (g) $y < x^2 + 4x - 7$ (h) $y \geq 2x^2 - 5x - 2$
3. Draw the graphs of the solution sets of the following systems of inequalities.
   (a) $y < 4x + 1$, $y < -2x + 2$
   (b) $y \leq 3x - 7$, $y \leq -x + 2$
   (c) $2x + 3y < 4$, $5x - 2y < 6$
   (d) $5x + 7y + 2 \geq 0$, $6x + 4y + 3 \leq 0$
   (e) $y \leq 2x + 1$, $y \leq -x + 2$, $y \geq \frac{2}{3}x - 4$
   (f) $y > x + 2$, $y < 3x + 4$, $y < -2x + 3$
   (g) $y \leq 4x + 3$, $y \leq -2x + 1$, $y \geq -x + 2$
   (h) $2x + 5y < 7$, $3x - 7y < 5$, $x + y > 0$
4. Draw graphs of the solution sets of the following systems of inequalities.
   (a) $y \leq 2x + 1$, $y \geq x^2$
   (b) $y \geq 3x - 1$, $x^2 + y^2 \leq 9$
   (c) $y \geq x^2 + 1$, $y \leq -2x^2 + 3$
   (d) $x \geq 0$, $y \geq 0$, $y \leq 2x + 3$, $y \leq -3x + 4$
   (e) $x \geq 2$, $y \geq 3$, $2x + 3y < 10$, $-3x + 4y < 20$
   (f) $x \geq 3$, $y \geq 4$, $x^2 + y^2 \leq 16$, $y \geq x^2$

*12. Graphs of Rational Functions; Solving Inequalities Graphically

From Section 10 we know that the graphs of $y = (x - a)^n$ and $y = x^n$ have the same shape. In general, if n is even the graph is tangent to the X-axis at $(a, 0)$ and if n is odd, it crosses the X-axis. The larger n is, the "flatter" the curve is at the point of tangency or the point of crossing. (See Figure 3–104.)

Example 1

Consider now the graph of

$$y = (x - 2)^2(x - 1)^3$$

For values of x between 1.9 and 2.1 this graph is between the graphs of

$$y = (.9)^3(x - 2)^2 \quad \text{and} \quad y = (1.1)^3(x - 2)^2$$

Figure 3–104

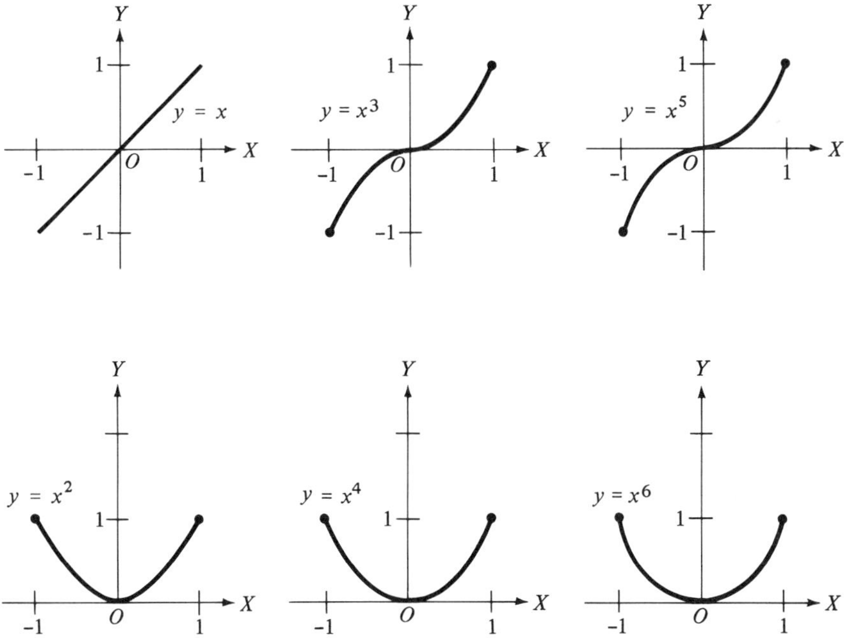

The graph of $y = (x - 2)^2$ is also between these and, therefore, at (2, 0) the graph must look something like the graph of $y = (x - 2)^2$ and be tangent to the X-axis with somewhat the same degree of flatness. (See Figure 3–105.) Between .9 and 1.1, this graph is between the graphs of

$$y = (-1.1)^2(x - 1)^3 \quad \text{and} \quad y = (-.9)^2(x - 1)^3$$

The graph of $y = (x - 1)^3$ is also between these and this graph must be approximated there by the graph of $y = (x - 1)^3$. Thus it crosses the X-axis at (1, 0).

In summary, then:

For values of x greater than 2, y must be positive.
For values of x between 1 and 2, y is also positive.
For values of x less than 1, y is negative.
As x gets larger so does $|y|$ and both increase without bound.

If we had more information as to the location of high points and low points, we could complete the graph. We can, however, sketch a rough graph using the above information. See Figure 3–105. □

Figure 3–105

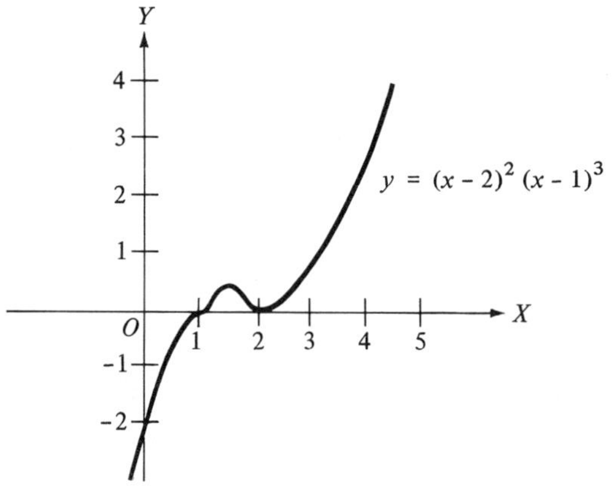

Consider now the general equation

$$y = c(x - a_1)^{n_1}(x - a_2)^{n_2} \cdots (x - a_k)^{n_k}$$

with $a_1 < a_2 < \cdots < a_k$ and each n_j is a positive integer.

If $1 \leq j \leq k$, the graph of this equation is approximated by the graph of

$$y = c(x - a_j)^{n_j}$$

for values of x close to a_j that are larger than a_{j-1} and smaller than a_{j+1}. This graph is tangent to the X-axis at $(a_j, 0)$ if n_j is even and crosses the X-axis there if n_j is odd. The graph of the original equation must do the same thing. (See Figure 3–106.)

Figure 3–106

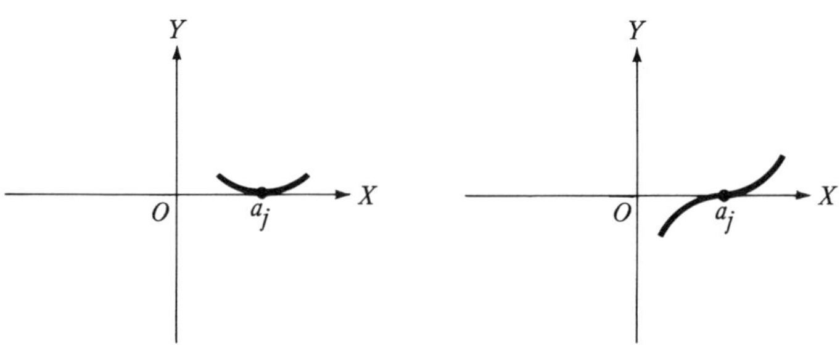

Thus, by examining the equation

$$y = c(x - a_1)^{n_1}(x - a_2)^{n_2}(x - a_3)^{n_3} \cdots (x - a_k)^{n_k}$$

we can determine the following:

(i) $y = 0$ if and only if $x = a_1, x = a_2, \ldots,$ or $x = a_k$.
(ii) y does not change sign for $x < a_1$ or $x > a_k$; in fact, y changes sign at no points other than possibly $a_1, \ldots, a_k$.
(iii) y changes sign at the point $(a_j, 0)$ if n_j is odd and does not change sign if n_j is even.
(iv) For values of x greater than a_k and less than a_1, the graph of this equation has a shape similar to the graph of

$$y = cx^{n_1+n_2+\cdots+n_k}$$

Thus as $|x|$ increases, so does $|y|$. If $c > 0$, then $y > 0$ for $x > a_k$. Furthermore, $y > 0$ for $x < a_1$ if $n_1 + \cdots + n_k$ is even, and $y < 0$ for $x < a_1$ if $n_1 + \cdots + n_k$ is odd. If $c < 0$, then all signs are changed; for example, $y < 0$ for $x > a_k$.

Example 2

Sketch the graph of

$$y = 3(x - 1)^2(x - 3)^3(x - 5)^3(x - 8)^2$$

First, $y = 0$ if and only if $x = 1, 3, 5,$ or 8. For $x > 8$, $y > 0$. For $x < 1$, $y > 0$ since $2 + 3 + 3 + 2$ is even.

The graph is tangent to the X-axis at (8, 0) and at (1, 0) and crosses it at (3, 0) and (5, 0). We can make the preliminary sketch as in Figure 3–107.

Figure 3–107

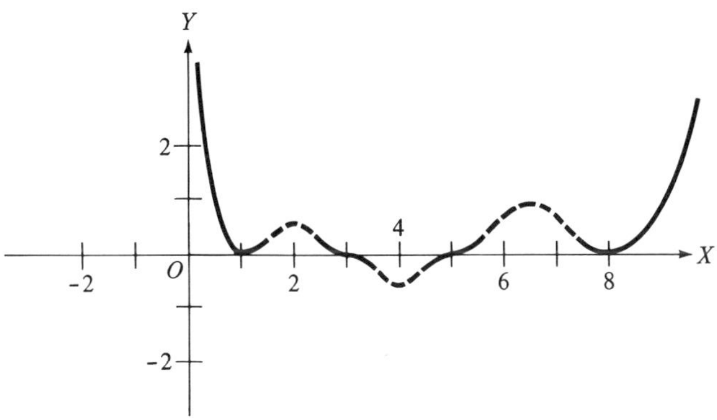

The dotted portions we know very little about and need calculus to determine them. We could, of course, compute some of the points. □

Example 3

$$y = 4(x + 2)^2(x^3)(x - 1)^5(x - 5)^3$$

$y = 0$ if and only if $x = -2, 0, 1,$ or 5. For $y > 5$, $y > 0$. For $x < -2$, $y < 0$ because $2 + 3 + 5 + 3$ is odd.

The graph crosses the X-axis at (5, 0), (1, 0), and (0, 0) and is tangent to it at $(-2, 0)$. (See Figure 3–108.) □

Figure 3–108

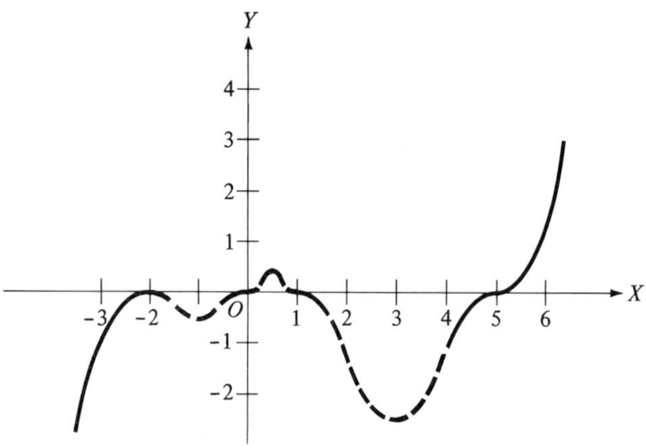

Since this technique tells us when $y > 0$ or $y < 0$, we can use it to solve polynomial inequalities.

Example 4

To solve

$$6(x + 2)^3(x - 1)(x - 4)^2(x - 5)^3 > 0$$

consider the graph of

$$y = 6(x + 2)^3(x - 1)^1(x - 4)^2(x - 5)^3$$

We need to determine the values of x for which $y > 0$. First, for $x > 5$, $y > 0$. For $y < -2$, $y < 0$ since $3 + 1 + 2 + 3$ is odd.

The graph crosses the X-axis at (5, 0), (1, 0), and $(-2, 0)$ and is tangent at (4, 0). The preliminary sketch is as in Figure 3–109. Thus $y > 0$ if and only if $-2 < x < 1$ or $x > 5$. □

Figure 3-109

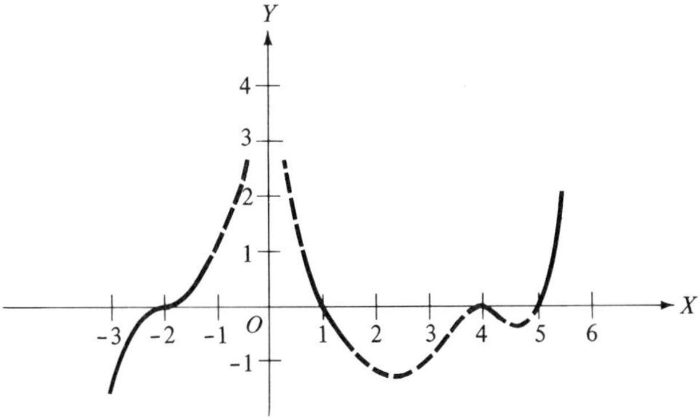

Example 5

To solve

$$3(x + 4)^2(x + 2)^3(x - 1)^4(x - 5) < 0$$

consider the equation

$$y = 3(x + 4)^2(x + 2)^3(x - 1)^4(x - 5)$$

For $x > 5$, $y > 0$. For $x < -4$, $y > 0$.

The graph crosses the X-axis at $(5, 0)$ and $(-2, 0)$. Thus $y \leq 0$ for $-2 \leq x \leq 5$ and $y < 0$ for $-2 < x < 1$ and $1 < x < 5$. Thus the solution set is the set of all x such that $-2 < x < 1$ or $1 < x < 5$. □

Example 6

To solve

$$-2(x + 1)^3(x - 3)^4(x - 7)^3(x - 10)^3 > 0$$

rewrite as

$$2(x + 1)^3(x - 3)^4(x - 7)^3(x - 10)^3 < 0$$

Consider the equation

$$y = 2(x + 1)^3(x - 3)^4(x - 7)^3(x - 10)^3$$

We wish to know where $y < 0$. For $x > 10$, $y > 0$. For $x < -1$, $y < 0$.

y changes sign at $(-1, 0)$ so $y > 0$ for $-1 < x < 3$.
y does not change sign at $(3, 0)$ so $y \geq 0$ for $-1 < x < 7$.
y changes sign at 7 so $y < 0$ for $7 < x < 10$.
y changes sign at 10 so $y \geq 0$ for $10 \leq x$.

Thus $y < 0$ for $x < -1$ or $7 < x < 10$. □

Notice how much faster these methods are than the case-by-case method from Chapter 2.

In general, the graph of

$$y = \frac{1}{(x - b)^m}$$

has the same shape as the graph of

$$y = \frac{1}{x^m}$$

it is just shifted $|b|$ units to the right if $b > 0$ and $|b|$ units to the left if $b < 0$. If m is even, then $y > 0$ for all $x \neq b$. If m is odd, then $y > 0$ for $x > b$ and $y < 0$ for $x < b$.

Example 7

$$y = \frac{1}{(x - 3)^2}$$

See Figure 3–110. □

Figure 3-110

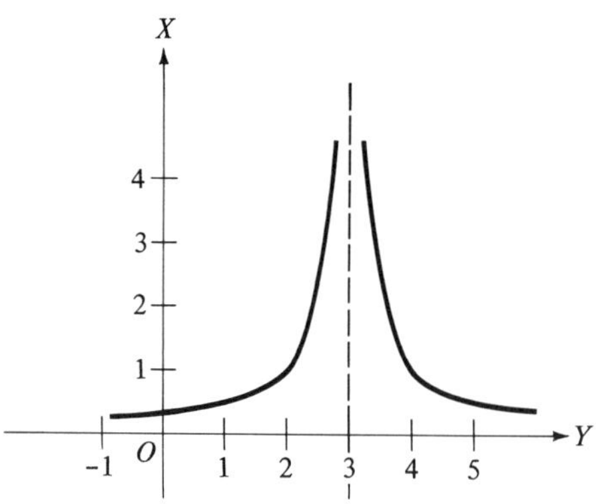

Example 8

$$y = \frac{1}{(x - 3)^3}$$

See Figure 3–111. □

Figure 3-111

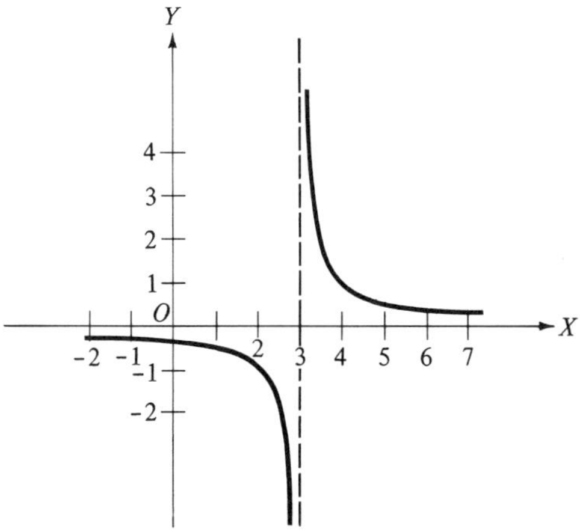

As x is chosen closer to 3, $|y|$ gets larger. □

In general, as we choose values of x closer to b and substitute in the equation

$$y = \frac{1}{(x - b)^m}$$

the absolute value of y gets larger. The vertical line $x = b$ is an *asymptote* to the graph in that the curve gets arbitrarily close to this line.

Consider now the graph of

$$y = \frac{1}{(x - b_1)^{m_1}(x - b_2)^{m_2} \cdots (x - b_t)^{m_t}}$$

with $b_1 < b_2 < \cdots < b_t$ and $m_1, \ldots, m_t$ positive integers.

If $1 \leq i \leq t$, for values of x between b_{i-1} and b_{i+1}, this graph is approximated by the graph of

$$y = \frac{c}{(x - b_i)^{m_i}}$$

The sign of the product

$$(x - b_1)^{m_1}(x - b_2)^{m_2} \cdots (x - b_{c-1})^{m_{c-1}}(x - b_{c+1})^{m_{c+1}} \cdots (x - b_t)^{m_t}$$

will be fixed for such values and will determine the sign of c. In this case x cannot assume the value b_i.

For $x > b_i$, y and c agree in sign.

For $x < b_i$, y and c agree if m_i is even and disagree if m_i is odd.

As we choose values of x closer to b_i, $|x - b_i|$ gets smaller and $|y|$ gets larger, that is, the line $x = b_i$ is an asymptote for the curve

$$y = \frac{c}{(x - b_i)^{m_i}}$$

and therefore, must be an asymptote for the curve

$$y = \frac{1}{(x - b_1)^{m_1}(x - b_2)^{m_2} \cdots (x - b_t)^{m_t}}$$

In general, each of the lines $x = b_1$, $x = b_2, \ldots, x = b_t$ is an asymptote for this curve. The numbers $b_1, \ldots, b_t$ are called the *poles* of the equation or graph.

To sketch the graph of this equation we first draw the asymptotes as dotted lines and sketch the portions of the curve near them. Notice:

(i) For $x > b_t$, $y > 0$.
(ii) For $x < b_1$, $y > 0$ if $m_1 + \cdots + m_t$ is even and $y < 0$ if $m_1 + \cdots + m_t$ is odd.
(iii) For $x < b_1$, this curve is approximated by

$$y = \frac{1}{(x - b_1)^{m_1}}$$

and as x gets smaller (and in fact $|x|$ gets larger) the graph approaches the X-axis. Similarly, for $x > b_t$, the curve is approximated by

$$y = \frac{1}{(x - b_t)^{m_t}}$$

and as x gets arbitrarily large, y becomes arbitrarily small.

Example 9

Draw the graph of

$$y = \frac{1}{(x + 2)(x - 1)^2(x - 2)^3}$$

The poles are -2, 1, and 2 and the asymptotes are $x = -2$, $x = 1$, and $x = 2$.

For $x > 2$, $y > 0$.
For $1 < x < 2$, $y < 0$.
For $-2 < x < 1$, $y < 0$.
For $x < -2$, $y > 0$.

See Figure 3–112. □

Figure 3-112

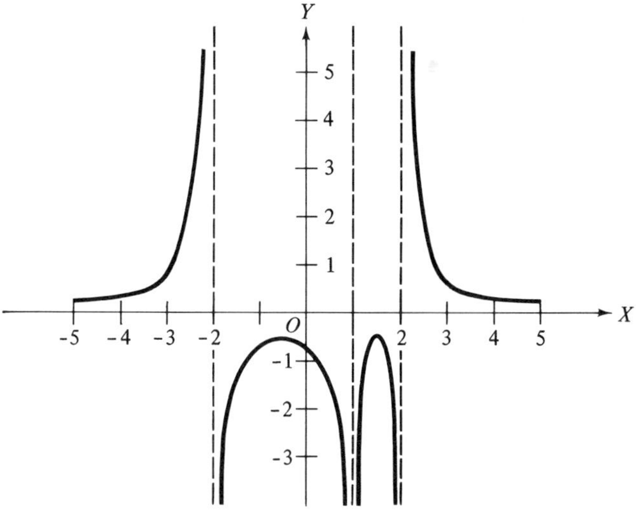

We do not have enough information to determine the high points of the sections of curve below the axis; calculus is needed for this. Thus we can only sketch these portions but we can, at least, determine what values of x yield positive or negative values of y.

Example 10

Now let us combine our ideas and consider the graph of

$$y = \frac{(x+2)^2(x-1)^3(x-4)^2}{12(x+1)^3(x-2)^2}$$

No simplification is possible in the expression. The equation has no meaning for $x = -1$ or $x = 2$ and $y = 0$ if and only if $x = -2$, $x = 1$, or $x = 4$. First, if $x > 4$, $y > 0$ because both the numerator and denominator are positive. If $x < -2$, then $y > 0$ because the numerator and denominator are both negative. The asymptotes are $x = -1$ and $x = 2$ since for values of x close to -1 or to 2, the graph is approximated by

$$y = \frac{1}{(x+1)^3} \quad \text{or} \quad y = \frac{1}{(x-2)^2}$$

The graph touches the X-axis only at $(-2, 0)$, $(1, 0)$, and at $(4, 0)$. At these points the graph is approximated by the graph of

$$y = \pm 3(x+2)^2(x-1)^3(x-4)^2$$

the + or − being determined by the sign of the denominator. Thus the graph is tangent to the X-axis at $(-2, 0)$ and $(4, 0)$ and crosses it at $(1, 0)$.

To draw the graph, first we draw the asymptotes. Since $y > 0$ for $x < -2$ and the graph is tangent to the X-axis at $(-2, 0)$, $y \geq 0$ for $x < -1$ and the graph begins as in Figure 3–113.

Figure 3–113

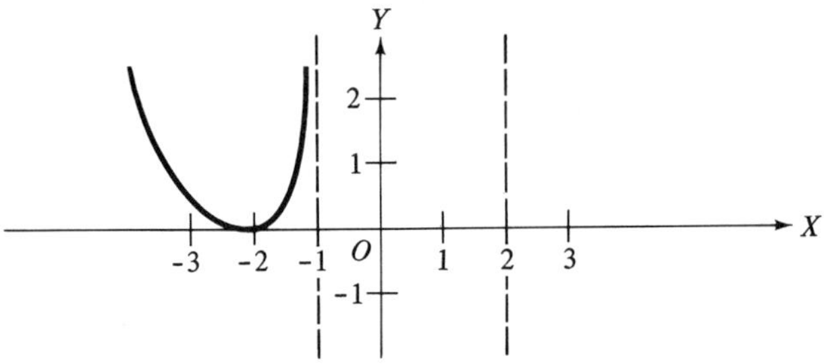

Since

$$y = \frac{1}{(x + 1)^3}$$

changes sign at -1, this curve does also and so it approaches the line $x = -1$ below the X-axis from the right. The curve crosses the X-axis at $(-1, 0)$. Since $x = 2$ is also an asymptote, we can now sketch the portion of the graph for $-1 < x < 2$. (See Figure 3–114.)

Figure 3–114

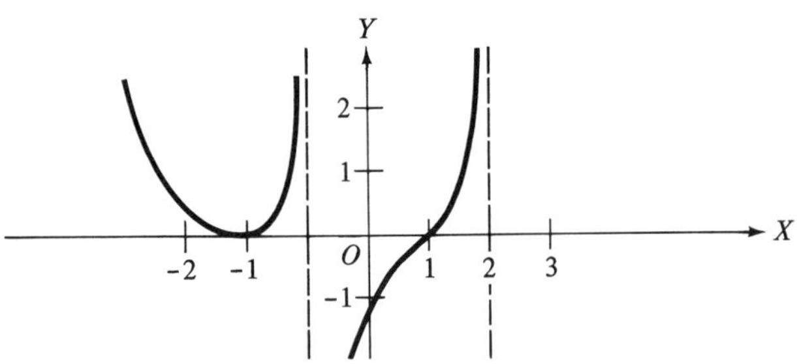

The curve does not change sign at $x = 2$ since for

$$y = \frac{1}{(x - 2)^2}$$

$y > 0$ for all values of x close to 2. The curve is tangent to the X-axis at $(4, 0)$ and we can now sketch the graph for $2 < x < 5$.

For large values of $|x|$, the curve has about the same shape as

$$y = 3x^2$$

Thus as $|x|$ gets larger, y increases without bound.

We now can sketch the rest of the graph. (See Figure 3–115.)

Figure 3–115

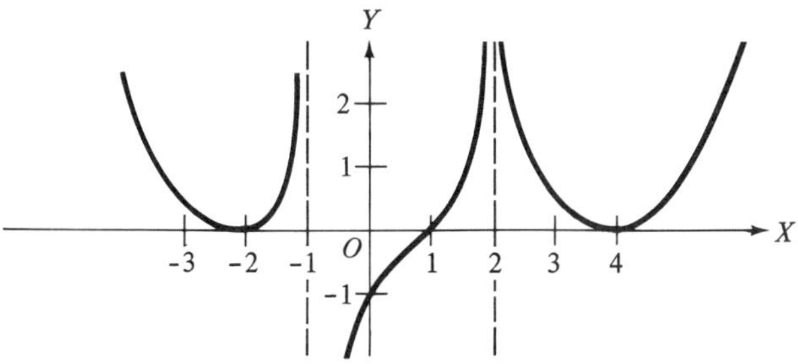

In general, the graph of

$$y = \frac{c(x - a_1)^{n_1}(x - a_2)^{n_2} \cdots (x - a_k)^{n_k}}{(x - b_1)^{m_1}(x - b_2)^{m_2} \cdots (x - b_t)^{m_t}} \qquad \text{with } a_i \neq b_j \text{ for all } i, j$$

has asymptotes $x = b_1, x = b_2, \ldots, x = b_t$, and touches the X-axis only at $(a_1, 0), (a_2, 0), \ldots, (a_k, 0)$.

For $x > a_k$ and $x > b_t$, y and c agree in sign.
For $x < a_1$ and $x < b_1$, y and c agree in sign if

$$n_1 + \cdots + n_k + m_1 + \cdots + m_t$$

is even and disagree if it is odd.

Also y changes sign at a_j if and only if n_j is odd.

Furthermore, y changes sign at b_i if and only if m_i is odd. To change sign at b_i means that y is positive for values of x immediately to one side of b_i and negative for values immediately to the other side. For very large values of $|x|$, the curve has about the same shape as

$$y = cx^{(n_1 + n_2 + \cdots + n_k) - (m_1 + m_2 \cdots m_t)}$$

If this exponent is positive, then $|y|$ increases without bound as $|x|$ gets large. If the exponent is negative, then as $|x|$ gets large, y gets close to 0. If it is 0, then y gets close to c.

Figure 3–116

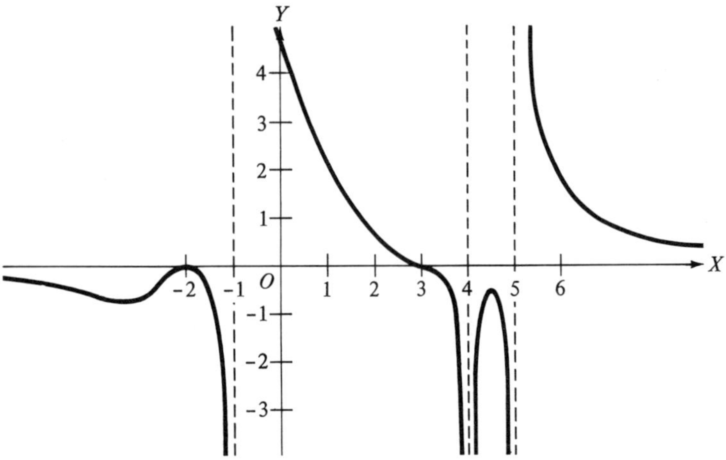

Example 11

Draw the graph of

$$y = \frac{4(x+2)^2(x-3)^3}{(x+1)^7(x-4)^2(x-5)}$$

The asymptotes are $x = -1$, $x = 4$, and $x = 5$, the curve changes sign at -1 and 5. The graph touches the X-axis only at $(-2, 0)$ and $(3, 0)$; it is tangent at $(-2, 0)$ and crosses at $(3, 0)$. As $|x|$ increases, y gets close to 0. (See Figure 3–116.) □

These techniques enable us to sketch rough graphs of some rational functions, rational functions being those that can be expressed as the quotient of polynomials.

Example 12

$$f(x) = \frac{x^2 + 3}{x^2 + 2x - 7}$$

is a rational function. □

Example 13

$$g(x) = \frac{4x^5 - 7x^3 + 3x^2 - 2x + 19}{6x^3 - 3x^2 + 18x - 91}$$

is a rational function. □

In order to apply the method we must first factor the numerator and denominator and, as we know, such complete factorizations do not always exist or even if they do exist, they are frequently difficult to find. Once found, however, they yield a great deal of information about the graph.

We can use these methods to solve polynomial inequalities involving quotients of polynomials.

Example 14

Determine all points where $y > 0$ if

$$y = \frac{4(x+3)^2(x-3)^3(x-5)^4(x-8)^3}{(x+2)^3(x-1)^4(x-2)^3(x-7)^3}$$

For $x < -3$, $y < 0$ because $2 + 3 + 4 + 3 + 3 + 4 + 3 + 3 = 25$. Furthermore, y does not change sign at $(-3, 0)$ because the exponent 2 is even. Thus for $x \leq -2$, $y \leq 0$. But y does change sign at $(-2, 0)$ since the exponent 3 is odd. Thus for $-2 < x < 1$, $y > 0$. Since the exponent on $(x - 1)$ is even, y does not change sign at 1 and therefore for $1 < x < 2$, $y > 0$. Since the exponent on $(x - 2)$ is odd, y changes sign at 2 and for $2 < x < 3$, $y < 0$. Again, y changes sign at 3 and for $3 < x < 5$, $y > 0$. Also y does not change sign at 5 and for $5 < x < 7$, $y > 0$. But y changes sign at 7 and for $7 < x < 8$, $y < 0$. Again y changes sign at 8 so for $8 < x$, $y > 0$. Thus, for values of x such that

$$-2 < x < 1, \qquad 1 < x < 2, \qquad 3 < x < 5, \qquad 5 < x < 7, \quad \text{and} \quad 8 < x$$

Figure 3-117

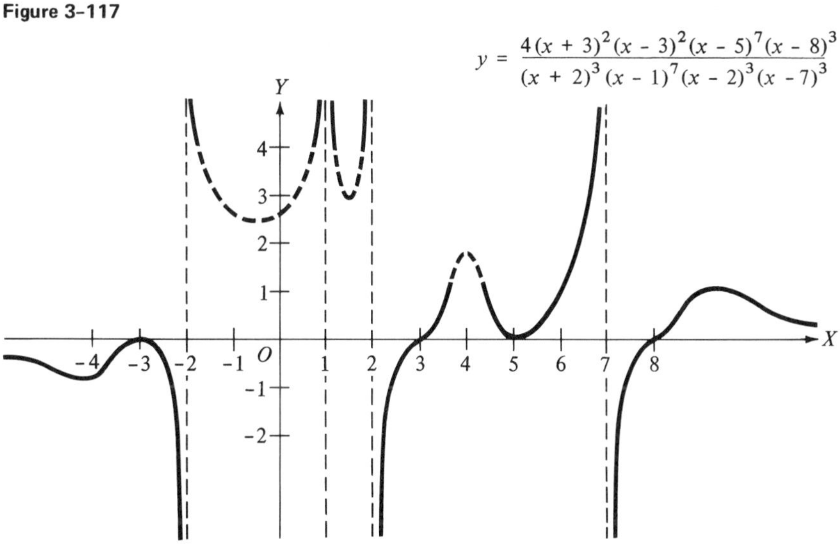

y is positive. Hence, the solution of

$$\frac{4(x+3)^2(x-3)^3(x-5)^4(x-8)^3}{(x+2)^3(x-1)^4(x-2)^3(x-7)^3} > 0$$

is

$$\{x: -2 < x < 1, 1 < x < 2, 3 < x < 5, 5 < x < 7, \text{ or } 8 < x\}$$

(See Figure 3–117.) □

EXERCISES

1. Draw graphs of the following equations.
   (a) $y = (x+2)^2(x-5)^3$
   (b) $y = (x+3)^2(x+1)(x-2)^2$
   (c) $y = 2(x+1)^3(x-2)^2(x-3)^2(x-7)^3$
   (d) $y = 5(x+4)^2(x^3)(x-1)^2(x-3)^2$
   (e) $y = (x+2)(x+1)^2(x-3)$
   (f) $y = 3(x+\frac{3}{2})^2(x-3)^3(x-4)^5(x-7)^3$
   (g) $y = x^2(x-7)^2(x-2)^3(x+3)^4$
   (h) $y = (x^2-2x-8)^2$
   (i) $y = x^4 - 13x^2 + 36$
2. Solve the following inequalities.
   (a) $6(x+3)^2(x+1)^2(x-2)(x-4)(x-5)^2 > 0$
   (b) $4.7(x-2.2)^2(x+4.7)(x+6.3)^2(x-7.2)^3 > 0$
   (c) $(x+1)(x-3)(x+4)(x+6) < 0$
   (d) $(x+\frac{1}{3})(x-\frac{4}{5})(x-\frac{7}{3})(x-\frac{1}{7}) > 0$
   (e) $(2x+1)(4x-3)(6x-1)(5x-4) < 0$
   (f) $(6x+5)^2(11x-2)^3(2x-7)^4(5x-3)^2 > 0$
3. Draw the graphs of the following functions.
   (a) $f(x) = \dfrac{4(x+2)^3(x-5)^2(x-8)}{(x+1)^2(x-3)^3}$
   (b) $g(x) = \dfrac{5(x+5)^2(x-1)^3}{(x+2)^3(x-7)^3}$
   (c) $k(x) = \dfrac{x^2-5x-14}{x^2+6x+5}$
   (d) $h(x) = \dfrac{(x+5)(x-7)}{(x+3)(x-9)}$
   (e) $p(x) = \dfrac{x^2+11x+18}{x^2+5x-6}$
   (f) $q(x) = \dfrac{x^4-13x^2+36}{x^4-26x^2+25}$
4. Solve the following inequalities.
   (a) $\dfrac{(x+2)(x-3)(x-5)}{(x+1)(x-4)} > 0$
   (b) $\dfrac{(x+3)^2(x-2)^3}{x^2(x-1)^3} < 0$
   (c) $\dfrac{6(x+4)^3(x-1)^2(x-5)^3}{(x+2)^2(x-4)^3(x-7)} > 0$
   (d) $\dfrac{x^2-7x+10}{x^2-3x-28} > 0$
   (e) $\dfrac{x^2-6x+9}{x^2+5x+4} > 0$
   (f) $\dfrac{x^2+6x+5}{x^2-10x-24} < 0$

4

Logarithms

The concept of a logarithm was first developed by Napier in order to speed and simplify computation. It is still very important in these days of computers not only for computational ease but also because of the analytic significance of logarithmic functions. In this chapter we will concentrate on computational techniques with common logarithms, but will, however, touch upon the more general concepts and logarithmic functions.

1. Common Logarithms

The computational simplicity achieved by using logarithms can be illustrated by using Table 4.1.

To compute the product of two numbers on the left we can use the table to rewrite them as powers of 2, add the exponents, and then rewrite the product by using the table.

Example 1

$$16 \cdot 32 = 2^4 \cdot 2^5 = 2^9 = 512 \ \square$$

Example 2

$$(\tfrac{1}{8}) \cdot 256 = 2^{-3} \cdot 2^8 = 2^5 = 32 \ \square$$

Table 4.1

$\frac{1}{8}$	2^{-3}
$\frac{1}{4}$	2^{-2}
$\frac{1}{2}$	2^{-1}
1	2^0
2	2^1
4	2^2
8	2^3
16	2^4
32	2^5
64	2^6
128	2^7
256	2^8
512	2^9
1024	2^{10}

Example 3

$$\left(\tfrac{1}{4}\right) \cdot 1024 = 2^{-2} \cdot 2^{10} = 2^8 = 256 \ \square$$

Certainly it is easier to add exponents than to multiply the numbers directly. We could use the table to simplify other computations.

Example 4

$$\sqrt[3]{512} = (512)^{1/3} = (2^9)^{1/3} = 2^3 = 8 \ \square$$

Example 5

$$4^4 = (2^2)^4 = 2^8 = 256 \ \square$$

Computational techniques with logarithms are generalizations of these methods.

The concept of a logarithm rests upon the following basic principles which are extensions of the properties and laws of rational exponents discussed in Chapter 1.

I. If b is a positive number and t is any real number, then b^t is a positive real number.

II. If t and s are real numbers and b is a positive real number, then $b^t b^s = b^{t+s}$, $(b^t)^s = b^{ts}$, $b^{-t} = 1/b^t$, and $b^t/b^s = b^{t-s}$.

III. If b and c are positive real numbers and t is any real number, then $(bc)^t = b^t c^t$ and $(b/c)^t = b^t/c^t$.

IV. If $b > 1$ and t and s are real numbers such that $t < s$, then $b^t < b^s$.

V. If $0 < b$, $b \neq 1$, and a is a positive real number, then there exists exactly one real number t such that $b^t = a$.

We will not prove any of the above five assertions. Principles I–IV can be shown to hold if t and s are rational numbers but a great deal of work with Axiom 12 must be done to establish general proofs. The proof of Principle V also depends on Axiom 12. Our purpose is to discuss the properties of logarithms and the benefits to be derived from their use. The effort involved in establishing the proofs would not, at this stage, contribute to the reader's understanding of the properties and, therefore, we will not undertake any of them.

If $a > 0$, then Principle V assures us that there is exactly one number t such that $a = 10^t$. This number t is defined to be the common logarithm of a.

The common logarithm of a is t if and only if $a = 10^t$.

We will omit the word "common" and just speak of the logarithm of a. It is also customary to abbreviate the phrase "logarithm of a" to "log a." Thus

$$\log a = t \quad \text{if and only if} \quad a = 10^t$$

Example 6

$\log 10 = 1$ because $10 = 10^1$ □

Example 7

$\log 1 = 0$ because $1 = 10^0$ □

Example 8

$\log 0.1 = -1$ because $.1 = 10^{-1}$ □

Example 9

$\log 1000 = 3$ because $1000 = 10^3$ □

Example 10

$\log \sqrt{10} = \frac{1}{2}$ because $\sqrt{10} = 10^{1/2}$ □

Example 11

$\log \sqrt[3]{100} = \frac{2}{3}$ because $\sqrt[3]{100} = \sqrt[3]{10^2} = 10^{2/3}$ □

Notice that for any number x,

$$\log 10^x = x$$

Also for any positive number a,

$$10^{\log a} = a$$

If $\log a = w$, then a is called the *antilogarithm* of w. We abbreviate the expression "antilogarithm of w" to "antilog w."

In the examples given after Table 4.1, we were able to do this multiplication by adding the corresponding exponents and a similar general result holds for logarithms. The fundamental principles used in computations with logarithms are given by the following propositions.

PROPOSITION 1

If a and b are positive real numbers, then

$$\log (ab) = \log a + \log b$$

Proof:

Let

$$\log a = t \quad \text{and} \quad \log b = s$$

Then

$$a = 10^t \quad \text{and} \quad b = 10^s$$

Hence

$$ab = 10^t 10^s = 10^{t+s}$$

and

$$\log (ab) = t + s = \log a + \log b$$ □

PROPOSITION 2

If a and b are positive real numbers, then

$$\log \left(\frac{a}{b}\right) = \log a - \log b$$

Proof:

Exercise.

PROPOSITION 3

If a is a positive real number and x is a real number, then

$$\log a^x = x \log a$$

Proof:

Let

$$\log a = t$$

that is,

$$a = 10^t$$

Then

$$a^x = (10^t)^x = 10^{tx}$$

and

$$\log a^x = \log 10^{tx} = tx = xt = x \log a \;\square$$

For the following examples, let us assume $\log 2 = .3010$ and $\log 3 = .4771$.

Example 12

$$\log 6 = \log (2 \cdot 3) = \log 2 + \log 3 = .3010 + .4771 = .7781 \;\square$$

Example 13

$$\log 16 = \log 2^4 = 4 \log 2 = 4(.3010) = 1.2040 \;\square$$

Example 14

$$\log \sqrt{2} = \log 2^{1/2} = \tfrac{1}{2} \log 2 = \tfrac{1}{2}(.3010) = .1505 \;\square$$

Example 15

$$\log \tfrac{2}{3} = \log 2 - \log 3 = .3010 - .4771 = -.1761 \;\square$$

Example 16

$$\log \sqrt[5]{3} = \log 3^{1/5} = \tfrac{1}{5} \log 3 = \tfrac{1}{5}(.4771) = .0954 \;\square$$

Example 17

$$\log 20 = \log (2 \cdot 10) = \log 2 + \log 10 = .3010 + 1 = 1.3010 \;\square$$

Example 18

$$\log 0.3 = \log 3 \cdot (0.1) = \log 3 + \log 0.1 = .4771 + (-1) = -.5229 \square$$

The student should realize that log 2 is not exactly .3010 nor is log 3 exactly .4771. These numbers are finite decimal approximations of the logarithms just as 1.414 is a finite decimal approximation of $\sqrt{2}$. Our computations will use 4-decimal-place approximations for common logarithms. More precise approximations are available but the principles governing their use are the same as those governing the use of 4-place approximations.

EXERCISES

In Exercises 1–25, assume that log 2 = .3010, log 3 = .4771, and log 7 = .8451. Compute the following logarithms.

1. $\log 21 =$
2. $\log \sqrt{7} =$
3. $\log 12 =$
4. $\log 30 =$
5. $\log 14 =$
6. $\log \sqrt[5]{210} =$
7. $\log (12)^6 =$
8. $\log 2000 =$
9. $\log .02 =$
10. $\log 700 =$
11. $\log 14^2 =$
12. $\log 5 =$ [*Hint*: $5 = \frac{10}{2}$]
13. $\log 4.2 =$
14. $\log (21)^4 =$
15. $\log 9 =$
16. $\log \frac{3}{7} =$
17. $\log \frac{7}{2} =$
18. $\log \sqrt{\frac{3}{2}} =$
19. $\log .0008 =$
20. $\log 9000 =$
21. $\log .9 =$
22. $\log 50 =$
23. $\log .5 =$
24. $\log .005 =$
25. $\log 1.5 =$
26. Prove Proposition 2.
27. If $\log a = p$, what is $\log 10a$? What is $\log 100a$? $\log 1000a$? $\log .01a$?
28. Use assertion IV and the fact that $10^0 = 1$ to describe numbers that have negative logarithms.
29. Describe all numbers a such that $0 < \log a < 1$.
30. Describe all numbers a such that $1 < \log a < 2$; all with logarithms between 2 and 3; all with logarithms between n and $n + 1$ where n is a positive integer.
31. Show, by example, that in general

    $$\log (a + b) \neq \log a + \log b$$

    [*Hint*: Let $a = 5$ and $b = 10$ and use your results from Exercises 7 and 22 above.]
32. Show, by example, that in general

    $$\frac{\log a}{\log b} \neq \log a - \log b$$

    [*Hint*: Let $a = 3$ and $b = 2$ and compute both numbers directly.]

2. Determining Logarithms by Using Tables

In order to use logarithms in computation, we need some way to determine the logarithm of a given number and vice versa, given a logarithm we need to be able to determine its antilogarithm. This is accomplished by using tables for logarithms of numbers between 1 and 10 and certain other facts.

In Exercise 29 of Section 1 you were asked to determine all numbers a such that $0 < \log a < 1$. This is done by noticing that if

$$1 < a < 10$$

then

$$10^0 < a < 10^1$$

and therefore

$$0 < \log a < 1$$

Conversely if

$$0 < \log a < 1$$

then

$$10^0 < 10^{\log a} < 10^1$$

and

$$1 < a < 10$$

Thus the logarithm of a number between 1 and 10 is a positive number smaller than 1 and, conversely, the antilogarithm of a positive number smaller than 1 is a number between 1 and 10.

Since any positive number can be expressed as the product of a power of 10 and a number between 1 and 10, the logarithm of any number can be determined if we can determine logarithms of numbers between 1 and 10 and logarithms of powers of 10. Since $\log 10^n = n$, our problem really is to determine logarithms of numbers between 1 and 10.

Example 1

$$\log 4235 = \log [10^3(4.235)] = \log 10^3 + \log 4.235 = 3 + \log 4.235 \quad \square$$

Example 2

$$\log .346 = \log [10^{-1}(3.46)] = \log 10^{-1} + \log 3.46 = (-1) + \log 3.46 \quad \square$$

Example 3

$$\log 74.53 = \log [10(7.453)] = \log 10 + \log 7.453 = 1 + \log 7.453 \quad \square$$

Example 4

$$\log .0057 = \log [10^{-3}(5.7)] = \log 10^{-3} + \log 5.7 = (-3) + \log 5.7 \quad \square$$

To determine the logarithms in Examples 1–4 we need to use tables to find the logarithms of 4.235, 3.46, 7.453, and 5.7. Logarithms of these numbers and other numbers between 1 and 10 are obtained from tables such as those provided in the back of the text. The tables available for logarithms of numbers between 1 and 10 are, in fact, finite decimal approximations and the number of decimal places varies with the degree of accuracy required. Thus one may find tables with 4-place approximations, 5-place, 6-place, or higher. For our purposes in this text, 4-place approximations will suffice.

The tables in the back provide logarithms of 2-decimal-place numbers between 1.00 and 10.00 and we use a technique, called *interpolation*, that enables us to approximate logarithms of 3-decimal-place numbers between 1 and 10. Logarithms of 2-place numbers between 1 and 10 are read from the table by finding the first two digits in the column at the left and the third digit by reading across the top. The logarithm is then on the same horizontal line as the first two digits and under the third digit. It is customary to eliminate the decimal points in such tables and it is *understood* that each logarithm in the table is preceded by a decimal point and that the digits in the left column are separated by a decimal point.

Example 5

Determine the logarithm of 5.23.

Find 52 in the left column and read under 3 on the top. The number is 7185. The 52 actually means 5.2 and the 7185 really means .7185. Thus $\log 5.23 = .7185$. □

Example 6

Notice that

$$\begin{aligned} \log 52.3 &= \log 10(5.23) \\ &= \log 10 + \log 5.23 \\ &= 1 + .7185 \\ &= 1.7185 \ \square \end{aligned}$$

Thus by knowing log 5.23 we can determine log 52.3 and in a similar fashion we could compute log 523, log .523, and in fact could compute $\log 10^n(5.23)$ for any integer n since, in fact, it will just be

$$n + \log 5.23 = n + .7185$$

Since n is an integer, the decimal part of each of these logarithms is the same.

The decimal part of the logarithm is called the *mantissa* and the integral part is called the *characteristic*.

To determine the logarithm of a number:

(i) Rewrite the number as a product of a power of 10 and a number between 1 and 10.
(ii) Use a table to determine the mantissa.
(iii) Determine the characteristic by reading the exponent of 10.

Example 7

$\log 735 = \log 10^2(7.35) = 2.8663$ □

Example 8

$$\begin{aligned}\log .0735 = \log 10^{-2}(7.35) &= (-2) + .8663\\ &= 8.8663 - 10\\ &= -1.1337\end{aligned}$$ □

For computational ease and simplicity, it is customary to write negative logarithms by using a positive decimal and a negative characteristic. We either write as we did earlier, $(-2) + .8663$, or in a form such as $8.8663 - 10$. By doing this we know that the characteristic is -2 and the mantissa is .8663 which is positive and can be found in the tables.

Example 9

If $\log x = 9.6590 - 10$, then $\log x = (-1) + .6590$.

The characteristic is -1 and the mantissa is .6590. From the table we see that .6590 is the mantissa of 4.57, that is, antilog $.6590 = 4.57$. Thus $x = 10^{-1}(4.57) = .457$. If we had expressed this as $\log x = -.3410$ we could not determine x immediately from the table. □

Once you develop some skill in applying these ideas, you may wish to eliminate some of the rewriting. Remember that the characteristic is always an integer. You can determine the characteristic of the logarithm of a number by observing the position of the decimal point in the number, and just *counting* the number of places we would have to move the decimal point of the original number in order to obtain a number between 1 and 10. If we move the point to the left, the characteristic of the logarithm is positive, if we move the point to the right, it is negative.

Example 10

The characteristic of log 93.2 is 1, since $93.2 = 10(9.32)$. □

Example 11

The characteristic of log 4750 is 3, since $4750 = 1000(4.750)$. □

Example 12

The characteristic of log .0692 is -2, since $.0692 = .01(6.92)$. □

We can use the table of mantissas for 3-digit numbers to approximate mantissas of 4-digit numbers by means of *interpolation* which is basically just a use of proportions.

Example 13

Determine log 75.56.

75.56 differs from 75.50 by $\frac{6}{10}$ of the difference between 75.50 and 75.60.

$$.10\left\{\begin{array}{l} .06\left\{\begin{array}{l}\log 75.50 = 1.8779\\ \log 75.56 = \ ?\end{array}\right\}\ ?\\ \log 75.60 = 1.8785\end{array}\right\}.0006$$

Since $1.8785 - 1.8779 = .0006$, we take $\frac{6}{10}$ of this difference, that is, $\frac{6}{10}(.0006)$ which is about .0004 and add this to 1.8779. Thus log 75.56 is approximately 1.8783.

$$.10\left\{\begin{array}{l} .06\left\{\begin{array}{l}\log 75.50 = 1.8779\\ \log 75.56 = 1.8783\end{array}\right\}.0004\\ \log 75.60 = 1.8785\end{array}\right\}.0006 \quad \square$$

Example 14

Determine log 6.534.

We know that log 6.534 is between log 6.530 and log 6.540.

$$.010\left\{\begin{array}{l} .004\left\{\begin{array}{l}\log 6.530 = .8149\\ \log 6.534 = \ ?\end{array}\right\}\ ?\\ \log 6.540 = .8156\end{array}\right\}.0007$$

Log 6.534 differs from log 6.530 by about $\frac{4}{10}$ of the difference between .8149 and .8156 which is .0007. Thus log 6.534 differs from log 6.530 by about .4(.0007) which is approximately .0003. Thus log 6.534 is approximately $.8149 + .0003 = .8152$.

$$.010\left\{\begin{array}{l} .004\left\{\begin{array}{l}\log 6.530 = .8149\\ \log 6.534 = .8152\end{array}\right\}.0003\\ \log 6.540 = .8156\end{array}\right\}.0007 \quad \square$$

Example 15

Determine x if $\log x = 9.5721 - 10$.

By consulting the table we find that the mantissa .5721 is between the mantissa .5717 which has antilog 3.730 and the mantissa .5729 which has antilog 3.740. Since the characteristic of the logarithm of x is -1, x is between .3730 and .3740.

$$\begin{array}{l} ? \\ .0010 \end{array} \left\{ \begin{array}{l} \log .3730 = 9.5717 - 10 \\ \log x \quad\ = 9.5721 - 10 \\ \log .3740 = 9.5729 - 10 \end{array} \right\} \begin{array}{l} .0004 \\ .0012 \end{array}$$

Since $\log x$ differs from $\log .3730$ by about $.0004/.0012 = 1/3$ of the difference between the logarithms of .3730 and .3740, x differs from .3730 by about 1/3 of the difference between .3730 and .3740, that is, by 1/3(.0010) which is approximately .0003. Thus x is about .3733.

$$\begin{array}{l} .0003 \\ .0010 \end{array} \left\{ \begin{array}{l} \log .3730 = 9.5717 - 10 \\ \log .3733 = 9.5721 - 10 \\ \log .3740 = 9.5729 - 10 \end{array} \right\} \begin{array}{l} .0004 \\ .0012 \end{array} \quad \square$$

Example 16

Determine antilog 2.7934.

If $x = \text{antilog } 2.7934$, then $\log x = 2.7934$. The mantissa of $\log x$ is .7934. From the table we find this between .7931, which is the mantissa for 6.21 and .7938 which is the mantissa for 6.22. Since the characteristic of $\log x$ is 2, x is between 621 and 622.

$$\begin{array}{l} ? \\ 1 \end{array} \left\{ \begin{array}{l} \log 621 = 2.7931 \\ \log x \quad = 2.7934 \\ \log 622 = 2.7938 \end{array} \right\} \begin{array}{l} .0003 \\ .0007 \end{array}$$

The difference between 621 and x is about 3/7 of 1, that is, about .4 and x is approximately 621.4. □

EXERCISES

1. Compute the following logarithms.
   (a) log 49.3 (b) log .0932 (c) log 73,900 (d) log 1080
   (e) log .763 (f) log 24.6 (g) log .847 (h) log 9450
   (i) log 62.4 (j) log .0779
2. Determine the following.
   (a) antilog 2.4099 (b) antilog 1.6355
   (c) antilog (9.7218 − 10) (d) antilog (8.2068 − 10)
   (e) antilog 4.8949 (f) antilog [(−2) + .8241]

(g) antilog $(8.9465 - 10)$ (h) antilog 3.8338
(i) antilog 2.7612 (j) antilog $(7.9886 - 10)$

3. Compute the following.
(a) log 59.34 (b) log 625.6 (c) log 9.232 (d) log .07139
(e) log 76.85 (f) log .9523 (g) log 24.67 (h) log 19.78
(i) log .003972 (j) log 45,970

4. Determine the following.
(a) antilog 2.6197 (b) antilog .4792
(c) antilog $(9.5773 - 10)$ (d) antilog 1.3226
(e) antilog $(8.4930 - 10)$ (f) antilog $(7.7654 - 10)$
(g) antilog 3.9225 (h) antilog 2.9412
(i) antilog $[(-3) + .6139]$ (j) antilog $(9.2134 - 10)$

5. Use the table to draw a graph of the equation $y = \log x$ for $0 < x \leq 10$.
6. Use the table to draw a graph of the equation $y = \text{antilog}\, x$ for $-1 \leq x \leq 1$.

3. Computations with Logarithms

The principles given in Propositions 1, 2, and 3 enable us to use logarithms to simplify computations that would be very tedious or time-consuming and also enable us to perform other computations which we have no other way of handling.

Example 1

We can use logarithms to multiply several numbers.

If $x = (4.73)(72.9)(36.2)(.923)$, then

$$\begin{aligned} \log x &= \log 4.73 + \log 72.9 + \log 36.2 + \log .923 \\ &= .6749 + 1.8627 + 1.5587 + 9.9652 - 10 \\ &= 14.0615 - 10 \\ &= 4.0615 \end{aligned}$$

Using interpolation, we find that x is approximately 11,520. Since the logarithms are approximate, our answer must be approximate. The symbol "$\doteq$" is frequently used to indicate that two numbers are approximately equal. In this case we would write

$x \doteq 11{,}520$ □

Strictly speaking, almost all logarithms obtained from a table are approximations and, therefore, all computations involving logarithms yield approximate answers and most statements such as "log 2 = .3010" should actually be "log 2 $\doteq$ .3010." It is not customary to emphasize the fact of the approximation in the latter case but it is good practice to indicate it when the answer to a problem has been obtained by using logarithms.

Example 2

If

$$y = \frac{(97.3)(6.12)}{72.9}$$

then

$$\begin{aligned}\log y &= \log 97.3 + \log 6.12 - \log 72.9\\ &= 1.9881 + .7868 - 1.8627\\ &= .9122\end{aligned}$$

Using the table, we see that $y \doteq 8.17$. □

More complicated multiplication and division problems can be handled by analogous methods.

We can use Proposition 3 ($\log a^x = x \log a$) to compute powers of numbers.

Example 3

Compute $(26.3)^7$.

If $y = (26.3)^7$, then

$$\begin{aligned}\log y &= \log (26.3)^7\\ &= 7(\log 26.3)\\ &= 7(1.4200)\\ &= 9.9400\end{aligned}$$

Thus $\log (26.3)^7 = 9.9400$. From the table, antilog $9.9400 = (10^9)(8.71)$. Thus $(26.3)^7 \doteq 8{,}710{,}000{,}000$. □

Example 4

Compute $\sqrt{493}$.

$$\begin{aligned}\log \sqrt{493} &= \log (493)^{1/2}\\ &= \tfrac{1}{2} \log 493\\ &= \tfrac{1}{2}(2.6928)\end{aligned}$$

Thus $\log \sqrt{493} = 1.3464$. Using the table we see that antilog 1.3464 is approximately 22.2. Thus $\sqrt{493} \doteq 22.2$. □

Example 5

Compute $\sqrt[5]{9762}$.

$$\begin{aligned}\log \sqrt[5]{9762} &= \log (9762)^{1/5}\\ &= \tfrac{1}{5} \log 9762\\ &= \tfrac{1}{5}(3.9895)\end{aligned}$$

Thus $\log \sqrt[5]{9762} = .7979$. Using the table we see that antilog .7979 is approximately 6.279. Thus $\sqrt[5]{9762} \doteq 6.279$. □

Example 6

Solve the equation $5^x = 92$.

Taking the logarithm of both sides we have

$$\begin{aligned}\log 5^x &= \log 92\\ x \log 5 &= \log 92\\ x &= \frac{\log 92}{\log 5}\\ x &= \frac{1.9638}{.6990}\end{aligned}$$

Dividing 1.9638 by .6990 we obtain approximately 2.81. Thus $x \doteq 2.81$. □

Sometimes it is useful to rewrite logarithms with negative characteristics in a form which is easier for use in our computation.

Example 7

Compute $\sqrt[3]{.416}$.

If

$$y = \sqrt[3]{.416}$$

then

$$\begin{aligned}y &= (.416)^{1/3}\\ \log y &= \tfrac{1}{3} \log .416\end{aligned}$$

Thus

$$\log y = \tfrac{1}{3}(9.6191 - 10)$$

Since 3 does not divide evenly into 10, we rewrite $9.6191 - 10$ as $29.6191 - 30$:

$$\begin{aligned}\log y &= \tfrac{1}{3}(29.6191 - 30)\\ \log y &= 9.8730 - 10\end{aligned}$$

Using the table we see that antilog 9.8730 − 10 is .7465. Thus

$\sqrt[3]{.416} \doteq .7465$ □

Example 8

Compute $\sqrt{\dfrac{12.3}{17.9}}$.

If

$$y = \sqrt{\frac{12.3}{17.9}}$$

$$y = \left(\frac{12.3}{17.9}\right)^{1/2}$$

$$\begin{aligned}\log y &= \tfrac{1}{2}\log\left(\frac{12.3}{17.9}\right)\\ &= \tfrac{1}{2}(\log 12.3 - \log 17.9)\\ &= \tfrac{1}{2}(1.0899 - 1.2529)\\ &= \tfrac{1}{2}(-.1630)\\ &= -.0815\end{aligned}$$

We now rewrite with a positive mantissa.

$$\begin{aligned}\log y &= 10 + (-.0815) - 10\\ &= 9.9185 - 10\end{aligned}$$

Using the table we see that antilog 9.9185 − 10 is .8288. Thus $y \doteq .8288$, that is,

$$\sqrt{\frac{12.3}{17.9}} \doteq .8288 \ \square$$

EXERCISES

1. Use logarithms to compute the following.

(a) $(493)(7.42)(.0812)(62.9)$

(b) $\dfrac{(7.42)(61.7)}{(8.92)(94.7)}$

(c) $\dfrac{(2.47)(69.45)}{(4.23)(783.4)}$

(d) $(8.629)(.435)(.008135)(6.293)(7.654)$

(e) $\dfrac{1}{4.53}$

(f) $\dfrac{1}{(84.65)(.124)(.008651)}$

(g) $\dfrac{(95.76)(493{,}400)(2.234)}{(4.976)(92{,}300)(712.5)}$

(h) $\dfrac{(56{,}350)(7949)(2761)}{(9876)(4539)(72{,}540)}$

2. Use logarithms to compute the following.
   (a) $(4.95)^3$ (b) $\sqrt{76.54}$ (c) $(9.725)^4$ (d) $\sqrt[3]{17}$
   (e) $\sqrt[4]{931}$ (f) $(1.06)^{20}$ (g) $\sqrt{4672}$ (h) $\sqrt[3]{713}$
   (i) $(1.005)^{12}$
3. Use logarithms to solve the following equations.
   (a) $4^x = 19$ (b) $6^x = 131$ (c) $22^x = 19$
   (d) $1.32^x = 5.47$ (e) $62.9^x = 123.9$
4. Compute.
   (a) $\sqrt[3]{.915}$ (b) $\sqrt{\dfrac{473}{929}}$ (c) $\sqrt[3]{\dfrac{(49.2)(76.1)(92.7)}{(81.6)(102)(62.3)}}$
   (d) $(.976)^{25}$ (e) $\sqrt[11]{.842}$

4. Compound Interest

The formula for simple interest is

$$i = Prt$$

where i is interest, P is the principal invested, r is the rate per unit of time, and t is the time. The amount A accumulated at the end of time t is given by

$$A = P + i$$

Since $i = Prt$, we have

$$A = P + Prt = P(1 + rt)$$

Thus the amount accumulated at 7 percent per year for 5 years is

$$A = P(1 + .35) = 1.35P$$

that is, the amount is 1.35 times the principal. At the end of one unit of time, the amount accumulated is just $P(1 + r)$. Let us call this amount A_1. If we now invest A_1, A_1 becomes the principal and at the end of the second unit of time, we have

$$A_2 = A_1(1 + r)$$

Since $A_1 = P(1 + r)$, we have

$$A_2 = (P(1 + r))(1 + r) \quad \text{and} \quad A_2 = P(1 + r)^2$$

If we invest A_2 for a third unit of time, we have

$$A_3 = A_2(1 + r) = [P(1 + r)^2](1 + r) = P(1 + r)^3$$

In general, if we start with an amount P at the rate r per unit time and reinvest our earnings at the end of each unit of time, the amount A_n at the end of n units of time is given by

$$A_n = P(1 + r)^n$$

The total earnings are called *compound interest.* We can use logarithms to simplify the computation of compound interest or the total amount including the principal and the interest.

Example 1

What is the total amount in the account at the end of 5 years if you invest \$5000 in a savings account at 6 percent per year for 5 years and the interest is credited to your account every 6 months?

In this case, the rate is $\frac{6}{2}$ percent for every 6 months, that is, $r = .03$. You have the money invested for 5 years which is 10 6-month units of time. Thus

$$\begin{aligned} A &= 5000(1 + .03)^{10} = 5000(1.03)^{10} \\ \log A &= \log 5000 + \log (1.03)^{10} \\ &= \log 5000 + 10 \log 1.03 \\ &= 3.6990 + 10(.0128) \\ &= 3.6990 + .1280 \\ &= 3.8270 \\ A &\doteq \$6714 \ \square \end{aligned}$$

Example 2

If you invest \$3200 at 8 percent per year compounded quarterly, what will be the amount at the end of 3 years?

The unit of time is 3 months and, therefore, the .08 rate per year is a rate of .02 per quarter. The money is invested for 12 quarters. Thus

$$\begin{aligned} A &= 3200(1 + .02)^{12} \\ &= 3200(1.02)^{12} \\ \log A &= \log 3200 + 12 \log 1.02 \\ &= 3.5051 + 12(.0086) \\ &= 3.5051 + .1032 \\ &= 3.6083 \\ A &\doteq \$4058 \ \square \end{aligned}$$

Example 3

If you invest money at 6 percent compounded annually, how long does it take for your money to double?

If you invest P at 6 percent per year, then

$$A = P(1 + .06)^n = P(1.06)^n$$

In order for the money invested to double, we need $A = 2P$. If

$$2P = P(1.06)^n$$
$$2 = (1.06)^n$$
$$\log 2 = n \log 1.06$$
$$n = \frac{\log 2}{\log 1.06}$$
$$n = \frac{.3010}{.0253} = 11.9$$

You must invest the money for 12 years. □

Notice that since $A = P(1 + r)^n$, if we had tables for $(1 + r)^n$, we could compute A just by multiplying P by the values from the tables. Such tables are available and are commonly used by persons involved in banking and investment.

EXERCISES

1. What is the total amount in your account at the end of 5 years if you invest $750 at 5 percent per year compounded annually?
2. What is the total amount accumulated in 2 years if you invest $3400 at 6 percent per year compounded quarterly?
3. How much would you have to invest at 8 percent per year compounded quarterly to yield $10,000 at the end of 6 years?
4. Suppose you invest $1000 a year in a savings account that pays 6 percent compounded semiannually. How much will you have at the end of 3 years? 5 years?
5. How long would it take your money to double if it is invested at 8 percent per year compounded quarterly?
6. What is the difference in earnings on $1000 invested at 6 percent compounded quarterly for 5 years and 6 percent compounded annually for 5 years? What are the earnings if the interest is compounded monthly?
7. What rate must your investment earn so that an investment of $1000 with interest compounded quarterly yields an amount of $1500 in 7 years?

*8. It seems the amount gets larger if the interest is compounded more frequently. Let the interest rate be 8 percent per year and compute the amount if we invest $1000 and the interest is compounded monthly. What is it if the interest is compounded 24 times a year? 48 times a year? 360 times a year?

5. Logarithms to Other Bases

When we defined the common logarithm of a number in Section 1, the number 10 was chosen because it is the base of our numeration system and therefore computations are much simpler than if some other number is used in the definition. In general, if b is a positive number different from 1, and N is a positive number,

the logarithm of N to the base b is x if and only if $N = b^x$

The logarithm to the base b of the number N is usually denoted by $\log_b N$. Thus we have

$$\log_b N = x \quad \text{if and only if} \quad N = b^x$$

Common logarithms are just logarithms to the base 10.

The laws which we established for common logarithms extend easily to more general logarithms. We list them here and leave the justification as exercises. You need only replace the "10" in the previous arguments by "b."

$$\log_b (xy) = \log_b x + \log_b y$$

$$\log_b \left(\frac{x}{y}\right) = \log_b x - \log_b y$$

$$\log_b a^x = x \log_b a$$

In calculus and other portions of higher mathematics, an important role is played by logarithms to a special base, the number e which is approximately 2.718. These logarithms are called *natural logarithms* and are frequently denoted by "$\ln x$" instead of "$\log_e x$."

Let b and c be any two positive numbers different from 1 and consider the relationship between $\log_b c$ and $\log_c b$.

$$\log_b c = x \quad \text{if and only if} \quad b^x = c$$

$$\log_c b = y \quad \text{if and only if} \quad c^y = b$$

Then

$$(b^x)^y = c^y = b$$

and

$$b^{xy} = b^1$$

Therefore

$$xy = 1$$

Thus

$$x = \frac{1}{y} \quad \text{and} \quad y = \frac{1}{x}$$

Since

$$y = \log_c b \quad \text{and} \quad x = \log_b c$$

$$\log_b c = \frac{1}{\log_c b}$$

How is this equation useful in dealing with general logarithms? Suppose we wish to compute $\log_b N$ for some number N but we have no table for logarithms to the base b. Let $N = 10^t$. Then

$$\begin{aligned}\log_b N &= \log_b 10^t \\ &= t(\log_b 10) \\ &= t\left(\frac{1}{\log_{10} b}\right)\end{aligned}$$

by the equation previously developed. Hence

$$\log_b N = t\left(\frac{1}{\log_{10} b}\right) = \frac{t}{\log_{10} b}$$

But since

$$N = 10^t$$

$$\log_{10} N = t$$

Thus we have

$$\log_b N = \frac{\log_{10} N}{\log_{10} b} = \frac{\log N}{\log b}$$

Example 1

$$\log_5 7 = \frac{\log 7}{\log 5} \doteq \frac{.8451}{.6990} \doteq 1.2090 \ \square$$

Example 2

$$\log_4 11 = \frac{\log 11}{\log 4} \doteq \frac{1.0414}{.6021} \doteq 1.7296 \ \square$$

Example 3

$$\ln 6 = \frac{\log 6}{\log e} \doteq \frac{.7782}{.4343} \doteq 1.7918 \ \square$$

From this discussion we see that any table of logarithms may be used to construct another logarithm table with a different base.

EXERCISES

1. Compute the following.
   (a) $\log_2 8 =$ (b) $\log_4 256 =$ (c) $\log_{1/2} 4 =$ (d) $\log_{100} 10 =$
   (e) $\log_{64} 4 =$ (f) $\log_5 625 =$ (g) $\log_{625} 5 =$ (h) $\log_{.1} 100 =$
   (i) $\log_{27} 3 =$ (j) $\log_{27} 81 =$
2. Compute the following.
   (a) $\log_3 10 =$ (b) $\log_7 10 =$ (c) $\log_{20} 10 =$ (d) $\log_8 10 =$
   (e) $\log_{15} 10 =$ (f) $\log_{17} 10 =$ (g) $\log_6 10 =$ (h) $\ln 10 =$
3. Compute the following.
   (a) $\log_3 5 =$ (b) $\log_7 8 =$ (c) $\log_{20} 15 =$ (d) $\log_8 2.7 =$
   (e) $\log_{15} .071 =$ (f) $\log_{17} 92.3 =$ (g) $\log_6 14 =$ (h) $\ln 6 =$
   (i) $\ln 4.2 =$ (j) $\ln .071 =$

6. Exponential and Logarithmic Functions

Once we give meaning to expressions such as $2^{\sqrt{3}}$, $5^{-4/3}$, and $17^{\sqrt[3]{11}}$, these expressions can be used in descriptions of functions. Functions such as

$$f(x) = 2^x, \qquad g(x) = 5^x, \quad \text{and} \quad h(x) = (.4)^x$$

are called *exponential functions.* In general, if a is any positive real number, then there is associated with a the exponential function $f(x) = a^x$. The graph of $g(x) = 10^x$ is given by Figure 4–1.

Figure 4–1

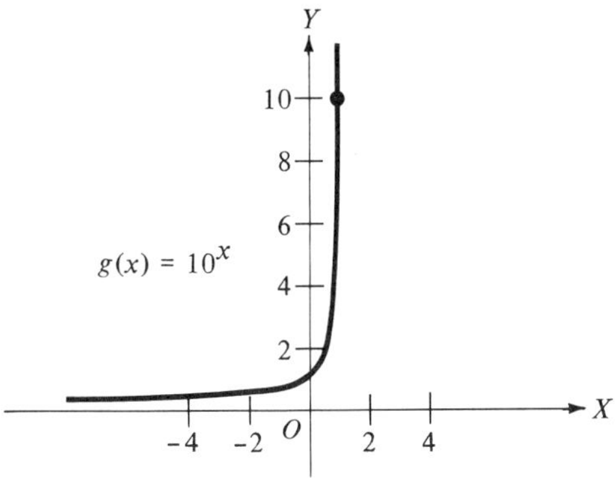

All exponential functions have certain properties in common and, in fact, are quite a bit alike. For example, if $a > 0$, and $f(x) = a^x$, then $f(0) = a^0 = 1$. Thus the point (0, 1) is on the graph of every exponential function.

Notice also that if $a > 0$, then $a^x > 0$ for each value of x and therefore the range of an exponential function is a set of positive numbers.

Exponential functions may be graphed by plotting a few key points and then sketching the rest of the curve. Once you have drawn several such graphs, the general patterns will become apparent and we leave the discovering of some of these as exercises.

If x is any positive number, then $\log x$ has meaning as defined in Section 1 and we can use this meaning to define a function, namely the logarithm function

$$f(x) = \log x$$

By plotting a few key points and sketching the rest of the curve we see that the graph of the logarithmic function is as in Figure 4–2.

Figure 4–2

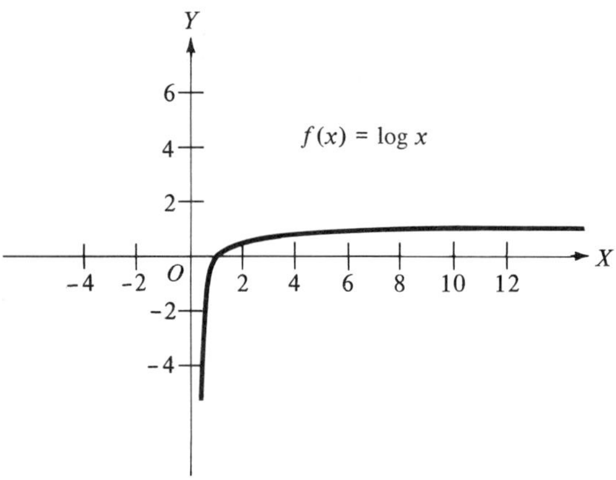

If $b > 0$, $b \neq 1$, then we can also draw the graph of $g(x) = \log_b x$. All logarithmic functions have certain common properties and you will be given an opportunity to discover some of them in the exercises.

EXERCISES

1. Draw graphs of the following functions.
   (a) $f(x) = 2^x$ (b) $g(x) = 3^x$ (c) $h(x) = (\frac{1}{2})^x$ (d) $k(x) = (.1)^x$
   (e) $p(x) = 20^x$ (f) $q(x) = 4^x$ (g) $r(x) = 8^x$ (h) $s(x) = (\frac{1}{4})^x$
2. Let $a > 0$, $a \neq 1$, and define $f(x)$ by $f(x) = a^x$.
   (a) What is the domain of f?
   (b) What is the range of f?

(c) What is $f(0)$?
(d) If $1 < a$ and $p < q$, what can be said about $f(p)$ and $f(q)$?
(e) If $0 < a < 1$ and $p < q$, what can be said about $f(p)$ and $f(q)$?

3. Draw graphs of the following functions.
(a) $F(x) = \log_2 x$ (b) $G(x) = \log_3 x$ (c) $H(x) = \log_{1/2} x$
(d) $K(x) = \log_{.1} x$ (e) $P(x) = \log_{20} x$ (f) $Q(x) = \log_4 x$
(g) $R(x) = \log_8 x$ (h) $S(x) = \log_{1/4} x$

4. Let $b > 0$, $b \neq 1$, and define $g(x)$ by $g(x) = \log_b x$.
(a) What is the domain of g?
(b) What is the range of g?
(c) What is $g(1)$?
(d) If $b > 1$ and $0 < p < q$, what can be said about $g(p)$ and $g(q)$?
(e) If $0 < b < 1$ and $0 < p < q$, what can be said about $g(p)$ and $g(q)$?

5. (a) Draw the following three graphs on one coordinate system: $f(x) = \log x$, $F(x) = 10^x$, and $y = x$.
(b) Draw the following three graphs on one coordinate system: $g(x) = \log_2 x$, $G(x) = 2^x$, and $y = x$.
(c) Draw the following three graphs on one coordinate system: $h(x) = \log_{1/2} x$, $H(x) = (\frac{1}{2})^x$, and $y = x$.
(d) From your drawings in parts (a), (b), and (c), what can you say in general about the graphs of $y = \log_b x$, $y = b^x$, and $y = x$?

6. Generalize the proofs of Propositions 1–3 of Section 1 for general logarithms.

7. Draw graphs of the following functions.
(a) $f(x) = \log(x + 3)$ (b) $g(x) = \log(x - 5)$ (c) $h(x) = \log(10x)$
(d) $k(x) = 1 + \log x$ (e) $r(x) = \log(5x)$ (f) $t(x) = \log(5x + 2)$
(g) $p(x) = \log(x^2)$ (h) $q(x) = 2 \log x$ (i) $m(x) = \log(x^3)$
(j) $n(x) = \log(x^4)$ (k) $z(x) = \log x$ (l) $v(x) = \log 5x$

8. Draw graphs of the following functions.
(a) $f(x) = 2^{x+1}$ (b) $g(x) = 2^{3x}$ (c) $h(x) = 2^{3x+1}$
(d) $k(x) = 2^{-x}$ (e) $m(x) = 2^{-x+1}$ (f) $n(x) = 2^{2x-3}$
(g) $p(x) = 2^{x/2}$ (h) $q(x) = 2^{x/2+1}$ (i) $r(x) = 2^{-x/3+2}$
(j) $t(x) = 3^{2x-5}$

7. Exponential Variation

We say that y *varies exponentially with* x if $y = ka^x$ for some number k and some positive number a.

Example 1

If $y = 3(10)^x$, then y varies exponentially with x. When $x = 1$, $y = 30$; when $x = 2$, $y = 300$. □

If y varies exponentially with x, we can determine k and a if we know two pairs of values of x and y.

Example 2

If $y = ka^x$ with (0, 4) and (3, 9) points on the graph, then

$$4 = ka^0 = k \quad \text{and} \quad k = 4$$

Also,

$$9 = 4a^3$$

$$\frac{9}{4} = a^3 \quad \text{and} \quad a = \left(\frac{9}{4}\right)^{1/3}$$

Thus

$$y = 4\left[\left(\frac{9}{4}\right)^{1/3}\right]^x$$

or

$$y = 4\left(\frac{9}{4}\right)^{x/3} \quad \square$$

Example 3

Bacterial growth is known to be exponential, that is, the amount A at time t varies exponentially with t. If the number of bacteria in a culture is 10,000 initially and 16,000 24 hours later, what will be the number of bacteria present after 40 hours?

Since A varies exponentially with time t, we have

$$A = ka^t$$

for some a and k. For $t = 0$, $A = 10{,}000$. Thus,

$$10{,}000 = kA^0 = k \quad \text{and} \quad k = 10{,}000$$

Also,

$$\begin{aligned} 16{,}000 &= 10{,}000a^{24} \\ 1.6 &= a^{24} \\ a &= (1.6)^{1/24} \end{aligned}$$

Thus

$$\begin{aligned} A &= 10{,}000[(1.6)^{1/24}]^t \\ &= 10{,}000(1.6)^{t/24} \end{aligned}$$

When $t = 40$,

$$\begin{aligned} A &= 10{,}000(1.6)^{40/24} \\ &= 10{,}000(1.6)^{5/3} \end{aligned}$$

To compute A,

$$\begin{aligned}\log A &= \log 10{,}000 + \log (1.6)^{5/3} \\ &= 4 + \tfrac{5}{3} \log 1.6 \\ &= 4 + \tfrac{5}{3}(.2041) \\ &= 4 + .3402 \\ &= 4.3402\end{aligned}$$

Since $\log A = 4.3402$, A is approximately 21,890.

Notice that even though we can express the relationship in the form

$$A = 10{,}000[(1.6)^{1/24}]^t$$

it is more convenient to express it as

$$A = 10{,}000(1.6)^{t/24}$$

and most relationships arising from exponential variation are expressed in such a fashion. □

EXERCISES

1. Determine the relationship between x and y if y varies exponentially with x, $y = 4$ when $x = 0$, and $y = 12$ when $x = 1$.
2. Determine the relationship between x and y if y varies exponentially with x, $y = 20$ when $x = 0$, and $y = 5$ when $x = 2$.
3. Determine the relationship between x and y if $y - 5$ varies exponentially with x, $y = 12$ when $x = 0$, and $y = 20$ when $x = 2$.
4. If the number of bacteria in a culture is 5000 at the beginning of an experiment, and is 8000 40 hours later, how many will there be 100 hours later? How long does it take the number to double?
5. Population in a nation varies (approximately) exponentially with time. If there are 14 million people in the country in 1950 and 20 million in 1960, what will be the approximate population in 1980? In the year 2000?
6. Consult a census report and repeat Exercise 5 above using the population of the United States in 1960 and 1970 to project the population in 2000.
7. The amount of radioactive substance remaining after time t varies exponentially with t. If 20 gm of a radioactive substance decays so that 15 gm remain after 2 years, how much will remain after 3 years?
8. The half-life of a radioactive substance is the length of time in which exactly half of a given amount decays (equivalently, half remains). Determine the half-life of the substance discussed in Exercise 7 above.
9. The half-life of radium is 1600 years. Give an equation that expresses the amount of radium remaining after t years.
10. Newton's law of cooling states that the difference between the temperature of a body and the temperature of the surrounding medium varies exponentially with time. Suppose the temperature of outside air is 10° above zero and a container of water set outside cools from 210° to 180° in 10 min. What will the temperature of the water be in 15 min? How long will it be until the water freezes?

5

Analytic Trigonometry

The study of trigonometry has its beginnings in the surveying and engineering works of ancient Egypt. The numerical aspects of the subject are still significant but have been overshadowed by the importance of trigonometric functions in calculus and in fields that use the calculus. In this chapter we introduce the fundamental principles of the subject and concentrate on the analytic aspects of the six elementary trigonometric functions. Numerical applications will be treated in Chapter 6.

1. The Trigonometric Functions

In geometry, an angle is considered to be two rays with a common end point. (See Figure 5–1.)

An angle is said to be in *standard position* if this common end point is at the origin of the coordinate system and one of the rays coincides with the origin and the positive portion of the X-axis. This ray is called the *initial* side and the other ray is called the *terminal* side. (See Figure 5–2.)

In trigonometry an angle in standard position is thought of as an amount of rotation of the initial side around the origin which ends in the terminal side. Positive rotation is in a counterclockwise direction and clockwise

Figure 5-1

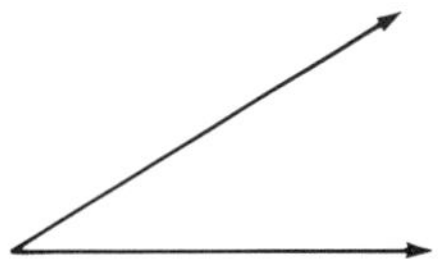

Figure 5-2

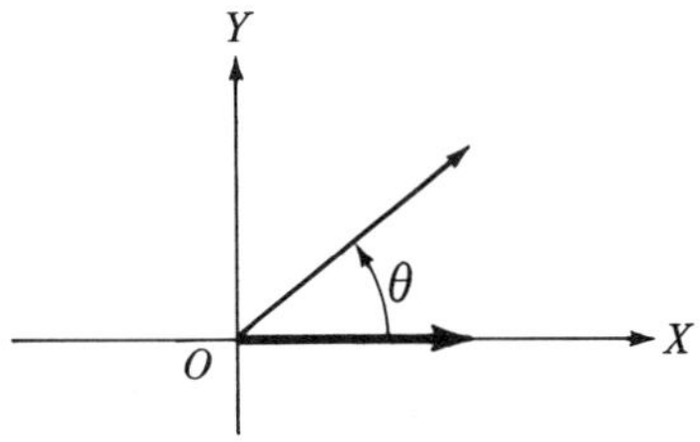

rotation is considered negative rotation. The measure of an angle is a measure of the amount of rotation. The customary degree measurement is adapted to this notion by treating one full rotation as 360°, two full rotations as 720°, and so on. Thus an angle of 325° means a rotation in a counterclockwise direction through 325/360 of one full rotation. An angle of -40° means rotation in a clockwise direction through $\frac{1}{9}$ of one full rotation. Notice that although the terminal side of an angle of 425° is the same as the terminal side of an angle of 65°, the amounts of rotation are not the same and, therefore, the angles are not the same.

Let θ be an angle with vertex at the origin, initial side along the positive X-axis; let (a, b) be any point on the terminal side of θ as illustrated in Figure 5–3 and let $r = \sqrt{a^2 + b^2}$, the distance of (a, b) from $(0, 0)$. The six elementary trigonometric functions of θ are defined by:

$$\text{sine } \theta = \frac{b}{r} \qquad \text{(abbreviated sin } \theta\text{)}$$

$$\text{cosine } \theta = \frac{a}{r} \qquad \text{(abbreviated cos } \theta\text{)}$$

$$\text{tangent } \theta = \frac{b}{a} \qquad \text{(abbreviated tan } \theta\text{)}$$

$$\text{cotangent } \theta = \frac{a}{b} \qquad \text{(abbreviated cot } \theta\text{)}$$

Figure 5–3

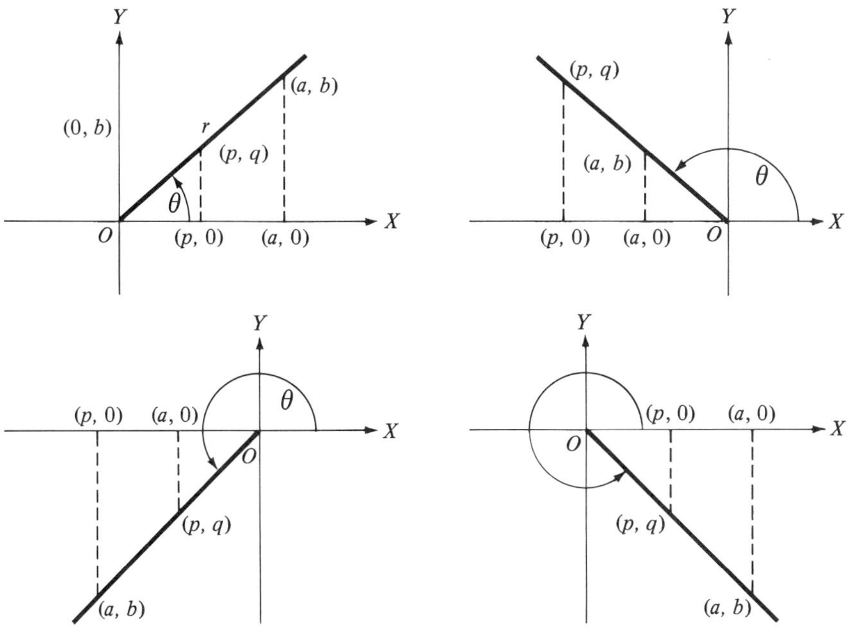

$$\text{secant } \theta = \frac{r}{a} \qquad \text{(abbreviated sec } \theta\text{)}$$

$$\text{cosecant } \theta = \frac{r}{b} \qquad \text{(abbreviated csc } \theta\text{)}$$

In order to be sure that we have, in fact, defined functions, we must be sure that the values do not depend on the particular point chosen but only on θ. If we let (p, q) be any other point on the terminal side of the angle, as in Figure 5–3, then the right triangle formed by (0, 0), (p, q), and $(p, 0)$ is similar to the triangle formed by (0, 0), (a, b), and $(a, 0)$. Since the triangles are similar, the corresponding sides are proportional. If we let $s = \sqrt{p^2 + q^2}$, the distance of (p, q) from the origin, we see that

$$\frac{|b|}{r} = \frac{|q|}{s} \qquad \frac{|a|}{r} = \frac{|p|}{s}$$

$$\frac{|b|}{|a|} = \frac{|q|}{|p|} \qquad \frac{|a|}{|b|} = \frac{|p|}{|q|}$$

$$\frac{r}{|a|} = \frac{s}{|p|} \qquad \frac{r}{|b|} = \frac{s}{|q|}$$

Since a and p agree in sign and b and q agree in sign, it follows that

$$\frac{b}{r} = \frac{q}{s} \qquad \frac{a}{r} = \frac{p}{s}$$

$$\frac{b}{a} = \frac{q}{p} \qquad \frac{a}{b} = \frac{p}{q}$$

$$\frac{r}{a} = \frac{s}{p} \qquad \frac{r}{b} = \frac{s}{q}$$

Notice that if we choose a point (x, y) on the terminal side of the angle, a unit length from the origin, as in Figure 5–4, then $x = \cos\theta$ and $y = \sin\theta$.

Figure 5-4

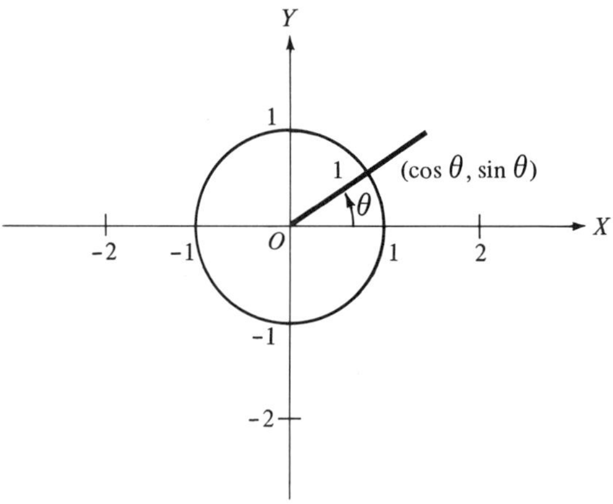

Not all trigonometric functions have meaning for all angles. For example, if we choose the point (0, 1) on the terminal side of a 90° angle, as illustrated in Figure 5–5, the definition would require the tangent to be 1/0 which is absurd. Thus an angle of 90° has no tangent. For the same reason, it has no secant.

Since the functions of an angle depend on the position of its terminal side, they are, in fact, functions of the measure of the angle; that is, if θ and α have the same measure, then the values of the trigonometric functions of θ are the same as those of α. Thus we can speak or write of sin 30° or tan 45° when, in fact, we really mean the sine of an angle measuring 30° or the tangent of an angle measuring 45°. We will not distinguish between an angle θ and an angle with measure θ.

Since any angle has a terminal side coinciding with the terminal side of an angle with measure between 0° and 360°, if we know the values of trigono-

Figure 5–5

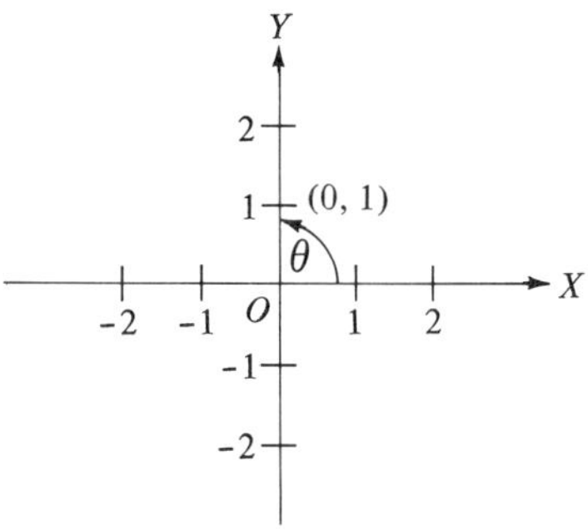

metric functions of angles between 0° and 360°, we know the values for all angles.

The usual method of coordinatizing the plane divides the plane into four regions called *quadrants* and these are numbered as in Figure 5–6. An angle is said to be in quadrant III if its terminal side is in the third quadrant and so on. An angle is said to be *quadrantal* if its terminal side is on one of the axes.

Figure 5–6

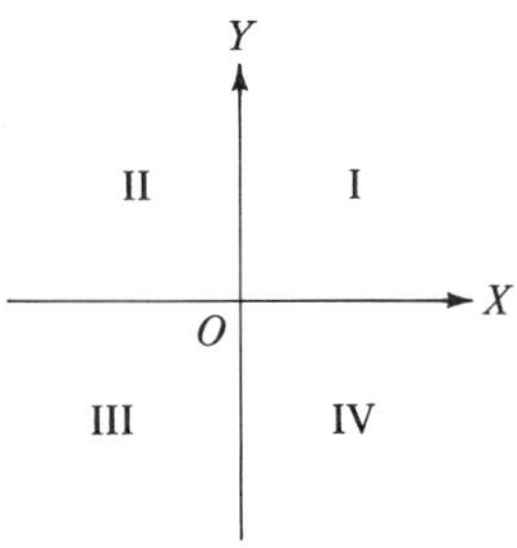

If the terminal side of the angle is in quadrant I, then all the trigonometric functions are positive because the coordinates of any point on the terminal side are positive. If the terminal side is in quadrant II and (a, b) is a point on the terminal side, then a is negative and b is positive. Thus

$$\sin \theta = \frac{b}{r} > 0$$

$$\cos \theta = \frac{a}{r} < 0$$

$$\tan\theta = \frac{b}{a} < 0$$

$$\cot\theta = \frac{a}{b} < 0$$

$$\sec\theta = \frac{r}{a} < 0$$

$$\csc\theta = \frac{r}{b} > 0$$

In summary, if θ is in the second quadrant, then $\sin\theta$ and $\csc\theta$ are positive and the other functions are negative. By similar reasoning using a diagram such as Figure 5–3, we can determine the signs of the trigonometric functions of angles in quadrants III and IV. This information is summarized in Table 5.1. By a plus sign, we mean that the function is positive for angles in that quadrant and a minus sign indicates that the function is negative for angles in that quadrant.

Table 5.1

Function	Quadrant I	II	III	IV
sine	+	+	−	−
cosine	+	−	−	+
tangent	+	−	+	−
cotangent	+	−	+	−
secant	+	−	−	+
cosecant	+	+	−	−

Since we know how to determine the signs of the trigonometric functions of any given angle, we could determine the values of the functions of any angle if we had a method of determining the absolute value of these functions. The following arguments show that if we know how to determine the values of trigonometric functions for angles between 0° and 90° we could then use this information to determine the values of the trigonometric functions of any angle whatsoever.

Case 1

Suppose the terminal side of an angle θ is in quadrant II and we may as well assume $90° < \theta < 180°$. If θ is not between 90° and 180°, it has the same terminal side as such an angle. (See Figure 5–7.) The triangle (0, 0), $(a, 0)$, (a, b) is congruent to the triangle (0, 0), $(-a, 0)$, $(-a, b)$, and $(-a, b)$

Figure 5–7

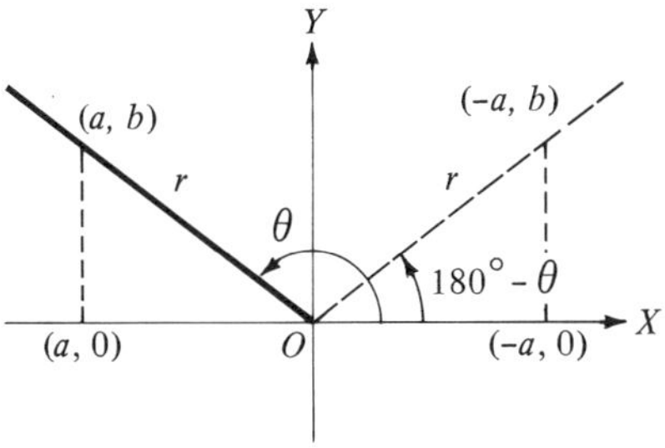

is a point on the terminal side of $180° - \theta$. Direct computation will show that the absolute values of the trigonometric functions of θ are the same as those of $180° - \theta$ which is between $0°$ and $90°$, that is, $0° < 180° - \theta < 90°$. The angle $180° - \theta$ is called a *reference* angle for the angle θ.

Example 1

$$\sin 110° = \sin 70°$$
$$\cos 110° = -\cos 70°$$
$$\tan 110° = -\tan 70° \quad \square$$

Case 2

Suppose the terminal side of θ is in quadrant III and $180° < \theta < 270°$. (See Figure 5–8.) In this case the triangle (0, 0), $(a, 0)$, (a, b) is congruent to the triangle (0, 0), $(-a, 0)$, $(-a, -b)$.

Figure 5–8

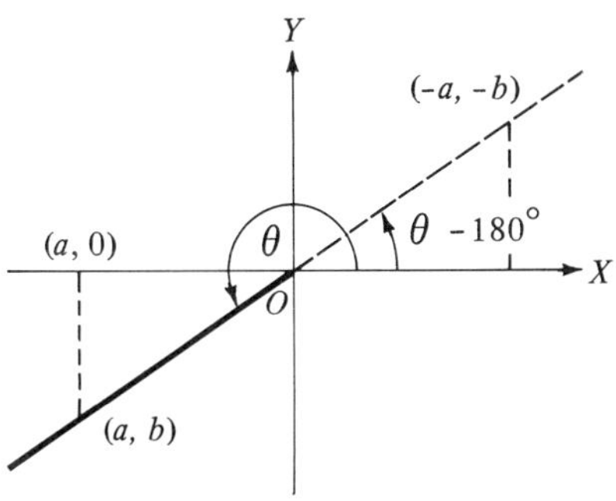

The terminal sides are on a line through the origin and we can determine the absolute values of the trigonometric functions of θ by determining the trigonometric functions of $\theta - 180°$ which is between $0°$ and $90°$. In this case, $\theta - 180°$ is the reference angle.

Example 2

$$\sin 230° = -\sin 50°$$
$$\cos 230° = -\cos 50°$$
$$\tan 230° = \tan 50° \quad \square$$

Case 3

Suppose the terminal side of θ is in quadrant IV and $270° < \theta < 360°$. (See Figure 5–9.) In this case the triangle $(0, 0)$, $(a, 0)$, (a, b) is congruent to the triangle $(0, 0)$, $(a, 0)$, $(a, -b)$.

Figure 5–9

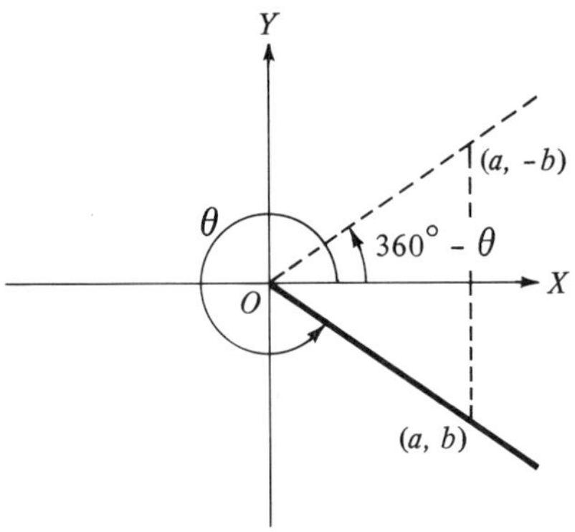

The absolute values of the trigonometric functions of θ can be found by determining the values of the trigonometric functions of $360° - \theta$ which, in this case, is between $0°$ and $90°$. In this case the reference angle is $360° - \theta$.

Example 3

$$\sin 305° = -\sin 55°$$
$$\cos 305° = \cos 55°$$
$$\tan 305° = -\tan 55° \quad \square$$

In each case, the reference angle is an acute angle between the terminal side and the X-axis. (See Figure 5–10.)

Once we place or picture the angle in standard position on the coordinate system, it is a rather easy matter to determine the reference angle.

Figure 5-10

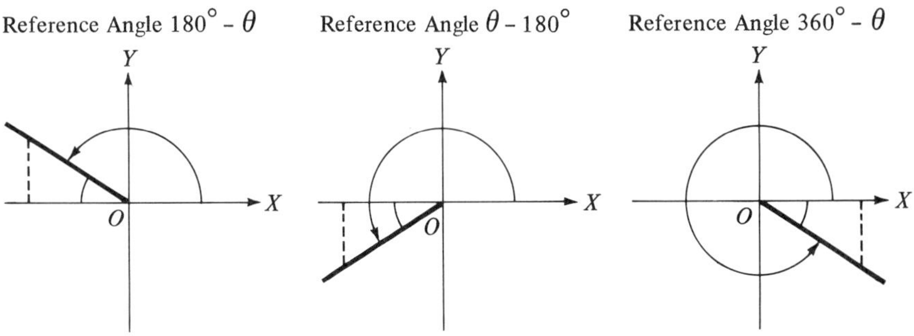

Example 4

The reference angle for 165° is 15°.
The reference angle for 235° is 55°.
The reference angle for 287° is 73°. □

Our knowledge of geometry enables us to determine the trigonometric functions of some special angles. First consider an angle of 45°. (See Figure 5–11.)

The point (1, 1) is on the terminal side of the angle and its distance from (0, 0) is $\sqrt{2}$. Thus

$$\sin 45^\circ = \frac{1}{\sqrt{2}} = \frac{\sqrt{2}}{2} \qquad \cos 45^\circ = \frac{\sqrt{2}}{2}$$

$$\tan 45^\circ = \frac{1}{1} = 1 \qquad \cot 45^\circ = 1$$

$$\sec 45^\circ = \frac{\sqrt{2}}{1} = \sqrt{2} \qquad \csc 45^\circ = \sqrt{2}$$

Consider now an equilateral triangle with sides having length 2 as in Figure 5–12. In Case 1, one vertex is at (0, 0), one side is along the positive X-axis, and the third vertex in quadrant I. In Case 2, one vertex is at (0, 0) and the positive X-axis bisects the opposite side.

In Case 1, we have an angle of 60° in standard position and, in Case 2, we have an angle of 30° in standard position. The Pythagorean theorem

Figure 5–11

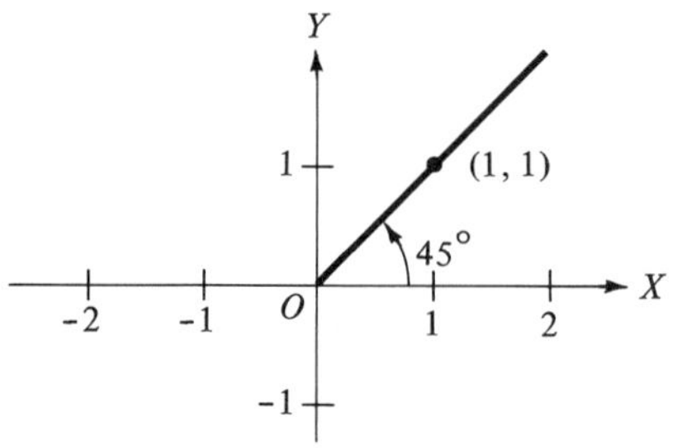

Figure 5–12

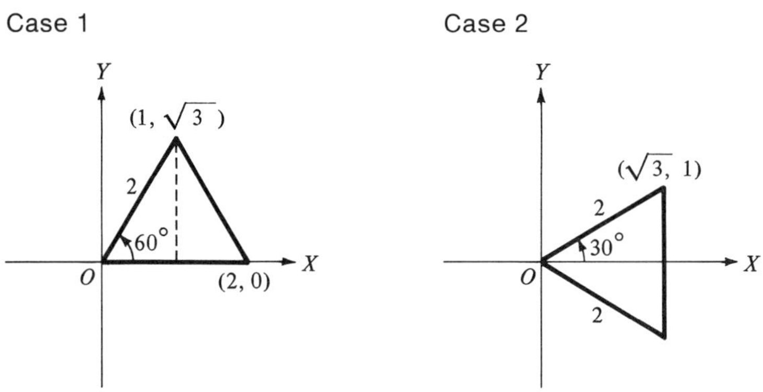

determines that the altitude of the triangle is $\sqrt{3}$ and the vertices have the coordinates as indicated in Figure 5–12.

Using Figure 5–12, we now see that

$$\sin 30^\circ = \frac{1}{2} \qquad \sin 60^\circ = \frac{\sqrt{3}}{2}$$

$$\tan 30^\circ = \frac{1}{\sqrt{3}} = \frac{\sqrt{3}}{3} \qquad \tan 60^\circ = \frac{\sqrt{3}}{1} = \sqrt{3}$$

We leave determining the rest of the values as an exercise.

EXERCISES

1. For some angles, certain trigonometric functions have no meaning; for example, as was pointed out in the text, tan 90° does not exist. Compute the values of the trigonometric functions of 0°, 90°, 180°, and 270° and determine which ones do not exist.
2. Determine the values of the trigonometric functions of 60° and 30°.

3. Compute the following.
   (a) $\sin 150^\circ$ (b) $\cos 225^\circ$ (c) $\tan -30^\circ$ (d) $\cot 120^\circ$
   (e) $\sin 300^\circ$ (f) $\sec 210^\circ$ (g) $\csc 315^\circ$ (h) $\cos -210^\circ$
   (i) $\tan 240^\circ$ (j) $\sec -60^\circ$
4. Find all values of x such that $-360^\circ < x < 360^\circ$ and
   (a) $\sin x = -1/2$ (b) $\cos x = 1/2$ (c) $\tan x = -1$
   (d) $\tan x = \sqrt{3}/3$ (e) $\sec x = 1$ (f) $\csc x = -2$
5. Can $\sin\theta = 1.5$? Explain your answer.
6. What is the range of $\sin\theta$? What is the range of $\cos\theta$? Justify your answers.
7. What is the range of $\sec\theta$? What is the range of $\csc\theta$? Justify your answers.
8. What is the range of $\tan\theta$? What is the range of $\cot\theta$? Justify your answers.

2. Radian Measure

Through most of your mathematical experience, you have measured angles in terms of degrees. One full rotation was 360°, one-fourth of a rotation or a right angle was 90°, and so on. Angles can also be measured in terms of rotations or revolutions and, for some purposes, such measurements can be most useful. In calculus and other higher mathematics, it is very useful to measure angles in terms of *radians*.

Let an angle be in standard position on a coordinate system and let a circle of radius 1 be drawn with center at the origin. (See Figure 5–13.) Such a circle is called a unit circle. Note that it has circumference 2π.

Figure 5–13

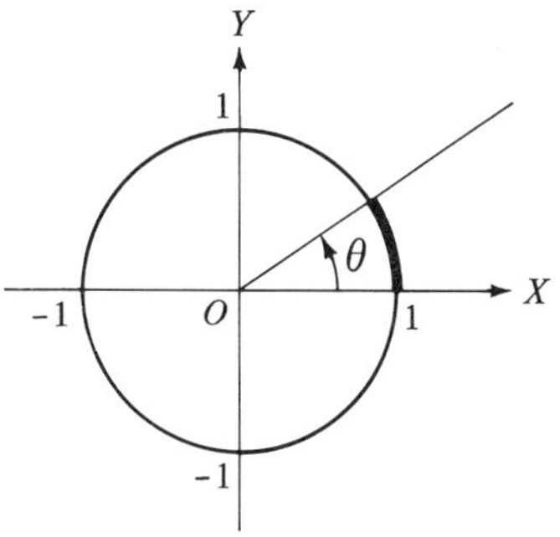

The *radian measure* of the angle is the length of the arc of the unit circle cut off by the two sides of the angle, the distance being measured from the initial side to the terminal side. If the rotation of the angle is positive, the radian measure is positive; if the rotation is negative, then the radian measure is negative. Notice that a 90° angle has radian measure $\pi/2$, and a 180°

angle has radian measure π. Thus we see that an angle whose degree measure is between 0° and 360° must have radian measure between 0 and 2π.

If we adopt the trigonometric convention of viewing angles as amounts of rotation we also can use radian measure. One full revolution traverses the circumference of the unit circle which is precisely 2π. Thus just as an angle of 450° means $1\frac{1}{4}$ revolutions, an angle of $5\pi/2$ radians also means $1\frac{1}{4}$ revolutions. Similarly, an angle of 8π radians means 4 complete revolutions. To change from radians to degrees, notice that 1° is $1/360$ of a revolution and 1 radian is $1/2\pi$ of a revolution. Thus

$$\text{I.}\quad \frac{\text{measure in degrees}}{360} = \frac{\text{measure in radians}}{2\pi}$$

Example 1

What is the radian measure of a 60° angle?

If we let r denote the radian measure we have

$$\frac{60}{360} = \frac{r}{2\pi}$$

Thus

$$\frac{1}{6} = \frac{r}{2\pi} \quad \text{and} \quad r = \frac{\pi}{3} \; \square$$

Example 2

What is the degree measure of an angle measuring $\pi/8$ radians?

If we let d denote the degree measure we have

$$\frac{d}{360} = \frac{\pi/8}{2\pi}$$

$$\frac{d}{360} = \frac{\pi}{16\pi} = \frac{1}{16}$$

$$d = \frac{360}{16} = 22.5$$

Thus the angle measures 22.5°. $\square$

If we solve Equation (I.) for each unit of measurement in terms of the other, we obtain the pair of equations

$$\text{II.}\quad \text{measure in degrees} = \frac{180\,(\text{measure in radians})}{\pi}$$

$$\text{III.} \quad \text{measure in radians} = \frac{\pi\,(\text{measure in degrees})}{180}$$

You should either memorize these two formulas or remember the basic relationship given in Equation (I.) above.

If r is any real number, it can be made to correspond to an angle with r as its radian measure in just about the same way as we discuss angles of 750° and so on. If $-2\pi < r < 2\pi$, then the directed distance r along the unit circle determines a point on that circle and this determines an angle. If $r \geq 2\pi$ or $r \leq -2\pi$, then the angle consists of more than 1 full revolution and we determine the point on the circle by determining the number of full revolutions plus the partial revolution.

Example 3

Just as 750° means 2 full revolutions plus 30°, $13\pi/6 = 2\pi + \pi/6$ means 2 full revolutions plus $\pi/6$ radians. □

Example 4

An angle of $16\pi/3$ radians is $4\pi + 4\pi/3$ which is 2 complete revolutions plus $4\pi/3$ radians. □

From the above discussion, we see that any real number can be used as the radian measure of an angle. It is customary to use radian measure to define the value of a trigonometric function of a real number.

By $\sin t$ we mean the sine of an angle with measure t radians. The standard notational convention is to use the degree symbol "°" when the angle measure is given in degrees and to assume radian measure when this symbol does not appear. Thus, for example, by $\cos 1.7$ we mean the cosine of an angle with measure 1.7 radians. We will use both types of measure in this text so that you will become more aware of the relationship between them and be able to handle both with ease.

In calculus and other higher mathematics, the notion of trigonometric functions of angles is suppressed and these functions are viewed just like any other functions of real numbers. For the balance of this chapter our primary interest is to study these functions, their behavior, and their interrelationships.

EXERCISES

1. Convert the following degree measurements into radians.
   (a) 30° (b) 120° (c) −45° (d) 270°
   (e) 750° (f) 150° (g) 225° (h) 75°
2. Convert the following radian measurements into degrees.
   (a) $\pi/4$ (b) $2\pi/3$ (c) $-\pi/12$ (d) $5\pi/4$ (e) $-7\pi/2$
   (f) $3\pi/2$ (g) $15\pi/4$ (h) $-3\pi/8$ (i) $\pi/6$ (j) $5\pi/6$

3. Complete the following table of values of trigonometric functions.

	0	$\pi/6$	$\pi/4$	$\pi/3$	$\pi/2$	$2\pi/3$	$3\pi/4$	$5\pi/6$	π	$7\pi/6$	$5\pi/4$	$4\pi/3$	$3\pi/2$	$5\pi/3$	$7\pi/4$	$11\pi/6$
sin																
cos																
tan																
cot																
sec																
csc																

4. Use your information from Exercise 3 to sketch graphs of the following equations where $0 \le x \le 2\pi$.
   (a) $y = \sin x$ (b) $y = \cos x$ (c) $y = \cot x$
   (d) $y = \sec x$ (e) $\tan x$ (f) $\csc x$

3. The Elementary Trigonometric Identities

There are two basic types of equations which concern us in mathematics, *conditional equations* and *identities*. Equations such as $2x + 3 = 5$, $x^2 + 1 = 10$, and $2x^2 - 3x - 7 = 0$ are examples of conditional equations. In general, conditional equations have relatively small solution sets; they do not hold for most real numbers and, in fact, most conditional equations you have seen have finite solution sets. In general, an equation is *conditional* if there are some real numbers which, when substituted for the variable, produce a meaningful but false equation. An *identity* is an equation which holds for all possible substitutions for the variable providing that the substitution yields a meaningful statement.

Example 1

The equation

$$x^2 - 4 = (x + 2)(x - 2)$$

is an identity because no matter what number is substituted for x, the sentence is true. □

Example 2

The equation

$$\frac{x^2 - 4}{x - 2} = x + 2$$

is also an identity because it holds for all substitutions which yield meaningful statements. If we substitute 2 for x, the left expression is meaningless. □

We are going to be concerned with trigonometric identities, that is, equations involving trigonometric functions which will hold for all values of the variable that yield meaningful statements. In this section we will establish some elementary identities which provide basic, fundamental relationships among the trigonometric functions.

Recall that if (a, b) is a point on the terminal side of θ and $r = \sqrt{a^2 + b^2}$, the distance of (a, b) from $(0, 0)$, then

$$\sin\theta = \frac{b}{r} \qquad \cos\theta = \frac{a}{r}$$

$$\tan\theta = \frac{b}{a} \qquad \cot\theta = \frac{a}{b}$$

$$\sec\theta = \frac{r}{a} \qquad \csc\theta = \frac{r}{b}$$

Example 3

The first identity we will establish is a relation among $\sin\theta$, $\cos\theta$, and $\tan\theta$.

$$\frac{\sin\theta}{\cos\theta} = \frac{b/r}{a/r} = \frac{b}{a} = \tan\theta$$

Thus we have established the identity

$$\frac{\sin\theta}{\cos\theta} = \tan\theta$$

Notice that if $\theta = \pi/2$, the expression on the left has no meaning. We have shown, however, that if $\cos\theta \neq 0$, then

$$\frac{\sin\theta}{\cos\theta} = \tan\theta \ \square$$

Example 4

$$\frac{1}{\cos\theta} = \frac{1}{a/r} = \frac{r}{a} = \sec\theta$$

Thus

$$\frac{1}{\cos\theta} = \sec\theta$$

Realize that if $\cos\theta = 0$, the above sentence is meaningless. □

Example 5

$$
\begin{aligned}
(\sin\theta)^2 + (\cos\theta)^2 &= \left(\frac{b}{r}\right)^2 + \left(\frac{a}{r}\right)^2 \\
&= \frac{b^2}{r^2} + \frac{a^2}{r^2} \\
&= \frac{a^2 + b^2}{r^2} \\
&= \frac{r^2}{r^2} \\
&= 1
\end{aligned}
$$

Thus

$$(\sin\theta)^2 + (\cos\theta)^2 = 1 \ \square$$

At this point, let us establish a notational convention. If $n \neq -1$, it is customary to denote "$(\sin\theta)^n$" by "$\sin^n\theta$," "$(\cos\theta)^n$" by "$\cos^n\theta$," and so on for other trigonometric functions. Thus the above identity becomes

$$\sin^2\theta + \cos^2\theta = 1$$

Example 6

Establish the identity

$$\tan^2\theta + 1 = \sec^2\theta$$

We have many routes to take but we must not start with the desired identity and manipulate. To do this would be to assume what we are trying to prove and our logic would be faulty. We may start with one side and make substitutions using known identities or the definitions to attempt to transform it into the other expression.

$$
\begin{aligned}
\tan^2\theta + 1 &= \left(\frac{\sin\theta}{\cos\theta}\right)^2 + 1 && \text{because } \tan\theta = \frac{\sin\theta}{\cos\theta} \\
&= \frac{\sin^2\theta}{\cos^2\theta} + 1 \\
&= \frac{\sin^2\theta}{\cos^2\theta} + \frac{\cos^2\theta}{\cos^2\theta} \\
&= \frac{\sin^2\theta + \cos^2\theta}{\cos^2\theta} \\
&= \frac{1}{\cos^2\theta} && \text{because } \sin^2\theta + \cos^2\theta = 1
\end{aligned}
$$

$$= \left(\frac{1}{\cos\theta}\right)^2$$

$$= \sec^2\theta \qquad \text{because } \frac{1}{\cos\theta} = \sec\theta$$

Thus $\tan^2\theta + 1 = \sec^2\theta$.

We would also establish this identity by beginning with the known identity

$$\sin^2\theta + \cos^2\theta = 1$$

Dividing both sides by $\cos^2\theta$ yields

$$\frac{\sin^2\theta + \cos^2\theta}{\cos^2\theta} = \frac{1}{\cos^2\theta}$$

$$\frac{\sin^2\theta}{\cos^2\theta} + \frac{\cos^2\theta}{\cos^2\theta} = \left(\frac{1}{\cos\theta}\right)^2$$

$$\left(\frac{\sin\theta}{\cos\theta}\right)^2 + 1 = \sec^2\theta \qquad \text{because } \frac{1}{\cos\theta} = \sec\theta$$

$$\tan^2\theta + 1 = \sec^2\theta \qquad \text{because } \frac{\sin\theta}{\cos\theta} = \tan\theta$$

We could also go back to the original definitions of the trigonometric functions and establish this identity using these definitions and some algebra.

$$\tan^2\theta + 1 = \left(\frac{b}{a}\right)^2 + 1$$

$$= \frac{b^2}{a^2} + 1$$

$$= \frac{b^2}{a^2} + \frac{a^2}{a^2}$$

$$= \frac{a^2 + b^2}{a^2}$$

$$= \frac{r^2}{a^2}$$

$$= \left(\frac{r}{a}\right)^2$$

$$= \sec^2\theta$$

Thus

$$\tan^2\theta + 1 = \sec^2\theta \quad \square$$

Any one of these methods is valid. The first and second, however, are more indicative of the methods to be used in establishing more complicated identities. In these two cases we used previously established information and did not have to go all the way back to the definitions.

We now list a number of fundamental identities and we leave the proofs of the new ones as exercises. These relationships are important and should be remembered.

1. $\dfrac{\sin\theta}{\cos\theta} = \tan\theta$

2. $\dfrac{\cos\theta}{\sin\theta} = \cot\theta$

3. $\dfrac{1}{\tan\theta} = \cot\theta$

4. $\dfrac{1}{\sin\theta} = \csc\theta$

5. $\dfrac{1}{\cos\theta} = \sec\theta$

6. $\sin^2\theta + \cos^2\theta = 1$
7. $\tan^2\theta + 1 = \sec^2\theta$
8. $1 + \cot^2\theta = \csc^2\theta$

Historically, the words *cosine*, *cotangent*, and *cosecant* are abbreviations for *complement's sine*, *complement's tangent*, and *complement's secant*; that is, the cosine of θ is the sine of the complement of θ, and so on. The complement of an angle is the difference between it and a right angle and therefore *the complement of θ is $\pi/2 - \theta$* if we are using radian measure. In order to show that the cosine of θ is the sine of the complement of θ we need to show that $\cos\theta = \sin(\pi/2 - \theta)$.

Case 1

Suppose θ is in quadrant I. Choose (a, b) on the terminal side of θ and consider the point (b, a) and the angle determined by the positive X-axis and the line through $(0, 0)$ and (b, a) as in Figure 5–14. This will be in quadrant I because a and b are both positive. The two rectangles are congruent and therefore the angle determined by the positive X-axis and the ray through (b, a) must be $\pi/2 - \theta$. Now by definition we have

$$\sin\left(\frac{\pi}{2} - \theta\right) = \frac{a}{\sqrt{b^2 + a^2}} = \frac{a}{r} = \cos\theta$$

Figure 5–14

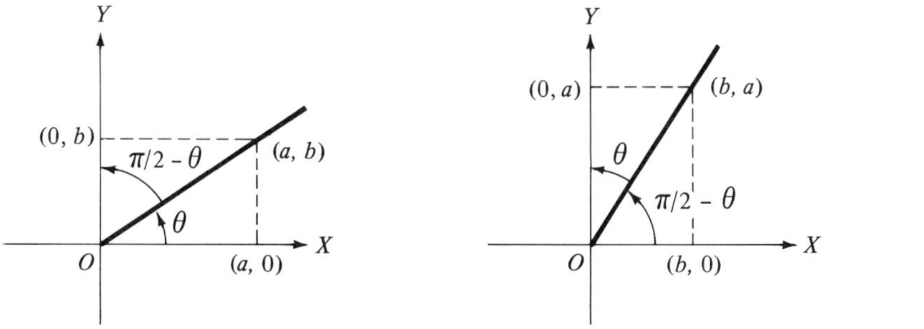

Also

$$\cos\left(\frac{\pi}{2} - \theta\right) = \frac{b}{\sqrt{b^2 + a^2}} = \frac{b}{r} = \sin\theta$$

Case 2

Suppose θ is in quadrant II and choose (a, b) on the terminal side of θ as in Figure 5–15. In this case $b > 0$ and $a < 0$ and if we now consider the point (b, a) and the angle it determines, the terminal side is in quadrant IV.

Since the rectangles are congruent and the angle determined by the positive X-axis and the line through (b, a) must be $\pi/2 - \theta$. (Notice that

Figure 5–15

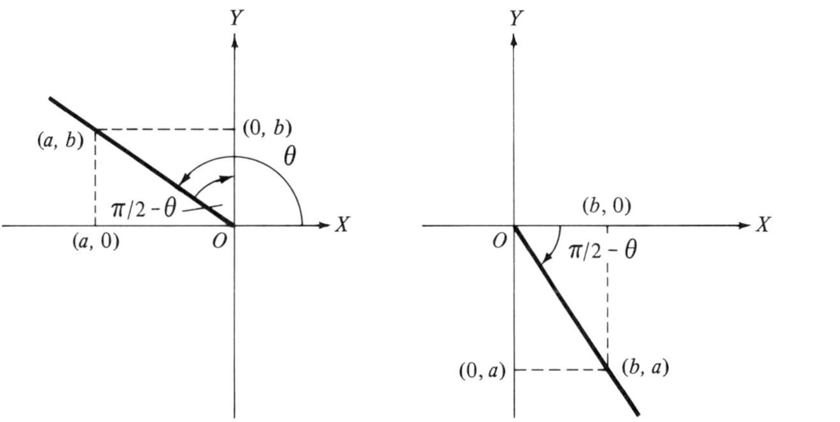

the rotation is clockwise for both.) Thus

$$\sin\left(\frac{\pi}{2} - \theta\right) = \frac{a}{\sqrt{b^2 + a^2}} = \frac{a}{r} = \cos\theta$$

and

$$\cos\left(\frac{\pi}{2} - \theta\right) = \frac{b}{r} = \sin\theta$$

If the terminal side is in quadrant III, then (b, a) is also in quadrant III and an argument similar to that for quadrant I holds. If the terminal side is in quadrant IV, then (b, a) is in quadrant II and the argument is much like that given when the terminal side is in quadrant II.

Thus we have two more fundamental identities:

9. $\sin(\pi/2 - x) = \cos x$
10. $\cos(\pi/2 - x) = \sin x$

We could proceed in a geometric fashion to establish similar identities for tangent and cotangent and for secant and cosecant. It is much simpler, however, if we use some of the identities previously established.

$$\tan\left(\frac{\pi}{2} - x\right) = \frac{\sin(\pi/2 - x)}{\cos(\pi/2 - x)} \qquad \text{since } \tan\theta = \frac{\sin\theta}{\cos\theta}$$

$$= \frac{\cos x}{\sin x} \qquad \text{by previously established identities}$$

$$= \cot x \qquad \text{because } \frac{\cos x}{\sin x} = \cot x$$

Thus

11. $\tan\left(\dfrac{\pi}{2} - x\right) = \cot x$

Similarly we can show

12. $\cot\left(\dfrac{\pi}{2} - x\right) = \tan x$

The identities

13. $\sec\left(\dfrac{\pi}{2} - x\right) = \csc x$

14. $\csc\left(\frac{\pi}{2} - x\right) = \sec x$

and be established in a similar fashion and we leave the proofs as exercises.

EXERCISES

1. Prove the identity

$$\frac{\cos\theta}{\sin\theta} = \cot\theta$$

2. Prove the identity

$$\frac{1}{\tan\theta} = \cot\theta$$

3. Prove the identity

$$\frac{1}{\sin\theta} = \csc\theta$$

4. Prove the identity

$$1 + \cot^2\theta = \csc^2\theta$$

5. Prove the identity

$$\cot\left(\frac{\pi}{2} - x\right) = \tan x$$

6. Prove the identity

$$\sec\left(\frac{\pi}{2} - x\right) = \csc x$$

7. Prove the identity

$$\csc\left(\frac{\pi}{2} - x\right) = \sec x$$

8. Prove the following identities.
   (a) $\cos x \csc x = \cot x$
   (b) $(\sin x)(\csc x - \sin x) = \cos^2 x$
   (c) $\csc x - \sin x = \cot x \cos x$
   (d) $(\cos t)(\tan t + \cot t) = \csc t$
9. Prove the following identity.

$$\sec^2 x \csc^2 x = \sec^2 x + \csc^2 x$$

10. Prove the following identity.

$$\frac{1 + \cos x}{\sin x} + \frac{\sin x}{1 + \cos x} = 2 \csc x$$

11. Suppose $\sin x = .4$ and $0 < x < \pi/2$. Use identities 1–8 to determine the values of the other trigonometric functions of x.
12. Generalize Exercise 11; that is, if $\sin x = t$ and $0 < x < \pi/2$, determine the other trigonometric functions of x.
13. Determine the other trigonometric functions of x if $\cos x = .3$ and $0 < x < \pi/2$.
14. Generalize Exercise 13 as you did Exercise 11.
15. If $\tan x = s$ and $0 < x < \pi/2$, determine all trigonometric functions of x.
16. If $\sin x = .6$ and $\pi/2 < x < \pi$, determine the values of the other trigonometric functions of x.
17. If $\cos x = -.2$ and $\pi < x < 3\pi/2$, determine the values of the other trigonometric functions of x.

18. If $\tan x = -5$ and $3\pi/2 < x < 2\pi$, determine the values of the other trigonometric functions of x.
19. If $\sin x = t$ and $-\pi/2 \le x \le \pi/2$, determine $\tan x$ and $\cos x$.
20. If $\cos x = t$ and $0 \le x \le \pi$, determine $\sin x$ and $\tan x$.
21. If $\tan x = t$ and $-\pi/2 < x < \pi/2$, determine $\sin x$ and $\cos x$.

4. More Trigonometric Identities

Trigonometric identities play an important role in integration techniques in calculus. The only way to develop skill with identities is to work with a great number of them and to do a great deal of practicing. There are, however, some general methods that we can follow. Instead of showing that one side can be transformed into the other, it may be more convenient to transform each side of the proposed identity separately into the same expression.

Example 1

Prove

$$\frac{\cos^2 x}{(1 - \sin x)^2} = (\sec x + \tan x)^2$$

Transforming the left side of the proposed identity we have:

$$\frac{\cos^2 x}{(1 - \sin x)^2} = \frac{1 - \sin^2 x}{(1 - \sin x)^2} = \frac{(1 - \sin x)(1 + \sin x)}{(1 - \sin x)^2} = \frac{1 + \sin x}{1 - \sin x}$$

Now we transform the right side of the proposed identity to the same expression:

$$\begin{aligned}(\sec x + \tan x)^2 &= \sec^2 x + 2 \sec x \tan x + \tan^2 x \\ &= \frac{1}{\cos^2 x} + 2\frac{1}{\cos x}\frac{\sin x}{\cos x} + \frac{\sin^2 x}{\cos^2 x} \\ &= \frac{1 + 2\sin x + \sin^2 x}{\cos^2 x} \\ &= \frac{(1 + \sin x)^2}{1 - \sin^2 x} \\ &= \frac{(1 + \sin x)^2}{(1 + \sin x)(1 - \sin x)} \\ &= \frac{1 + \sin x}{1 - \sin x}\end{aligned}$$

Thus

$$\frac{\cos^2 x}{(1 - \sin x)^2} = \frac{1 + \sin x}{1 - \sin x} \quad \text{and} \quad (\sec x + \tan x)^2 = \frac{1 + \sin x}{1 - \sin x}$$

Therefore

$$\frac{\cos^2 x}{(1 - \sin x)^2} = (\sec x + \tan x)^2 \ \square$$

Sometimes it is helpful to express all functions involved in terms of sines and cosines, and work primarily with the identities involving these functions. We did this in the latter part of Example 1.

Example 2

Prove

$$\frac{\cot x + \csc y}{\tan y + \tan x \sec y} = \cot x \cot y$$

$$\begin{aligned}
\frac{\cot x + \csc y}{\tan y + \tan x \sec y} &= \frac{\dfrac{\cos x}{\sin x} + \dfrac{1}{\sin y}}{\dfrac{\sin y}{\cos y} + \dfrac{\sin x}{\cos x}\,\dfrac{1}{\cos y}} \\
&= \frac{\dfrac{\cos x \sin y + \sin x}{\sin x \sin y}}{\dfrac{\sin y \cos x + \sin x}{\cos x \cos y}} \\
&= \frac{\dfrac{1}{\sin x \sin y}}{\dfrac{1}{\cos x \cos y}} \\
&= \frac{\cos x \cos y}{\sin x \sin y} \\
&= \cot x \cot y \ \square
\end{aligned}$$

Example 3

Prove

$$\tan \theta + \cot \theta = \sec \theta \csc \theta$$

Rewriting the left side using sines and cosines we have

$$\begin{aligned}\tan\theta + \cot\theta &= \frac{\sin\theta}{\cos\theta} + \frac{\cos\theta}{\sin\theta}\\ &= \frac{\sin^2\theta + \cos^2\theta}{\cos\theta\sin\theta}\\ &= \frac{1}{\cos\theta\sin\theta}\\ &= \frac{1}{\cos\theta}\frac{1}{\sin\theta}\\ &= \sec\theta\csc\theta \quad \square\end{aligned}$$

To prove identities:

(i) Look for relationships between the two sides that may lead easily to being able to transform one side to the other.
(ii) Attempt to simplify one side or both sides in the hope that the simplified expressions may be easier to work with.
(iii) If other methods fail, it may be useful to rewrite one or both sides using only two functions such as sine and cosine. Simplify the resulting expressions and attempt the needed transformations.

EXERCISES

Prove the following identities.

1. $\dfrac{1 - \csc\theta}{\sin\theta - 1} = \csc\theta$
2. $\cos^4 x - \sin^4 x = \cos^2 x - \sin^2 x$
3. $\dfrac{\cot^3 x - \tan^3 x}{\cot x - \tan x} = \tan^2 x + \csc^2 x$
4. $\dfrac{1}{\csc y - \cot y} = \dfrac{\sin y}{1 - \cos y}$
5. $\sin^2 x + \tan^2 y + \cos^2 x = \sec^2 y$
6. $(1 + \cos x)(1 + \sec x) = 2 + \cos x + \sec x$
7. $\dfrac{1 + \sin x}{1 - \sin x} - \dfrac{1 - \sin x}{1 + \sin x} = 4\tan x\sec x$
8. $\dfrac{\sin x\tan x}{1 - \cos x} - \sec x = 1$
9. $(\sec x - \tan x)(1 + \sin x) = \cos x$
10. $\dfrac{\tan^2 x}{\cos^2 y} - \dfrac{\tan^2 y}{\cos^2 x} = \tan^2 x - \tan^2 y$

11. $\dfrac{\cos x + \sec y}{\sec x + \cos y} = \dfrac{\cos x}{\cos y}$

12. $\dfrac{1}{1 + \cos x} + \dfrac{1}{1 - \cos x} = 2 \csc^2 x$

13. $\tan^2 x - \sin^2 x = \tan^2 x \sin^2 x$

14. $\dfrac{2 \sin x \cos x}{\cos^2 x - \sin^2 x} = \dfrac{2 \tan x}{1 - \tan^2 x}$

15. $\tan x - \tan^2 x = \dfrac{\csc x - \sec x}{\csc x \cot x}$

16. $\dfrac{\sin x}{1 - \cos x} - \dfrac{1 + \cos x}{\sin x} = 0$

17. $\dfrac{1}{\sec x + \tan x} = \dfrac{1 - \sin x}{\cos x}$

18. $\dfrac{\sin x}{\cot x + \csc x} - \dfrac{\sin x}{\cot x - \csc x} = 2$

19. $(\sec x + \tan y)^2 - (\tan x - \sec y)^2 = 2(\tan x \sec y + \sec x \tan y)$

20. $(\cos x \cos y + \sin x \sin y)^2 + (\sin x \cos y - \cos x \sin y)^2 = 1$

5. Graphs of Trigonometric Functions

We know that we can determine all values of the six trigonometric functions once the values are determined for angles between 0° and 90° or between 0 and $\pi/2$. Table 5.2 provides values of these functions for acute angles measured in degrees or by the corresponding radian measures. More precise tables are provided in the back of the book. If $0° < \theta < 45°$, the value of the function is read by finding θ on the left and reading *below* the function; if $45° < \theta < 90°$, the value is read by finding θ on the right and reading *above* the function. The table can be so constructed because of the relationships between co-functions; that is, because $\sin 52° = \cos 38°$ and so on.

Example 1

$$\sin 28° = .4695$$
$$\tan 57° = 1.5399$$
$$\cot 1.1345 = 0.4663 \quad \square$$

As with logarithms, the values given are 4-place approximations of the actual values but are sufficient for our purposes.

Table 5.2 Trigonometric Functions

degrees	radians	sin	cos	tan	cot		
0	0.0000	0.0000	1.0000	0.0000		1.5708	90
1	0.0175	0.0175	0.9998	0.0175	57.290	1.5533	89
2	0.0349	0.0349	0.9994	0.0349	28.636	1.5359	88
3	0.0524	0.0523	0.9986	0.0524	19.081	1.5184	87
4	0.0698	0.0698	0.9976	0.0699	14.301	1.5010	86
5	0.0873	0.0872	0.9962	0.0875	11.430	1.4835	85
6	0.1047	0.1045	0.9945	0.1051	9.5144	1.4661	84
7	0.1222	0.1219	0.9925	0.1228	8.1443	1.4486	83
8	0.1396	0.1392	0.9903	0.1405	7.1154	1.4312	82
9	0.1571	0.1564	0.9877	0.1584	6.3138	1.4137	81
10	0.1745	0.1736	0.9848	0.1763	5.6713	1.3963	80
11	0.1920	0.1908	0.9816	0.1944	5.1446	1.3788	79
12	0.2094	0.2079	0.9781	0.2126	4.7046	1.3614	78
13	0.2269	0.2250	0.9744	0.2309	4.3315	1.3439	77
14	0.2443	0.2419	0.9703	0.2493	4.0108	1.3265	76
15	0.2618	0.2588	0.9659	0.2679	3.7321	1.3090	75
16	0.2793	0.2756	0.9613	0.2867	3.4874	1.2915	74
17	0.2967	0.2924	0.9563	0.3057	3.2709	1.2741	73
18	0.3142	0.3090	0.9511	0.3249	3.0777	1.2566	72
19	0.3316	0.3256	0.9455	0.3443	2.9042	1.2392	71
20	0.3491	0.3420	0.9397	0.3640	2.7475	1.2217	70
21	0.3665	0.3584	0.9336	0.3839	2.6051	1.2043	69
22	0.3840	0.3746	0.9272	0.4040	2.4751	1.1868	68
23	0.4014	0.3907	0.9205	0.4245	2.3559	1.1694	67
24	0.4189	0.4067	0.9135	0.4452	2.2460	1.1519	66
25	0.4363	0.4226	0.9063	0.4663	2.1445	1.1345	65
26	0.4538	0.4384	0.8988	0.4877	2.0503	1.1170	64
27	0.4712	0.4540	0.8910	0.5095	1.9626	1.0996	63
28	0.4887	0.4695	0.8829	0.5317	1.8807	1.0821	62
29	0.5061	0.4848	0.8746	0.5543	1.8040	1.0647	61
30	0.5236	0.5000	0.8660	0.5774	1.7321	1.0472	60
31	0.5411	0.5150	0.8572	0.6009	1.6643	1.0297	59
32	0.5585	0.5299	0.8480	0.6249	1.6003	1.0123	58
33	0.5760	0.5446	0.8387	0.6494	1.5399	0.9948	57
34	0.5934	0.5592	0.8290	0.6745	1.4826	0.9774	56
35	0.6109	0.5736	0.8192	0.7002	1.4281	0.9599	55
36	0.6283	0.5878	0.8090	0.7265	1.3764	0.9425	54
37	0.6458	0.6018	0.7986	0.7536	1.3270	0.9250	53
38	0.6632	0.6157	0.7880	0.7813	1.2799	0.9076	52
39	0.6807	0.6293	0.7771	0.8098	1.2349	0.8901	51
40	0.6981	0.6428	0.7660	0.8391	1.1918	0.8727	50
41	0.7156	0.6561	0.7547	0.8693	1.1504	0.8552	49
42	0.7330	0.6691	0.7431	0.9004	1.1106	0.8378	48
43	0.7505	0.6820	0.7314	0.9325	1.0724	0.8203	47
44	0.7679	0.6947	0.7193	0.9657	1.0355	0.8029	46
45	0.7854	0.7071	0.7071	1.0000	1.0000	0.7854	45
		cos	sin	cot	tan	radians	degrees

We can use Table 5.2 or the table determined in Exercise 3, Section 2, to draw the graphs of the six elementary trigonometric functions. (See Figures 5–16 and 5–17). Note that we need only values of the functions

Figure 5-16

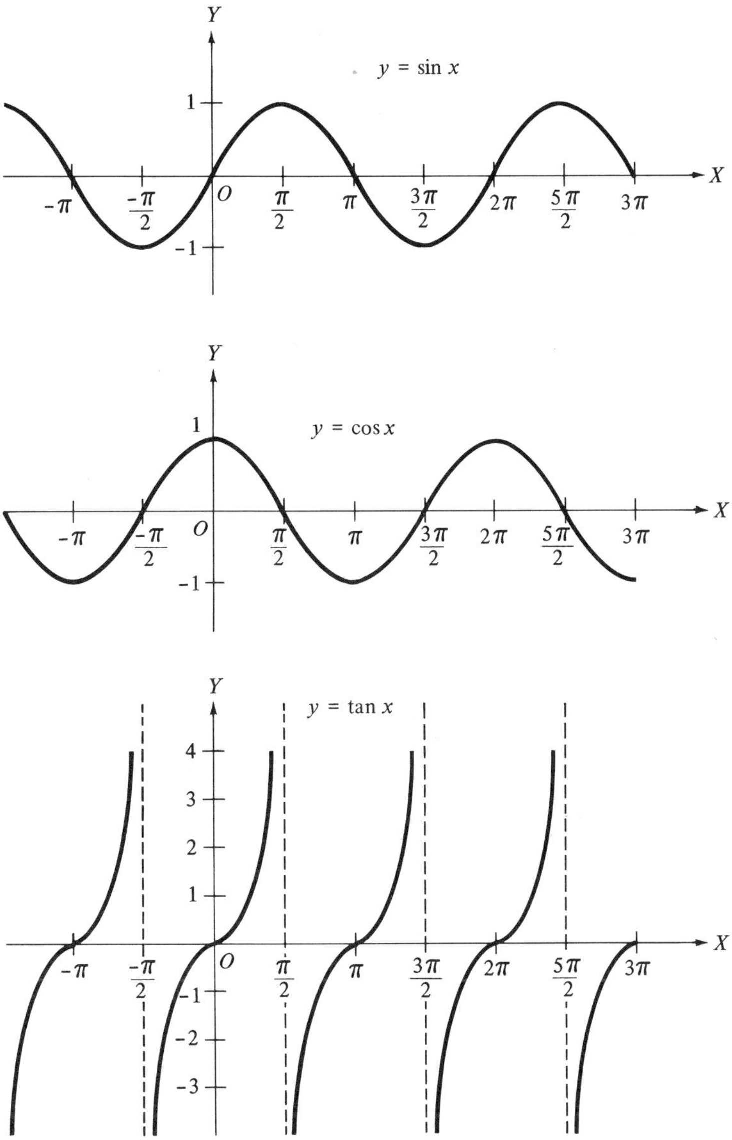

Figure 5-17

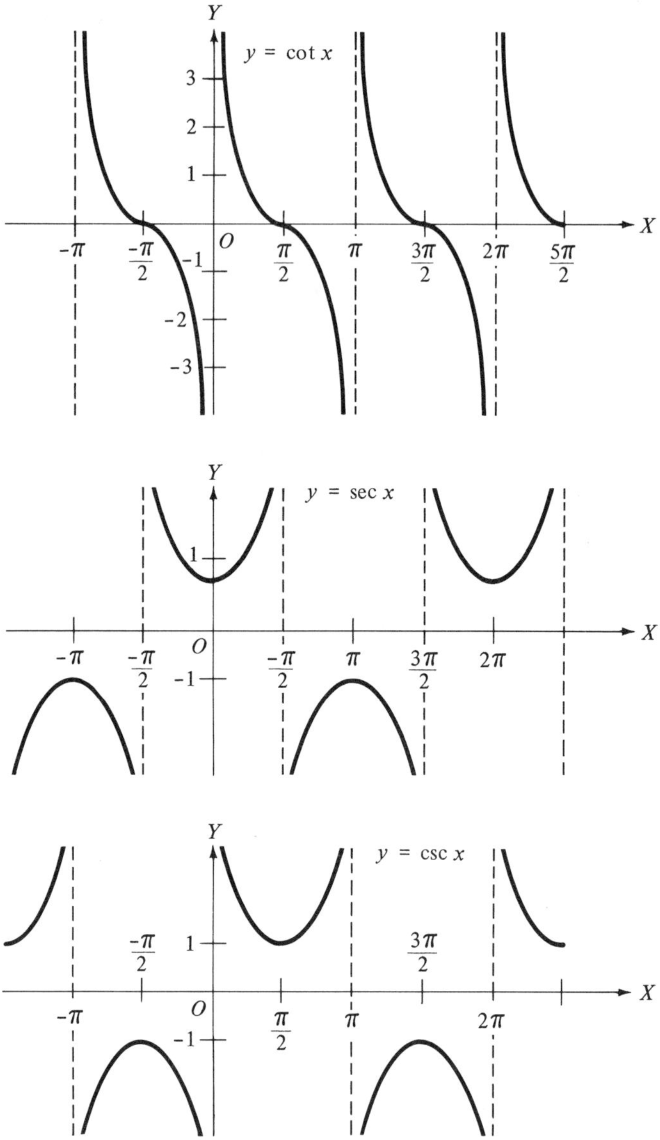

for values of x between 0 and 2π because the pattern repeats for $2\pi \leq x \leq 4\pi$, $4\pi \leq x \leq 6\pi$, $-2\pi \leq x \leq 0$, and so on. When dealing with graphs of trigonometric functions, we will always express the angle in terms of radians.

The graph of $y = \cos x$ seems to be symmetric with respect to the Y-axis, that is, it appears to have the same shape to the right of the Y-axis as it has to the left. This might lead us to expect that $\cos \theta = \cos(-\theta)$ for any number θ and this is, in fact, the case. We will establish it for positive angles t between 0 and 2π. If $0 < t < \pi/2$ or $3\pi/2 < t < 2\pi$, the terminal side is in quadrant I or IV. If $\pi/2 < t < 3\pi/2$, the terminal side is in quadrant II or III. The two cases are illustrated by Figure 5–18.

Figure 5–18

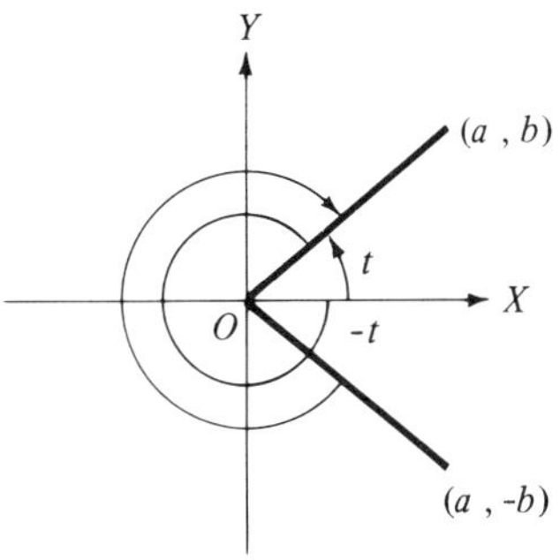

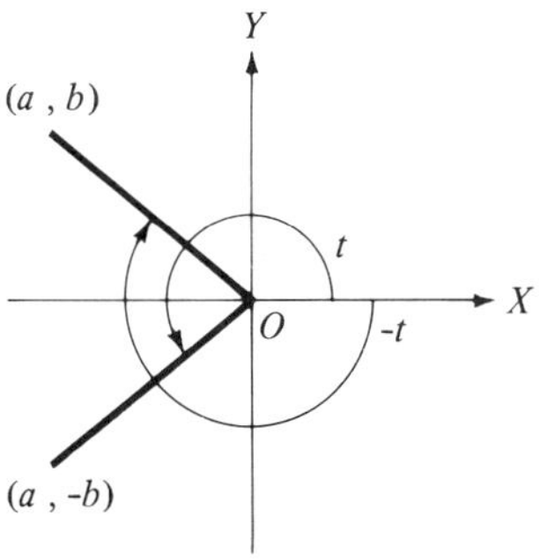

If we pick a point (a, b) on the terminal side of an angle with measure t, then by similar triangles, the point $(a, -b)$ must be on the terminal side of an angle with measure $-t$. Thus we see that

$$\cos t = \frac{a}{\sqrt{a^2 + b^2}} \quad \text{and} \quad \cos(-t) = \frac{a}{\sqrt{a^2 + b^2}}$$

Therefore $\cos t = \cos(-t)$.

Furthermore, notice that

$$\sin t = \frac{b}{\sqrt{a^2 + b^2}} \quad \text{and} \quad \sin(-t) = \frac{-b}{\sqrt{a^2 + b^2}}$$

Therefore $\sin(-t) = -\sin t$.

These two identities are also important and should be remembered.

15. $\sin(-x) = -\sin x$
16. $\cos(-x) = \cos x$

What relationship exists between $\tan x$ and $\tan(-x)$? Using a fundamental identity and those just developed we have the following:

$$\begin{aligned}\tan(-x) &= \frac{\sin(-x)}{\cos(-x)} \\ &= \frac{-\sin x}{\cos x} \\ &= -\frac{\sin x}{\cos x} \\ &= -\tan x\end{aligned}$$

Thus we have shown

17. $\tan(-x) = -\tan x$

The following three identities can now be established using 15, 16, and 17.

18. $\cot(-x) = -\cot x$
19. $\sec(-x) = \sec x$
20. $\csc(-x) = -\csc x$

We leave the proofs as exercises.

In Chapter 3, we saw that we could use the graph of $y = f(x)$ to determine the graphs of functions of the form $g(x) = f(x) + c$, $h(x) = f(x - b)$, and $p(x) = kf(x)$.

(i) The graph of $g(x) = f(x) + c$ is just the graph of f shifted c units up if c is positive and c units down if c is negative. We can draw the graph by drawing the graph of f as if the line $y = c$ were the X-axis.

(ii) The graph of $h(x) = f(x-b)$ is just the graph of $f(x)$ shifted b units to the right if b is positive and $|b|$ units to the left if b is negative. The number b is called the *phase shift*. We can draw the graph of $h(x)$ by drawing the graph of f as if the line $x = b$ were the Y-axis.

(iii) The graph of $p(x) = kf(x)$ is obtained from the graph of f by multiplying each of the y-coordinates of the graph of f by the number k.

We now use these techniques to draw the graphs of some functions obtained from trigonometric functions.

Example 2

To draw the graph of $y = 3 \cos x$, we can obtain all y-coordinates by multiplying each y-coordinate of the graph of $y = \cos x$ by 3. Both curves are illustrated in Figure 5–19. □

Figure 5–19

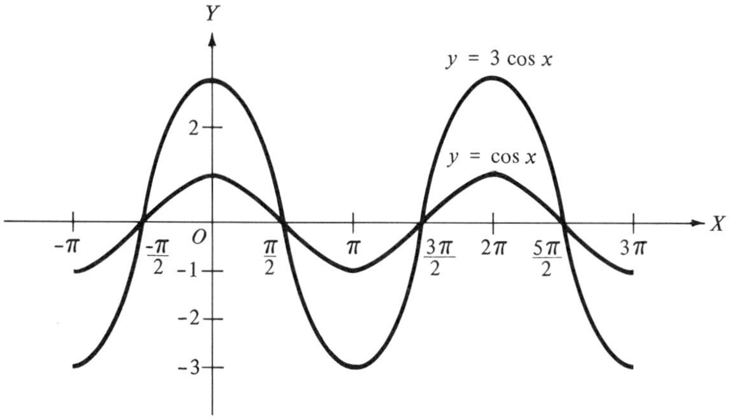

Example 3

Draw the graph of $h(x) = \sin x + 2$.

The curve $y = \sin x + 2$ is just 2 units above $y = \sin x$ as in Figure 5–20. □

Figure 5–20

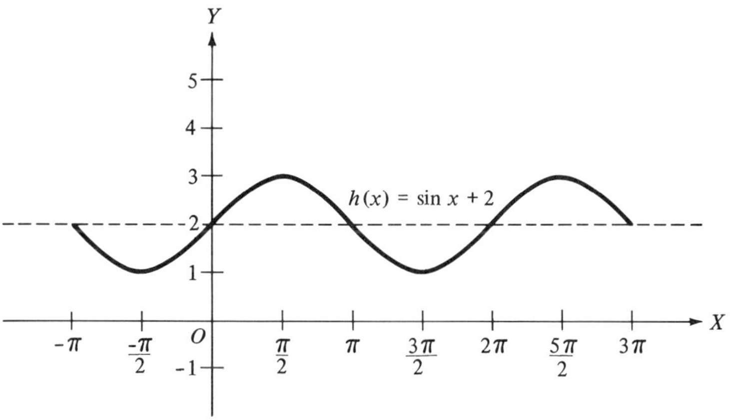

Example 4

The graph of $y = \cos(x + \pi/4)$ is just the graph of $y = \cos x$ shifted $\pi/4$ units to the left. The phase shift is $-\pi/4$. The graph is given in Figure 5–21. □

We can use identities 15–20 to draw graphs of functions of the form $f(-x)$ where f is a trigonometric function.

Figure 5-21

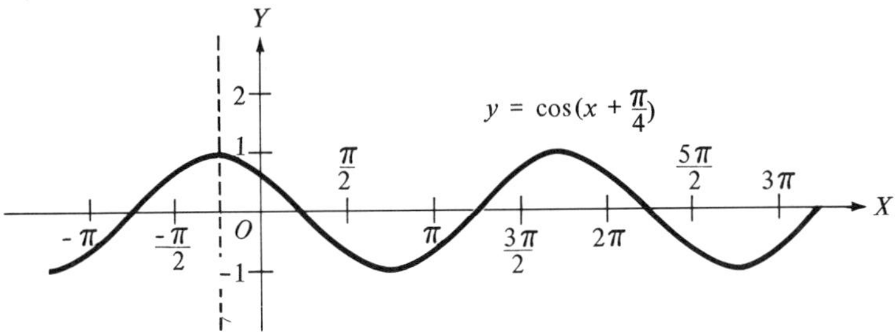

Example 5

Draw the graph of $y = \sin(-x)$.

From identity 15, we know that $\sin(-x) = -\sin x$. Thus $y = \sin(-x) = -\sin x = (-1) \sin x$ and we can draw this graph by multiplying the y-coordinates of $y = \sin x$ by (-1). Geometrically, this reflects the graph about the X-axis. The graph is given in Figure 5–22. □

Figure 5-22

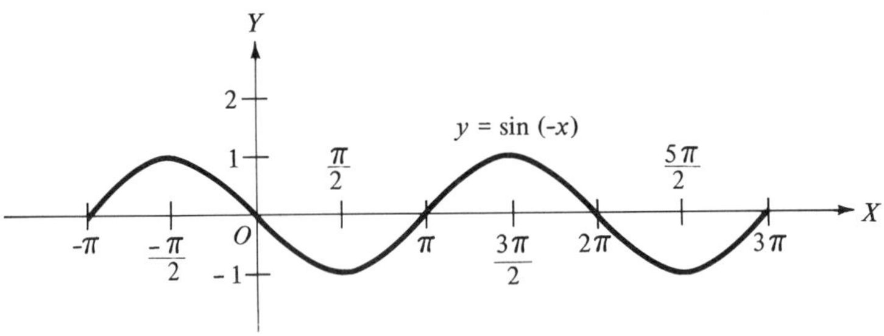

EXERCISES

1. Use Table 5.2 to determine the following.
   (a) $\sin 24^\circ$ (b) $\cos 61^\circ$ (c) $\tan 47^\circ$
   (d) $\cot 19^\circ$ (e) $\sec 73^\circ$ (f) $\csc 39^\circ$

2. Use Table 5.2 to determine the following.
   (a) $\sin 147^\circ$ (b) $\cos 93^\circ$ (c) $\tan 259^\circ$ (d) $\cot 427^\circ$
   (e) $\sin -53^\circ$ (f) $\cos 922^\circ$ (g) $\sec -43^\circ$ (h) $\csc 419^\circ$
   (i) $\tan -110^\circ$ (j) $\sin .6283$ (k) $\cos 1.3434$ (l) $\tan 3.7001$
3. Prove identities 18, 19, and 20.
4. Draw the graphs of $y = \sin x$ and $y = \csc x$ on the same coordinate system.
5. Draw the graphs of $y = \cos x$ and $y = \sec x$ on the same coordinate system.
6. Draw the graphs of $y = \tan x$ and $y = \cot x$ on the same coordinate system.
7. Draw the graphs of the following equations.
   (a) $y = \tan(-x)$ (b) $y = \cot(-x)$
   (c) $y = \sec(-x)$ (d) $y = \csc(-x)$
8. Draw the graphs of the following equations.
   (a) $y = \sin(x - \pi/4)$ (b) $y = \cos(x + \pi/3)$ (c) $y = \tan(x + \pi/2)$
   (d) $y = \csc(x - \pi)$ (e) $y = \sec(x + \pi/8)$ (f) $y = \cot(x - 3\pi/4)$
9. Draw the graphs of the following equations.
   (a) $y = \sin(x + \pi)$ (b) $y = \cos(x - \pi/4)$ (c) $y = \cot(x - \pi/8)$
   (d) $y = \sec(x + \pi/6)$ (e) $y = \csc(x - \pi/2)$ (f) $y = \tan(x + 5\pi/4)$
10. Draw the graphs of the following equations.
   (a) $y = 3 \sin x$ (b) $y = -2 \cos x$ (c) $y = \frac{1}{2} \sin x$
   (d) $y = \frac{2}{3} \csc x$ (e) $y = 4 \tan x$ (f) $y = \frac{1}{4} \cot x$
11. Draw the graphs of the following equations.
   (a) $y = 4 \cos x$ (b) $y = 5 \sin x$ (c) $y = \frac{3}{4} \tan x$
   (d) $y = 2 \sec x$ (e) $y = 3 \cot x$ (f) $y = \frac{2}{3} \cos x$
12. Draw graphs of the following equations.
   (a) $y = \cos x - 3$ (b) $y = \cos(x - 3)$ (c) $y = \sin x + 4$
   (d) $y = \sin(x + 4)$ (e) $y = \tan x + 1$ (f) $y = \tan(x + 1)$
   (g) $y = \csc x - 2$ (h) $y = \csc(x - 2)$
13. Draw graphs of the following equations.
   (a) $y = \sin x - 5$ (b) $y = \sin(x - 5)$ (c) $y = \cos x + 2$
   (d) $y = \cos(x + 2)$ (e) $y = \cot x - 2$ (f) $y = \cot(x - 2)$
   (g) $y = \sec x + 1$ (h) $y = \sec(x + 1)$

*14. Draw graphs of the following equations.
   (a) $y = 2 \cos x + 5$ (b) $y = 3 \sin(x - \pi/4)$
   (c) $y = 3 \cos(x - \pi/4) + 1$ (d) $y = -2 \sin(x + \pi/3) + 2$

6. Graphs of Functions Constructed from Trigonometric Functions

Quite complicated functions can be constructed using algebraic expressions involving trigonometric functions.

Example 1

$$f(x) = \sin 4x + 2 \cos 3x \;\square$$

Example 2

$$g(x) = \sin^2 x + \cos x \;\square$$

Example 3

$h(x) = -5 \sin 2x - 9$ □

Example 4

$p(x) = \tan x^4$ □

We will not attempt to develop techniques for graphing all such functions. Such an objective is far too ambitious for a course at this level. We will, however, develop techniques that will enable us to graph all equations of the form

$$y = cf(ax - b) + d \qquad \text{with } c \neq 0$$

where f is one of the elementary trigonometric functions.

Let k be a positive number. A function $f(x)$ is said to be *periodic with period* k if $f(k + x) = f(x)$ for all x for which $f(x)$ has meaning and k is the smallest positive number such that this equation holds. From the definitions we know that if g is any of the six elementary trigonometric functions, then

$$g(2\pi + x) = g(x)$$

From the table previously constructed in the exercises of Section 2, we see that for the sine, cosine, secant, and cosecant functions, 2π is the smallest positive number k such that $\sin(k + x) = \sin x$, $\cos(k + x) = \cos x$, $\sec(k + x) = \sec x$, and $\csc(k + x) = \csc x$. Thus each of these functions has period 2π. This same table indicates that the tangent and cotangent functions have period π and we can prove that this is, in fact, true. We use Figure 5–23 to illustrate the case for $0 < \theta < \pi/2$. The reader should provide diagrams for the other cases.

Figure 5–23

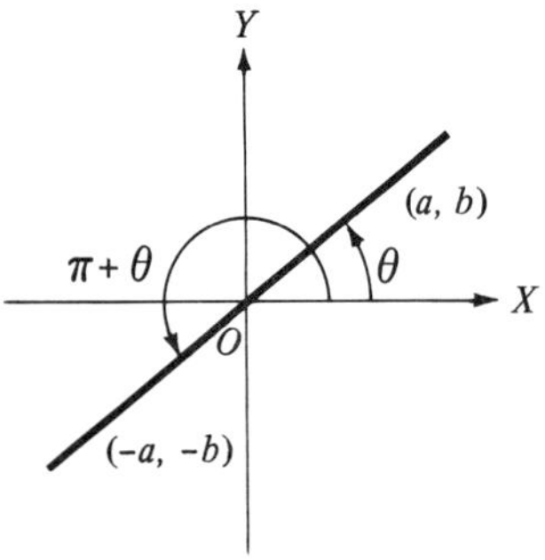

For all cases, if (a, b) is on the terminal side of θ, then $(-a, -b)$ is on the terminal side of $\pi + \theta$ and therefore

$$\tan(\pi + \theta) = \frac{-b}{-a} = \frac{b}{a} = \tan\theta$$

Also

$$\cot(\pi + \theta) = \frac{-a}{-b} = \frac{a}{b} = \cot\theta$$

From the tables we see that there is no number k, $0 < k < \pi$, such that $\tan(k + x) = \tan x$ or $\cot(k + x) = \cot x$. Thus these functions are periodic with period π.

The fact that $\sin x$ is periodic with period 2π enables us to determine the period of a function of the form $g(x) = \sin ax$.

Example 5

Draw the graph of $f(x) = \sin 4x$.

In order for $4x$ to assume all values between 0 and 2π, x must assume all values between 0 and $\pi/2$ since $0 \leq 4x \leq 2\pi$ if and only if $0 \leq x \leq \pi/2$. Thus the period cannot be less than $\pi/2$. On the other hand,

$$\begin{aligned} f\left(x + \frac{\pi}{2}\right) &= \sin\left[4\left(x + \frac{\pi}{2}\right)\right] \\ &= \sin(4x + 2\pi) \\ &= \sin 4x \\ &= f(x) \end{aligned}$$

Thus f is periodic with period $\pi/2$.

The graph is given in Figure 5–24 and looks much like the graph of $\sin x$. It just repeats over intervals of length $\pi/2$ instead of 2π. □

Figure 5–24

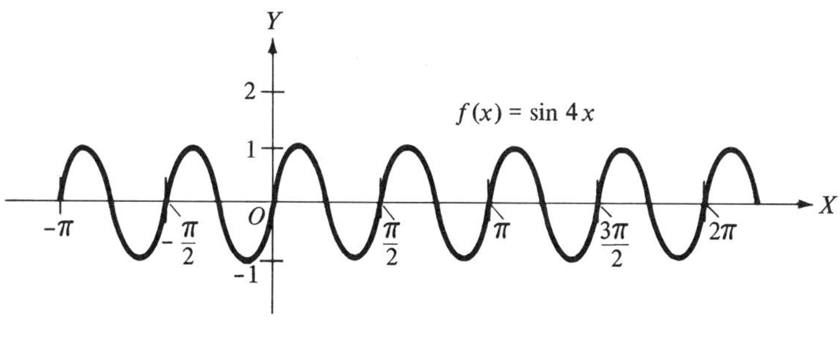

There was nothing special about the number 4 in the above example. Let a be a positive number and consider the function $g(x) = \sin ax$. Since $0 \leq x \leq 2\pi/a$ if and only if $0 \leq ax \leq 2\pi$, x must assume all values between 0 and $2\pi/a$ in order for ax to assume all values between 0 and 2π and, therefore, the period is not less than $2\pi/a$. Furthermore,

$$\begin{aligned} g\left(x + \frac{2\pi}{a}\right) &= \sin\left[a\left(x + \frac{2\pi}{a}\right)\right] \\ &= \sin(ax + 2\pi) \\ &= \sin ax \\ &= g(x) \end{aligned}$$

Thus g is periodic with period $2\pi/a$.

To draw the graph of such a function, we draw a graph much like that of the sine function only it repeats over intervals of length $2\pi/a$ instead of length 2π.

The argument given above can be generalized for any periodic function.

THEOREM

If f is a periodic function with period k and a is a positive number, then the function $g(x) = f(ax)$ is periodic with period k/a.

A formal proof of this assertion can be given by adapting the argument given above. We will omit it.

Example 6

$h(x) = \cos 3x$ is periodic with period $2\pi/3$. □

Example 7

$p(x) = \tan 5x$ is periodic with period $\pi/5$. □

Example 8

$q(x) = \sin \frac{1}{4}x$ is periodic with period $2\pi/\frac{1}{4} = 8\pi$. □

In order to graph a function of the form $g(x) = f(ax)$ where f is a trigonometric function, we just "squeeze" or "broaden" the graph of f according to the period of g.

Example 9

Draw the graph of $g(x) = \cos \frac{1}{2}x$.

Figure 5–25

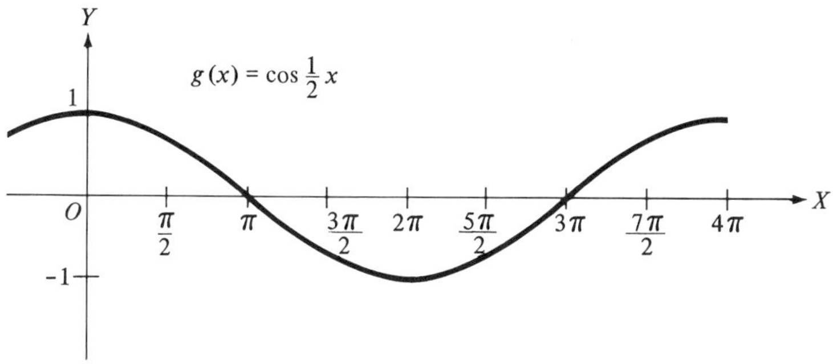

The period of this function is $2\pi/\frac{1}{2} = 4\pi$. The graph is much like the graph of the cosine function and is given in Figure 5–25. □

If f is a periodic function, the amplitude of f is $\frac{1}{2}(M - m)$ where M is the maximum value assumed by f and m is the minimum value assumed by f (providing such maximum and minimum values exist).

Example 10

The amplitude of $\sin x$ is 1 since the maximum value in the range is 1 and the minimum value in the range is -1. □

Example 11

The amplitude of $\cos x$ is 1. □

Example 12

The tangent function has no amplitude. □

Example 13

The function $f(x) = 3 \sin x$ has amplitude 3. □

We can now use the techniques of this section along with those of Section 5 to draw the graphs of more complicated functions.

Example 14

Draw the graph of $y = 4 \sin 2x$.

This graph can be drawn by multiplying each y-coordinate of the graph of $y = \sin 2x$ by 4. The graph of $y = \sin 2x$ is much like the graph

Figure 5–26

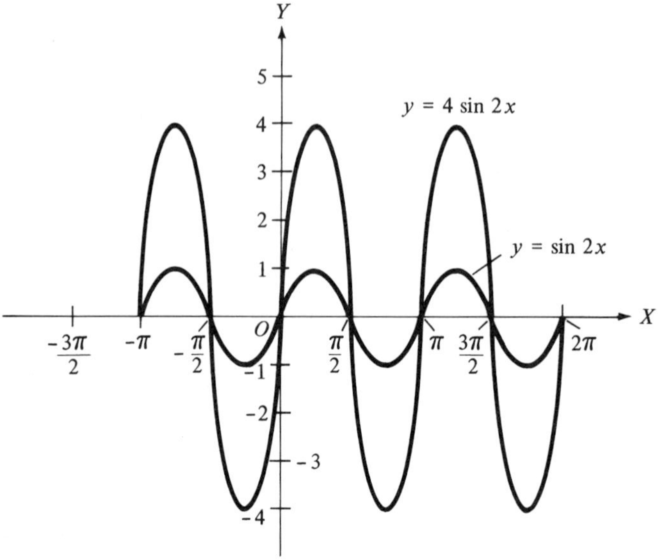

of $y = \sin x$ only the period is $2\pi/2 = \pi$. The amplitude of $y = 4 \sin 2x$ is 4. The graphs of both curves are given in Figure 5–26. □

Example 15

Draw the graph of $y = \sin(-4x)$.

From the identities we know that $\sin(-4x) = -\sin 4x = (-1)\sin 4x$. Therefore the graph of $y = \sin(-4x)$ can be obtained from the graph of $y = \sin 4x$ by multiplying each y-coordinate by (-1). The graph is given in Figure 5–27. □

Figure 5–27

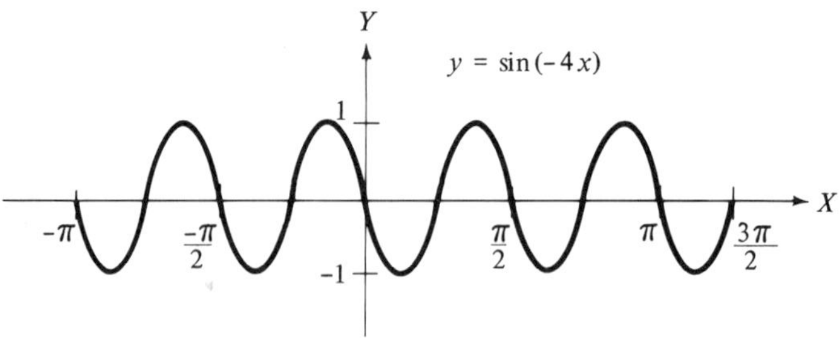

Example 16

Draw the graph of $y = \sin(2x - \pi)$.

First we rewrite this in the form $y = \sin[2(x - \pi/2)]$. We know that this graph has the same shape as the graph of $y = \sin 2x$, it is just shifted $\pi/2$ units to the right; that is, the phase shift is $\pi/2$. We can, as before, draw the line $x = \pi/2$ and now draw the graph of $y = \sin 2x$ as if $x = \pi/2$ were the Y-axis. This function has period π and amplitude 1. The graph is given in Figure 5–28. □

Figure 5–28

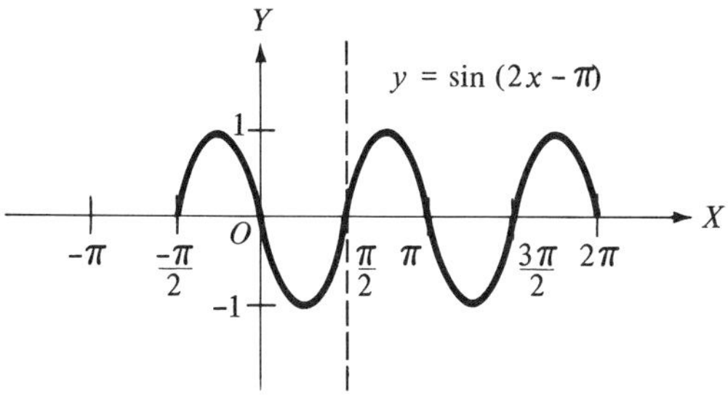

Example 17

Draw the graph of $y = \cos 3x + 2$.

This graph is just the graph of $y = \cos 3x$ shifted upward by 2. We can draw the line $y = 2$ and draw the graph of $y = \cos 3x$ as if $y = 2$ were the X-axis. This function has period $2\pi/3$ and amplitude 1. The graph is given in Figure 5–29. □

Figure 5–29

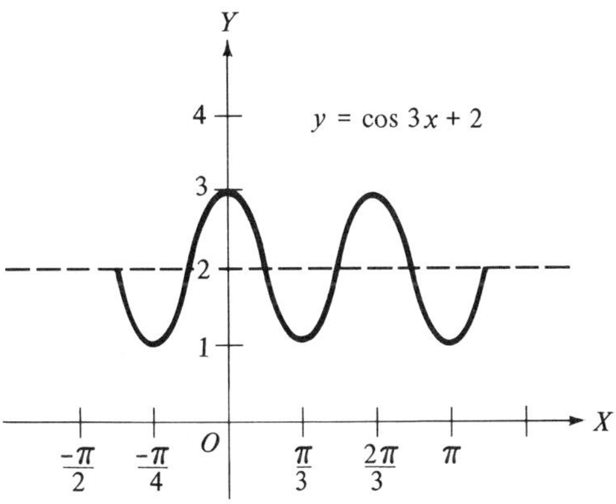

Example 18

Draw the graph of $y = 3 \sin (2x - \pi/2) + 1$.

This is more complicated than the other examples but if we use all the techniques available, we can do it.

First, this graph is the graph of $y = 3 \sin (2x - \pi/2)$ shifted up by 1. Thus we can draw the graph of $y = 3 \sin (2x - \pi/2)$ as if $y = 1$ were the X-axis.

Second, the graph of $y = 3 \sin (2x - \pi/2) = 3 \sin 2(x - \pi/4)$ has the same shape as the graph of $y = 3 \sin 2x$, it is just shifted $\pi/4$ units to the right; that is, the phase shift is $\pi/4$. We can draw the line $x = \pi/4$ and then draw the graph of $y = 3 \sin 2x$ as if $x = \pi/4$ were the Y-axis and $y = 1$ were the X-axis. We have reduced the problem to one much like Example 14. The period is π and the amplitude is 3. The graph is given by Figure 5–30. □

Figure 5-30

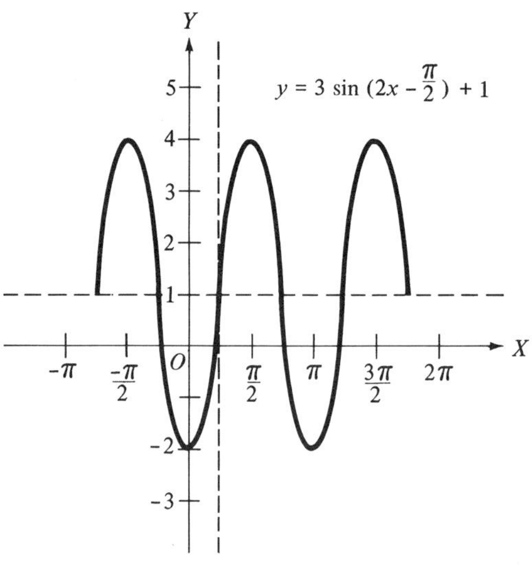

Example 19

Draw the graph of $y = 2 \cos (3x - \pi) + 1$.

First we rewrite as $y = 2 \cos [3(x - \pi/3)] + 1$. This graph is just the graph of $y = 2 \cos [3(x - \pi/3)]$ shifted upward by 1 and the phase shift is $\pi/3$. We draw the lines $y = 1$ and $x = \pi/3$ and then draw the graph of $y = 2 \cos 3x$ as if these lines were the axes. The period is $2\pi/3$ and the amplitude is 2. The graph is given in Figure 5–31. □

Figure 5–31

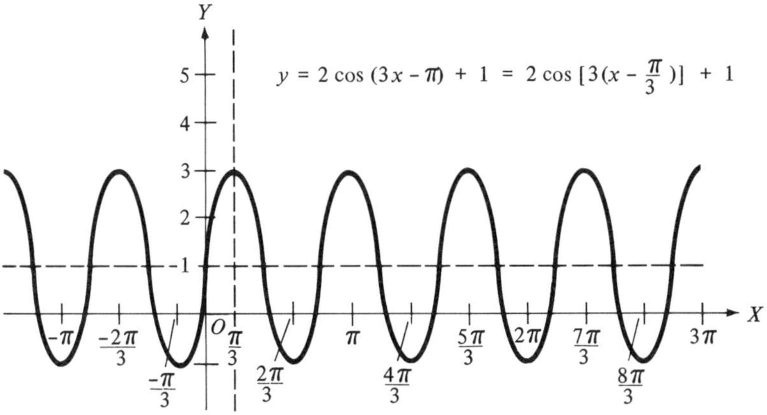

In summary: Let $g(x) = cf(ax - b) + d$ where f is a periodic function with period k ($c \neq 0$ and $a \neq 0$). To draw the graph of g, first rewrite in the form $g(x) = cfa(x - b/a) + d$. Draw the lines $y = d$ and $x = b/a$. (These are not part of the graph, they are just guidelines. It is customary to draw them as dotted lines.) Draw the graph of $y = cf(ax)$ as if these lines were were the axes. The period of $y = cf(ax)$ is $|k/a|$. The amplitude is $|c|$ times the amplitude of $f(x)$ providing $f(x)$ has an amplitude.

EXERCISES

1. Prove the following identities. [*Hint*: See the proof that $\tan(\pi + \theta) = \tan\theta$.]
   (a) $\sin(\pi + \theta) = -\sin\theta$ (b) $\cos(\pi + \theta) = -\cos\theta$
   (c) $\sec(\pi + \theta) = -\sec\theta$ (d) $\csc(\pi + \theta) = -\csc\theta$
2. Draw graphs of the following functions.
   (a) $f(x) = \sin 2x$ (b) $g(x) = \cos\frac{1}{4}x$ (c) $h(x) = \tan 3x$
   (d) $p(x) = \sec 4x$ (e) $q(x) = \cot\frac{1}{3}x$ (f) $r(x) = \csc\frac{4}{3}x$
3. Draw graphs of the following functions.
   (a) $f(x) = \cos 2x$ (b) $g(x) = \tan\frac{1}{3}x$ (c) $h(x) = \sin\frac{1}{4}x$
   (d) $p(x) = \csc 2x$ (e) $q(x) = \sec\frac{1}{2}x$ (f) $r(x) = \cot 2x$
4. Draw graphs of the following functions.
   (a) $f(x) = 3\sin 2x$ (b) $g(x) = 2\cos\frac{1}{4}x$ (c) $h(x) = -2\tan 3x$
   (d) $p(x) = \frac{1}{3}\sec 4x$ (e) $q(x) = 2\cot\frac{1}{3}x$ (f) $r(x) = 2\csc\frac{4}{3}x$
5. Draw graphs of the following functions.
   (a) $f(x) = 3\cos 2x$ (b) $g(x) = 2\tan\frac{1}{3}x$ (c) $h(x) = -3\sin\frac{1}{4}x$
   (d) $p(x) = \frac{3}{4}\csc 2x$ (e) $q(x) = 3\sec\frac{1}{2}x$ (f) $r(x) = \frac{2}{3}\cot 2x$
6. Draw graphs of the following functions.
   (a) $f(x) = \sin(2x - \pi)$ (b) $g(x) = \cos(\frac{1}{4}x + \pi/8)$
   (c) $h(x) = \tan(3x - \pi)$ (d) $p(x) = \sec(4x + 2\pi)$
   (e) $q(x) = \cot(\frac{1}{3}x - \pi/2)$ (f) $r(x) = \csc(\frac{4}{3}x - \pi)$

7. Draw graphs of the following functions.
   (a) $f(x) = \cos(2x + \pi)$
   (b) $g(x) = \tan(\frac{1}{3}x - \pi/3)$
   (c) $h(x) = \sin(\frac{1}{4}x + \pi/2)$
   (d) $p(x) = \csc(2x - \pi/2)$
   (e) $q(x) = \sec(\frac{1}{2}x + \pi/3)$
   (f) $r(x) = \cot(2x - \pi/2)$

7. Addition and Multiple-Angle Formulas

If α and β are two angles, it is sometimes useful to express function values such as $\sin(\alpha + \beta)$, $\cos(\alpha - \beta)$, or $\tan(\alpha + \beta)$ in terms of trigonometric functions of α and β. We shall use the definitions and geometry to develop a formula for $\cos(\alpha - \beta)$ and use previously developed identities to derive others. We will prove the formula for the case $0 \leq \alpha - \beta < \beta$. The proof can be generalized using similar techniques.

Choose a unit circle with center at the origin on the coordinate system and let $P_0 = (1, 0)$, $P_1 = (\cos(\alpha - \beta), \sin(\alpha - \beta))$, $P_2 = (\cos\beta, \sin\beta)$, and $P_3 = (\cos\alpha, \sin\alpha)$ as in Figure 5–32. Since the length of the arc from P_2 to P_3 is the same as the length of the arc from P_0 to P_1, the distance from P_2 to P_3 is the same as the distance from P_0 to P_1.

The distance from P_2 to P_3 is

$$\sqrt{(\cos\alpha - \cos\beta)^2 + (\sin\alpha - \sin\beta)^2}$$

The distance from P_0 to P_1 is

$$\sqrt{[\cos(\alpha - \beta) - 1]^2 + \sin^2(\alpha - \beta)}$$

Figure 5–32

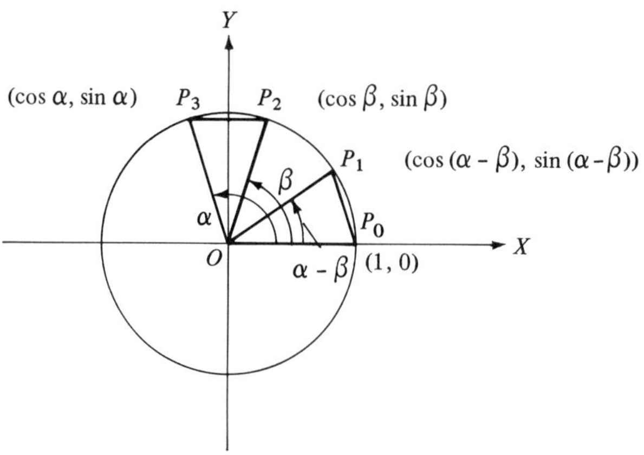

Thus

$$\sqrt{(\cos\alpha - \cos\beta)^2 + (\sin\alpha - \sin\beta)^2} = \sqrt{(\cos(\alpha - \beta) - 1)^2 + \sin^2(\alpha - \beta)}$$

Squaring both sides yields

$$(\cos\alpha - \cos\beta)^2 + (\sin\alpha - \sin\beta)^2 = (\cos(\alpha - \beta) - 1)^2 + \sin^2(\alpha - \beta)$$

Thus

$$\cos^2\alpha - 2\cos\alpha\cos\beta + \cos^2\beta + \sin^2\alpha - 2\sin\alpha\sin\beta + \sin^2\beta$$
$$= \cos^2(\alpha - \beta) - 2\cos(\alpha - \beta) + 1 + \sin^2(\alpha - \beta)$$
$$(\cos^2\alpha + \sin^2\alpha) + (\cos^2\beta + \sin^2\beta) - 2\cos\alpha\cos\beta - 2\sin\alpha\sin\beta$$
$$= \cos^2(\alpha - \beta) + \sin^2(\alpha - \beta) + 1 - 2\cos(\alpha - \beta)$$
$$2 - 2\cos\alpha\cos\beta - 2\sin\alpha\sin\beta = 2 - 2\cos(\alpha - \beta)$$

and therefore

$$\cos\alpha\cos\beta + \sin\alpha\sin\beta = \cos(\alpha - \beta)$$

Thus we have established the following identity:

21. $\cos(x - y) = \cos x \cos y + \sin x \sin y$

The formula for $\cos(x + y)$ can now be established using identity 21 and other previously established identities.

$$\begin{aligned} \cos(x + y) &= \cos[x - (-y)] \\ &= \cos x \cos(-y) + \sin x \sin(-y) \quad \text{by identity 21} \\ &= \cos x \cos y + \sin x(-\sin y) \\ &= \cos x \cos y - \sin x \sin y \end{aligned}$$

Thus we have the following identity:

22. $\cos(x + y) = \cos x \cos y - \sin x \sin y$

The next four identities can be established using these formulas and other identities.

23. $\sin(x + y) = \sin x \cos y + \cos x \sin y$
24. $\sin(x - y) = \sin x \cos y - \cos x \sin y$
25. $\tan(x + y) = \dfrac{\tan x + \tan y}{1 - \tan x \tan y}$
26. $\tan(x - y) = \dfrac{\tan x - \tan y}{1 + \tan x \tan y}$

Proof of Identity 23:

$$\begin{aligned}\sin(x+y) &= \cos\left[\frac{\pi}{2}-(x+y)\right]\\ &= \cos\left[\left(\frac{\pi}{2}-x\right)-y\right]\\ &= \cos\left(\frac{\pi}{2}-x\right)\cos y+\sin\left(\frac{\pi}{2}-x\right)\sin y\\ &= \sin x\cos y+\cos x\sin y\end{aligned}$$

Proof of Identity 25:

$$\begin{aligned}\tan(x+y) &= \frac{\sin(x+y)}{\cos(x+y)}\\ &= \frac{\sin x\cos y+\cos x\sin y}{\cos x\cos y-\sin x\sin y}\\ &= \frac{\dfrac{\sin x\cos y+\cos x\sin y}{\cos x\cos y}}{\dfrac{\cos x\cos y-\sin x\sin y}{\cos x\cos y}}\\ &= \frac{\dfrac{\sin x\cos y}{\cos x\cos y}+\dfrac{\cos x\sin y}{\cos x\cos y}}{\dfrac{\cos x\cos y}{\cos x\cos y}-\dfrac{\sin x\sin y}{\cos x\cos y}}\\ &= \frac{\dfrac{\sin x}{\cos x}+\dfrac{\sin y}{\cos y}}{1-\dfrac{\sin x}{\cos x}\dfrac{\sin y}{\cos y}}\\ &= \frac{\tan x+\tan y}{1-\tan x\tan y}\end{aligned}$$

We leave the proofs of identities 24 and 26 as exercises.

If we let $x = y$ in identities 22, 23, and 24, we obtain the following *double-angle formulas.*

27. $\cos 2x = \cos^2 x - \sin^2 x$
28. $\sin 2x = 2\sin x\cos x$
29. $\tan 2x = \dfrac{2\tan x}{1-\tan^2 x}$

Notice that

$$\begin{aligned}\cos 2x &= \cos^2 x - \sin^2 x \\ &= \cos^2 x - (1 - \cos^2 x) \\ &= 2\cos^2 x - 1\end{aligned}$$

Furthermore,

$$\begin{aligned}\cos 2x &= \cos^2 x - \sin^2 x \\ &= (1 - \sin^2 x) - \sin^2 x \\ &= 1 - 2\sin^2 x\end{aligned}$$

Let us denote these identities by 27′ and 27″:

27′. $\cos 2x = 2\cos^2 x - 1$
27″. $\cos 2x = 1 - 2\sin^2 x$

These formulas can be used to derive the *half-angle formulas*:

30. $\left|\sin \frac{y}{2}\right| = \sqrt{\frac{1 - \cos y}{2}}$

31. $\left|\cos \frac{y}{2}\right| = \sqrt{\frac{1 + \cos y}{2}}$

32. $\left|\tan \frac{y}{2}\right| = \sqrt{\frac{1 - \cos y}{1 + \cos y}}$

Proof of Identity 30:

If we let $y/2 = x$ in identity 27″ we have

$$\cos y = 1 - 2\sin^2 \frac{y}{2}$$

Thus

$$2\sin^2 \frac{y}{2} = 1 - \cos y$$

$$\sin^2 \frac{y}{2} = \frac{1 - \cos y}{2}$$

and

$$\left|\sin \frac{y}{2}\right| = \sqrt{\frac{1 - \cos y}{2}}$$

The proof of identity 31 is similar to the above, just use identity 27′. The proof of identity 32 follows from identities 30 and 31 and the fact that

$$\tan\frac{y}{2} = \frac{\sin y/2}{\cos y/2}$$

We leave the proofs as exercises.

The following identities are sometimes useful to students of integral calculus. The first four can be derived from identities 21–24. We leave the verifications as exercises.

33. $2\sin x\cos y = \sin(x+y) + \sin(x-y)$
34. $2\cos x\sin y = \sin(x+y) - \sin(x-y)$
35. $2\cos x\cos y = \cos(x+y) + \cos(x-y)$
36. $2\sin x\sin y = \cos(x-y) - \cos(x+y)$

The next four identities can be derived from the last four by suitable substitutions; that is, let $a = x + y$ and $b = x - y$. Again we leave the verifications as exercises.

37. $\sin a + \sin b = 2\cos\frac{a-b}{2}\sin\frac{a+b}{2}$

38. $\sin a - \sin b = 2\cos\frac{a+b}{2}\sin\frac{a-b}{2}$

39. $\cos a + \cos b = 2\cos\frac{a+b}{2}\cos\frac{a-b}{2}$

40. $\cos a - \cos b = 2\sin\frac{a+b}{2}\sin\frac{a-b}{2}$

EXERCISES

1. Prove identity 24.
2. Prove identity 26.
3. Prove identity 31.
4. Prove identity 32.
5. Prove identities 33–36.
6. Prove identities 37–40.
7. Use the identities in this section along with your knowledge of the values of the trigonometric functions of 30°, 45°, and 60° to compute the following.
   (a) $\cos 15^\circ$ [*Hint*: $\cos 15^\circ = \cos(60^\circ - 45^\circ)$] (b) $\sin 75^\circ$
   (c) $\cos 75^\circ$ (d) $\tan 15^\circ$
   (e) $\tan 75^\circ$ (f) $\sin 22.5^\circ$
   (g) $\cos 22.5^\circ$ (h) $\tan 22.5^\circ$
8. Prove the following identities.
   (a) $(\sin x + \cos x)^2 = 1 + \sin 2x$ (b) $2\sin x\csc 2x = \sec x$

(c) $\sec^2 \frac{x}{2} = \frac{2}{1 + \cos x}$

(d) $\frac{\sin 2y}{1 + \cos 2y} = \tan y$

(e) $\cos^4 x - \sin^4 x = \cos 2x$

9. Prove the following identities.

(a) $\cot 2x = \frac{\cot^2 x - 1}{2 \cot x}$

(b) $\sec 2x = \frac{\sec^2 x}{2 - \sec^2 x}$

(c) $\csc 2x = \frac{1 + \tan^2 x}{2 \tan x}$

(d) $\tan \frac{x}{2} = \frac{1 - \cos x}{\sin x} = \frac{\sin x}{1 + \cos x}$

(e) $\sin^2 2y - \sin^2 y = \cos^2 y - \cos^2 2y$

(f) $\sin 3x = 3 \sin x - 4 \sin^3 x$

10. Prove the following identities.

(a) $\frac{\cos 2y + \cos y + 1}{\sin 2y + \sin y} = \cot y$

(b) $\frac{\cos 2\theta - 1 + \sin 2\theta}{\cos 2\theta + 1 - \sin 2\theta} = \tan \theta$

8. Inverse Trigonometric Functions

We know that the equation

$$\sin s = .5$$

has many solutions; for example, $\pi/6$, $5\pi/6$, and $13\pi/6$. Notice, however, that there is only one solution between $-\pi/2$ and $\pi/2$. In fact, if $-1 \leq t \leq 1$, there is exactly one number s such that $-\pi/2 \leq s \leq \pi/2$ and $\sin s = t$. (See Figure 5–33.)

Figure 5-33

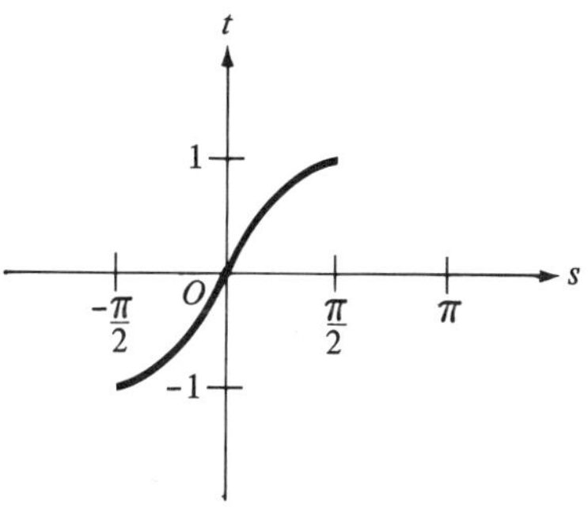

We can use this fact to define a function with domain $[-1, 1]$ and range $[-\pi/2, \pi/2]$. This function is called the *Arcsine function*, abbreviated Arcsin, and defined by:

$$s = \text{Arcsin}\, t \text{ if and only if } \sin s = t \text{ with } -1 \le t \le 1 \text{ and } -\pi/2 \le s \le \pi/2$$

Example 1

$$\text{Arcsin}\ .5 = \frac{\pi}{6}$$

$$\text{Arcsin}\ 0 = 0$$

$$\text{Arcsin}\ 1 = \frac{\pi}{2}$$

$$\text{Arcsin}\ -\frac{\sqrt{3}}{2} = -\frac{\pi}{3} \quad \square$$

Notice that if $-1 \le t \le 1$, then

$$\sin(\text{Arcsin}\, t) = t$$

and if $-\pi/2 \le s \le \pi/2$

$$\text{Arcsin}(\sin s) = s$$

Thus the Arcsine function is the inverse of the sine function if we limit the domain of the sine function to $[-\pi/2, \pi/2]$.

The graph of the Arcsine function is given in Figure 5–34.

Figure 5–34

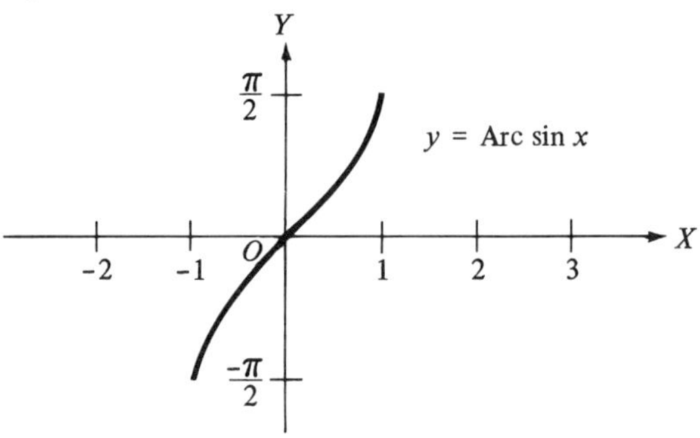

In a similar fashion we can define inverses for other trigonometric functions if we limit the domain under consideration. The Arccosine function, abbreviated Arccos, is defined by:

$$s = \text{Arccos}\, t \text{ if and only if } \cos s = t \text{ with } -1 \le t \le 1 \text{ and } 0 \le s \le \pi$$

Example 2

$$\text{Arccos } 1 = 0$$

$$\text{Arccos } .5 = \frac{\pi}{3}$$

$$\text{Arccos } -.5 = \frac{2\pi}{3}$$

$$\text{Arccos } -1 = \pi \ \square$$

As expected, if $0 \le s \le \pi$, then

$$\text{Arccos } (\cos s) = s$$

Furthermore, if $-1 \le t \le 1$, then

$$\cos (\text{Arccos } t) = t$$

The graph of $y = \text{Arccos } x$ is given in Figure 5–35.

Figure 5–35

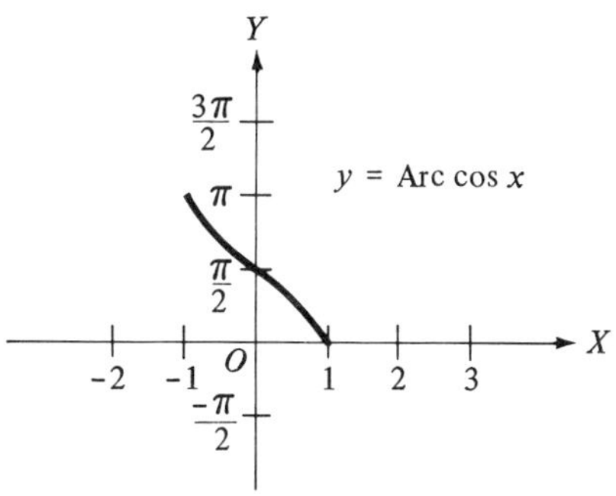

The Arctangent function, abbreviated Arctan, is defined by:

$s = \text{Arctan } t$ if and only if $\tan s = t$ with t any real number and $-\pi/2 < s < \pi/2$

As expected,

$$\text{Arctan } (\tan s) = s \text{ if } -\frac{\pi}{2} < s < \frac{\pi}{2}$$

and

$$\tan (\text{Arctan } t) = t \text{ for any number } t$$

The Arccotangent function, abbreviated Arccot, is defined by:

$s =$ Arccot t if and only if $\cot s = t$ with t any real number and $0 < s < \pi$

The graphs of these functions are given in Figures 5–36 and 5–37.

Figure 5–36

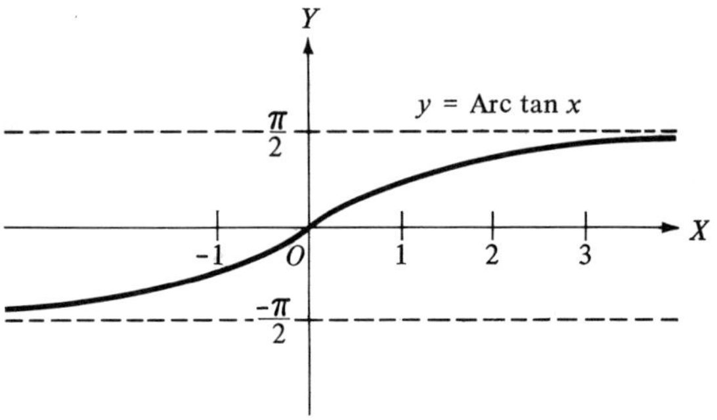

Figure 5–37

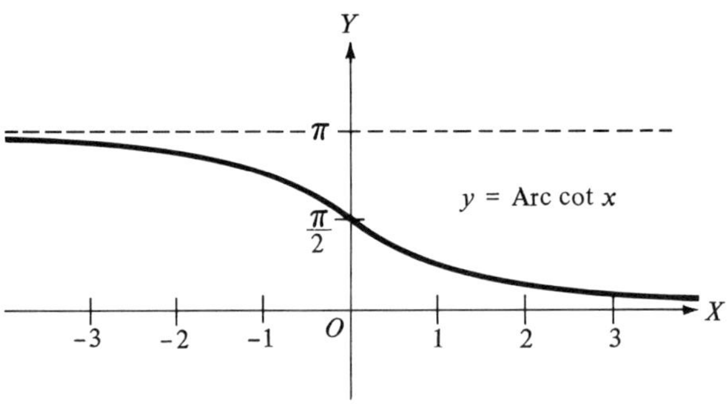

We will not attempt to define Arcsecant and Arccosecant functions. The definitions are not universally agreed upon and we do not need them. For example, if $\sec \theta = 2.2$, then $\cos \theta = 1/2.2$ and instead of considering Arcsec 2.2 we could just use Arccos 1/2.2.

In the exercises of Section 3 you were asked to determine the value of one trigonometric function of an angle given a value of another trigonometric function of the same angle. This notion generalizes to determining numbers such as sin (Arctan 2), cos (Arcsin .3), and so on.

Example 3

Determine sin (Arccos .6).

If

$$\begin{aligned} s &= \text{Arccos } .6 \\ \cos s &= .6 \\ \cos^2 s &= .36 \\ 1 - \sin^2 s &= .36 \\ \sin^2 s &= .64 \\ |\sin s| &= .8 \end{aligned}$$

Since $0 \leq s \leq \pi$, $\sin s \geq 0$. Thus $\sin s = .8$ and sin (Arccos .6) = .8. □

Example 4

Determine tan (Arcsin −.3).

If

$$\begin{aligned} s &= \text{Arcsin } -.3 \\ \sin s &= -.3 \\ \sin^2 s &= .09 \\ 1 - \cos^2 s &= .09 \\ \cos^2 s &= .91 \\ \frac{\sin^2 s}{\cos^2 s} &= \frac{.09}{.91} = \frac{9}{91} \\ \tan^2 s &= \frac{9}{91} \\ |\tan s| &= \sqrt{\frac{9}{91}} \end{aligned}$$

But $\sin s = -.3$ and $-\pi/2 \leq s \leq \pi/2$ imply that $\tan s$ is negative. Thus $\tan s = -\sqrt{9/91}$. □

We can also use a drawing to facilitate such computations.

Example 5

Determine sin (Arctan .7).

If $s = \text{Arctan } .7$, then $0 < s < \pi/2$ and s is in quadrant I. We draw the angle as in Figure 5–38 and since $\tan s = .7$, the point (1, .7) is on the terminal side since $\tan s = .7/1 = .7$. By the Pythagorean theorem, the distance of the point from the origin is

$$\sqrt{1 + .49} = \sqrt{1.49}$$

Figure 5-38

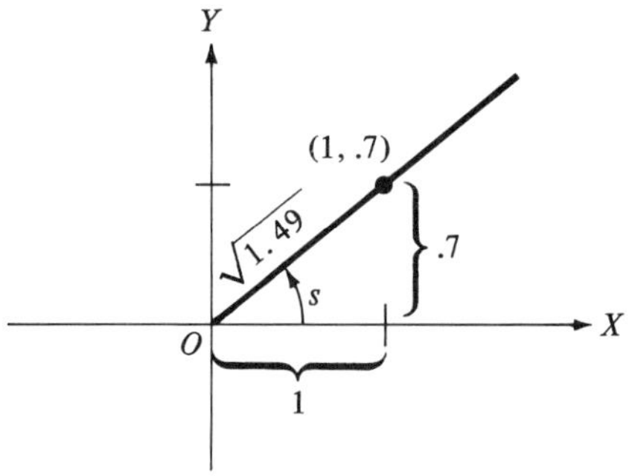

Thus

$$\sin s = \frac{.7}{\sqrt{1.49}}$$

that is,

$$\sin(\text{Arctan } .7) = \frac{.7}{\sqrt{1.49}} = \frac{7}{\sqrt{149}} \quad \square$$

We remarked earlier that the Arcsine, Arccosine, Arctangent, and Arccotangent functions are the inverses of the sine, cosine, tangent, and cotangent functions. This relationship is with respect to the operation of composition of functions; not with respect to multiplication. It is, however, traditional to use the exponent "-1" to denote these functions. Thus, in some books and tables of functions, you may see expressions such as $\sin^{-1} x$ and $\tan^{-1} x$ instead of Arcsin x or Arctan x.

Remember:

$\sin^{-1} x$ means Arcsin x

$\tan^{-1} x$ means Arctan x

$\cos^{-1} x$ means Arccos x

$\cot^{-1} x$ means Arccot x

This is why, when we introduced the notation "$\sin^k x$", we did not allow $k = -1$. Remember that $\sin^{-1} x$ does not mean $1/\sin x$.

EXERCISES

1. Determine the following. (You may wish to use Table 5.2.)
   (a) Arccos .5 (b) Arcsin -1 (c) Arccos .8660
   (d) Arctan 1 (e) Arccos .3420 (f) Arctan -3.474
   (g) Arcsin $-.7986$ (h) Arccot -2.3559

2. Determine the following.
   (a) Arcsin .8660 (b) Arccos .7071 (c) Arctan -1.732
   (d) Arccos $-.7071$ (e) Arccot $-.2309$ (f) Arcsin $-.6561$
   (g) Arccos .8192 (h) Arccot 4.7046
3. Determine the following.
   (a) Arcsin 1 $-$ Arcsin (-1)
   (b) Arccot 0 $-$ Arccot (-1)
   (c) Arctan .9 $-$ Arctan .6 (approximately)
   (d) Arcsin .5 $-$ Arcsin $(-.5)$
   (e) Arccos .5 $-$ Arccos (-1)
   (f) Arcsin .81 $-$ Arcsin .39 (approximately)
4. Compute.
   (a) sin (Arccos .5) (b) cos (Arcsin .75) (c) sin (Arccos .9)
   (d) tan (Arccot $\frac{7}{11}$) (e) sin (Arctan $-\frac{5}{6}$) (f) tan (Arcsin .4)
5. Compute.
   (a) cos (Arcsin .7071) (b) tan (Arcsin .8) (c) cot (Arcsin $-.5$)
   (d) sin (Arccos $-.9$) (e) cos (Arctan $\frac{7}{8}$) (f) sec (Arccos $\frac{10}{13}$)
6. Compute.
   (a) Arcsin (cos $\pi/3$) (b) Arctan (cot $\pi/8$) (c) Arccos (2 sin $\pi/3$)
   (d) Arcsin (2 cos $\pi/6$) (e) Arctan (sin π) (f) Arccot (cos $\pi/4$)
7. Compute.
   (a) Arcsin .5 $+$ Arccos .5
   (b) Arcsin $-.7$ $+$ Arccos $-.7$
   (c) Arcsin t $+$ Arccos t for any t, $-1 \leq t \leq 1$
   (d) Arctan 1 $+$ Arccot 1
   (e) Arctan .8 $+$ Arccot .8
   (f) Arctan t $+$ Arccot t for any t in R
8. Express the following in terms of x.
   (a) sin (Arccos x) (b) cos (Arcsin x) (c) tan (Arcsin x)
   (d) cot (Arcsin x) (e) sec (Arcsin x) (f) csc (Arcsin x)
   (g) sin (Arctan x) (h) sin (Arccot x) (i) cos (Arctan x)
   (j) tan (Arccos x)
9. Adapt the techniques discussed in Sections 5 and 6 and draw the graphs of the following equations.
   (a) $y = 2$ Arcsin x (b) $y =$ Arcsin $x + 2$
   (c) $y =$ Arcsin $2x$ (d) $y =$ Arctan $x - 3$
   (e) $y = 4$ Arccot x (f) $y = 3$ Arcsin $(x - 2) + 1$

6

Numerical Trigonometry

There are important applications of trigonometry in surveying, engineering, physics, and other technical and scientific areas. In this chapter we will study some of the basic techniques and formulas and will consider several applications.

1. Tables of Trigonometric Functions

We saw in Chapter 5 that in order to determine values of the six elementary trigonometric functions of any angle, we need only values of these functions for angles between 0° and 90° or between 0 and $\pi/2$. Thus tables are provided only for functions of acute angles. Furthermore, because of the co-function relationships, we only need values of functions for angles between 0° and 45° or between 0 and $\pi/4$.

Example 1

$\sin 62^\circ = \cos 28^\circ$

$\tan 47^\circ = \cot 43^\circ$

$\cos 57^\circ = \sin 33^\circ$ □

In general, tables providing the needed degree of precision are commonly available for values of the sine, cosine, tangent, and cotangent functions for angles between 0° and 90° or between 0 and $\pi/2$. The secant and cosecant functions are seldom used in numerical computations and when involved can be replaced by 1/cosine θ and 1/sine θ, respectively. The tables in this text, however, provide values for all six trigonometric functions.

Most tables are constructed as was the one given in Chapter 5, so that values of functions of angles up to 45° or $\pi/4$ are read by finding the angle measure on the *left* of the table and reading the value on that line, *under* the column headed by the name of the function under discussion. For acute angles greater than 45° or $\pi/4$, find the angle measure on the *right* of the table and read the value in the column *above* the name of the function under discussion.

The table provided in Chapter 5 gave values only for exact degree measurements and the corresponding radian measures. The tables in the back of the book provide 4-place approximations for angles measured to the nearest 10 min.

Example 2

To find sin 32°, find "32°" on the left side of the table and read .5299 in the column headed by "sin." □

Example 3

To find sin 63°, find "63°" on the right side of the table and read .8910 in the column above "sin." □

Example 4

tan .2793 = .2867 □

Since we can interpolate with trigonometric tables just as we did with logarithmic tables, our tables can be used to approximate function values for angles measured to the nearest minute.

Example 5

Determine sin 43°14′.

$$10'\left\{ 4'\left\{ \begin{array}{l} \sin 43^\circ 10' = .6841 \\ \sin 43^\circ 14' \end{array} \right. \right. \qquad \left. \left. \begin{array}{l} \sin 43^\circ 10' = .6841 \\ \sin 43^\circ 14' \end{array} \right\} t \right.$$

$$\begin{array}{ll} 4' \left\{ \begin{array}{l} \sin 43^\circ 10' = .6841 \\ \sin 43^\circ 14' \end{array} \right\} t & \\ 10' \quad \sin 43^\circ 20' = .6862 & .0021 \end{array}$$

$$\frac{4}{10} = \frac{t}{.0021}$$

Thus t is approximately .0008. Since the difference between .6841 and sin 43°14′ is about .0008,

$$\sin 43^\circ 14' = .6849 \quad \square$$

Example 6

If tan x = .2137, what is x in degree measurement?

$$10\left\{ t\left\{ \begin{array}{l} \tan 12^\circ 00' = .2126 \\ \tan x \quad\;\; = .2137 \end{array} \right. \atop \tan 12^\circ 10' = .2156 \right. \quad \begin{array}{l} .0011 \\ .0030 \end{array}$$

$$\frac{t}{10} = \frac{11}{30}$$

$$t = \frac{11}{3}$$

Thus t is approximately 4 and x is approximately 12°04′. □

Since using logarithms can frequently simplify our computations, it is useful to know the logarithms of values of trigonometric functions. We could look up the value of the function and then look up the logarithm of the number obtained. However, since such information is used quite often, tables are available which combine these two steps and provide logarithms of values of trigonometric functions. We include such a table in the back of the text. Notice, however, that the entries in the table are not quite complete. For example, although the table indicates that

$$\log \sin 9^\circ 10' = 9.2022$$

it really means that

$$\log \sin 9^\circ 10' = 9.2022 - 10$$

To determine the logarithm of a trigonometric function using the table, you must subtract 10 from each entry in the table. Interpolation is handled as before.

Example 7

Determine log cos 15°13′.

$$10\left\{ 3\left\{ \begin{array}{l} \log \sin 15^\circ 10' = 9.4177 - 10 \\ \log \sin 15^\circ 13' = \qquad ? \end{array} \right. \atop \log \sin 15^\circ 20' = 9.4223 - 10 \right. \quad \begin{array}{l} t \\ .0046 \end{array}$$

$$\frac{3}{10} = \frac{t}{.0046}$$

$$t = .00138 \doteq .0014$$

Thus

$\log \sin 15°13' \doteq 9.4191 - 10$ □

Example 8

Find x if $\log \sin x = 9.8463 - 10$.

$$10\left\{ t\left\{ \begin{array}{l} \log \sin 44°30' = 9.8457 - 10 \\ \log \sin x \qquad = 9.8463 - 10 \\ \log \sin 44°40' = 9.8469 - 10 \end{array} \right. \right. \begin{array}{l} \left.\right\} 6 \\ \left.\right\} 12 \end{array}$$

$$\frac{t}{10} = \frac{6}{12}$$

$$t = 5$$

Thus t is 5 and x is approximately $44°35'$. □

Example 9

Determine an approximation for $435 \sin 47°10'$.

If

$$\begin{aligned} x &= 435 \sin 47°10' \\ \log x &= \log 435 + \log \sin 47°10' \\ &= 2.6385 + 9.8653 - 10 \end{aligned}$$

Thus

$\log x = 2.5038$

and

$x \doteq 319.0$ □

EXERCISES

1. Determine the following.
   (a) $\sin 15°20'$ (b) $\cos 49°30'$ (c) $\tan 67°20'$ (d) $\cot 19°50'$
   (e) $\sin 52°40'$ (f) $\cos 11°30'$ (g) $\sin .5265$ (h) $\tan .6254$
2. Solve for x in terms of degrees. You may assume $0° \le x \le 90°$.
   (a) $\sin x = .2700$ (b) $\cos x = .8465$ (c) $\tan x = 1.868$
   (d) $\cot x = 2.583$ (e) $\sin x = .3934$ (f) $\cos x = .5783$
3. Determine the following.
   (a) $\sin 31°17'$ (b) $\cos 43°52'$ (c) $\tan 14°38'$ (d) $\cot 68°24'$
   (e) $\sin 57°51'$ (f) $\cos 71°06'$ (g) $\sin .2362$ (h) $\cot 1.4758$
4. Solve for x in terms of degrees. You may assume $0° \le x \le 90°$.
   (a) $\sin x = .2776$ (b) $\cos x = .4871$ (c) $\tan x = .5030$
   (d) $\cot x = 2.425$ (e) $\sin x = .8452$ (f) $\cos x = .3372$

5. Solve for x in terms of radians. You may assume $0 \le x \le \pi/2$.
   (a) $\sin x = .2108$ (b) $\cos x = .9576$ (c) $\tan x = .4592$
   (d) $\cot x = 2.660$ (e) $\sin x = .8220$ (f) $\cos x = .5625$
6. Determine the following.
   (a) $\log \sin 32°14'$ (b) $\log \cos 42°37'$ (c) $\log \tan 72°23'$
   (d) $\log \cot 38°47'$ (e) $\log \sec 29°32'$ (f) $\log \csc 51°48'$
7. Determine x in terms of degrees.
   (a) $\log \sin x = 9.4273 - 10$ (b) $\log \cos x = 9.1235 - 10$
   (c) $\log \tan x = .3010$ (d) $\log \cot x = .5312$
   (e) $\log \sec x = .2156$ (f) $\log \csc x = .4127$
8. Solve for x. [*Hint*: Use logarithms.]

   (a) $x = \dfrac{192}{\sin 14°13'}$ (b) $x = \dfrac{57.3}{\sin 73°19'}$

   (c) $x = \dfrac{94.2}{\sin 113°31'}$ (d) $\dfrac{x}{\sin 54°10'} = \dfrac{73.5}{\sin 47°40'}$

   (e) $\dfrac{82.4}{\sin x} = \dfrac{92.7}{\sin 34°15'}$ (f) $\dfrac{71.42}{\sin x} = \dfrac{64.19}{\sin 51°37'}$
9. Determine a 4-place approximation for x.

   (a) $x = 17.3 \cos 51°17'$ (b) $x = \dfrac{\sin 31°14'}{\cos 19°15'}$

   (c) $x = \sin^5 19°10'$ (d) $x = \sqrt{\tan 76°18'}$

   (e) $x = \dfrac{14 \sin 142°}{11 \cos 256°}$ (f) $x = \tan^2 32°15' + \cot^2 19°13'$

   (g) $6.62 \cos x = 4.36$ (h) $284.7 \cos x = 168.7$
10. Solve for x.
    (a) $14.3x + \sin 37°10' = 73.2$
    (b) $97.2x = \sin 56°12'$
    (c) $x^2 + x \sin 32°10' - 12 = 0$
    (d) $11.3x^2 - x \cos 69°47' - \tan 11°13' = 0$
    (e) $\tan^2 x + \sec^2 x = 5.23$
    (f) $4 \sin^2 x - \sin x - 2 = 0$

2. The Trigonometry of Right Triangles

If a right triangle ABC with right angle at C is placed with vertex B at the origin, vertex C on the X-axis, and vertex A above the X-axis, as in Figure 6–1, we see that

$$\sin B = \frac{b}{c} \qquad \csc B = \frac{c}{b}$$

$$\cos B = \frac{a}{c} \qquad \sec B = \frac{c}{a}$$

$$\tan B = \frac{b}{a} \qquad \cot B = \frac{a}{b}$$

Figure 6–1

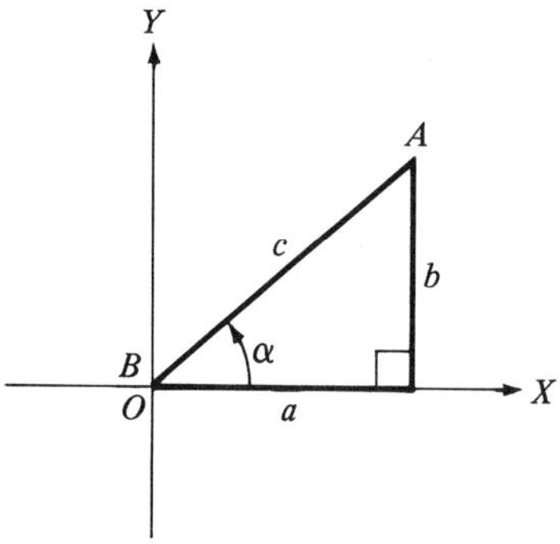

In general, if α is one of the acute angles of a right triangle, then

$$\sin \alpha = \frac{\text{side opposite } \alpha}{\text{hypotenuse}} \qquad \csc \alpha = \frac{\text{hypotenuse}}{\text{side opposite } \alpha}$$

$$\cos \alpha = \frac{\text{side adjacent to } \alpha}{\text{hypotenuse}} \qquad \sec \alpha = \frac{\text{hypotenuse}}{\text{side adjacent to } \alpha}$$

$$\tan \alpha = \frac{\text{side opposite } \alpha}{\text{side adjacent to } \alpha} \qquad \cot \alpha = \frac{\text{side adjacent to } \alpha}{\text{side opposite } \alpha}$$

We can use these facts and the tables of values of trigonometric functions to approximate unknown parts of right triangles.

Example 1

Find the unknown parts of the right triangle given in Figure 6–2.

First, the other acute angle must be exactly 55° because the sum of the angles of a triangle is 180°. Since

$$\sin 35^\circ = \frac{12}{x}$$

$$x = \frac{12}{\sin 35^\circ}$$

Thus

$$\log x = \log 12 - \log \sin 35^\circ$$

$$\log x = 1.0792 - (9.7586 - 10)$$

$$\log x = 1.3206$$

Figure 6–2

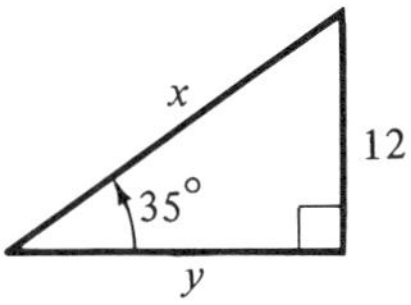

Hence

$x \doteq 20.92$

Furthermore,

$$\cot 35^\circ = \frac{y}{12}$$

$$\begin{aligned} y &= 12 \cot 35^\circ \\ \log y &= \log 12 + \log \cot 35^\circ \\ \log y &= 1.0792 + .1548 \\ \log y &= 1.2340 \\ y &\doteq 17.14 \ \square \end{aligned}$$

If the above problem arose from measurement, then the approximations of 20.92 and 17.14 would be unrealistic in light of the fact that the original data were correct to only two digits. We would approximate the two unknown sides by 21 and 17. If the original data were correct to 3 digits, say $35°0'$ and 12.0, we would approximate by 20.9 and 17.1. In general, we cannot expect to obtain better accuracy in the solutions than we had in the original data.

Example 2

Find the acute angles of the right triangle given in Figure 6–3.

Figure 6–3

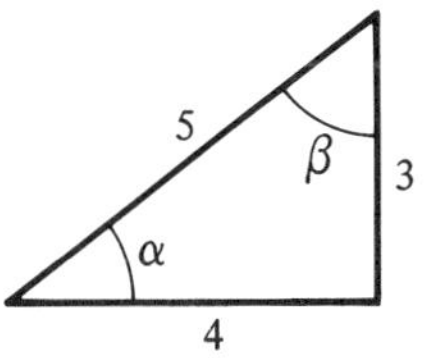

$\text{Sin } \alpha = \frac{3}{5} = .6$

Using the table and interpolating we see that α is approximately $36°52'$ and therefore β must be approximately $53°08'$.

Example 3

Find the acute angles and the third side of a right triangle with hypotenuse 7 and one side 4.

First we draw a triangle and label accordingly (Figure 6–4).

Figure 6–4

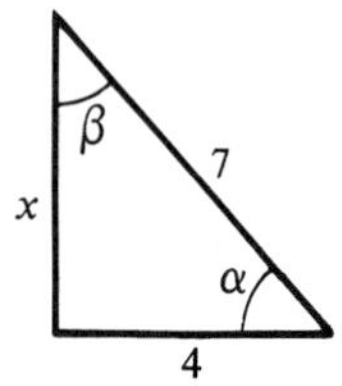

Since $\cos \alpha = \frac{4}{7}$, $\cos \alpha$ is approximately .5714.

From the table, we see that α is approximately $55°09'$.

Since $\sin \alpha = x/7$, $7 \sin \alpha = x$.

Reading from the table and interpolating we have

$$7(.8206) = x$$

and

$$x \doteq 5.7442$$

Since $\alpha = 55°09'$, $\beta = 34°51'$.

In this case we used trigonometry instead of the Pythagorean theorem to find the third side of a right triangle. We could verify our result by checking to see if

$$(5.7442)^2 + 4^2 \doteq 7^2 \square$$

There are many physical and practical applications of the techniques illustrated in Examples 1–3.

Example 4

Suppose we wish to determine the height of a tree. It is very difficult or impossible to actually measure the distance from the top of the tree to the ground. However, we can measure distances along the ground and with sighting devices, we can measure certain angles.

Suppose, from a point 100 ft from the base of the tree, we find that the acute angle between the ground and the line of sight from the ground to the tree is about 37°. (See Figure 6–5.) If we let h denote the height

Figure 6–5

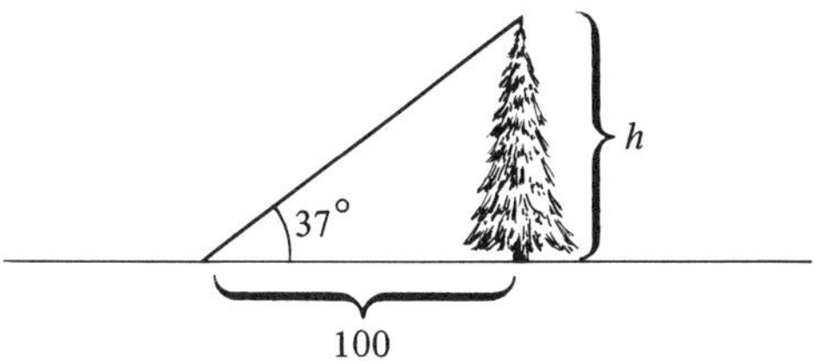

of the tree, then

$$\frac{h}{100} = \tan 37°$$

From a table, we find that $\tan 37° = .7536$. Thus

$$\frac{h}{100} = .7536 \quad \text{and} \quad h = 75.36$$

The tree is about 75 ft high. □

The angle between the line of sight of a point and the horizontal is called the *angle of elevation* if the object is above the horizontal and is called the *angle of depression* if it is below the horizontal.

Example 5

If two forces are applied at right angles to each other, the resultant force is in the direction of the diagonal of the rectangle so determined. (See Figure 6–6.)

Figure 6–6

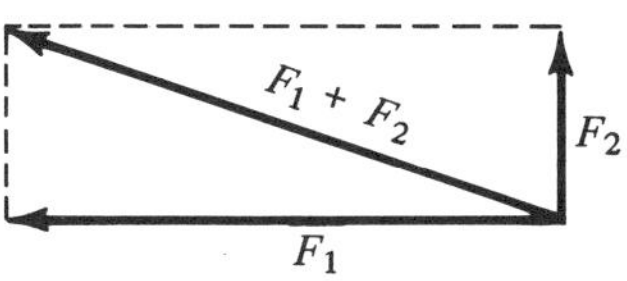

Suppose a plane is heading east at a speed of 500 mph and the wind is blowing from the north with a speed of 41 mph. What is the actual course and speed of the plane?

First we draw a diagram as in Figure 6–7. The actual course is indicated by the direction of the diagonal of the rectangle. The tangent of θ is 41/500, that is, $\tan\theta = .0820$. From the table, we see that θ is about $4°42'$. Thus the actual course is about $94°42'$. (Course headings are measured clockwise from due north.)

Figure 6–7

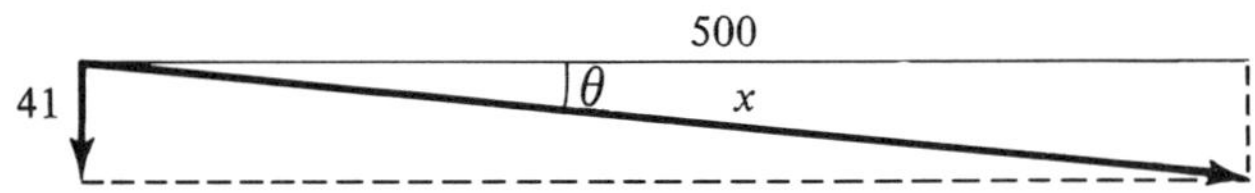

The speed can be found by noting that $500/x = \cos\theta$. Thus $x = 500/\cos\theta = 500/\cos 4°42' = 500/.9966 = 501.7$. Although the course is changed quite a bit by the wind, the speed is changed very little. □

EXERCISES

1. If the hypotenuse of a right triangle is 42 in and one acute angle is $37°10'$, what are the lengths of the other two sides of the triangle?
2. Determine the acute angles of a right triangle with sides 5, 12, and 13.
3. Determine the acute angles of a right triangle with hypotenuse 15.3 and one side 9.72.
4. Determine the acute angles of a right triangle having sides 42.7 and 19.8.
5. If one acute angle of a right triangle is $64°15'$ and the side adjacent to this angle is 17.2 in., what are the lengths of the other two sides?
6. If a flagpole casts a shadow 32 ft long and the line of sight from the tip of the shadow to the top of the pole makes an angle of 28° with the ground, how high is the flagpole? [*Hint*: First draw a picture.]
7. A boy is flying a kite and has let out 300 ft of string. If the string makes an angle of 35° with the ground, how high is the kite?
8. A boat is heading due north at a speed of 30 mph but the wind is blowing 8 mph due east. What is the actual course the boat is traveling? (Direction is measured clockwise from due north using 0° as north.) What is the actual speed of the boat? What course should be followed to go due north if the boat is, as before, to cruise at 30 mph?
9. Suppose that a skydiver jumps out of a plane over a point 800 ft from you and, at the instant he jumps out, the angle of elevation of the plane is 76°. Furthermore, suppose that 10 sec later, his angle of elevation is 70°. If you assume that his fall is vertical, at what average speed did he fall during the 10 sec?
10. Suppose a ship is sighted from the top of a 120-ft tower. At 10:15 the angle of

depression was 11° and at 10:25, the angle of depression was 18°. Assuming that the ship is moving directly toward the tower, how fast is it moving?

11. From one side of a river we can see a tree on the other shore. Let two points A and B be chosen on our side such that
    (i) A is on the shore.
    (ii) A, B, and the tree form a right triangle with right angle at A.
    (iii) A and B are 100 ft apart.
    (iv) The line of sight from B to the tree makes an angle of 74° with the line of sight from B to A.

    How wide is the river? (Draw a picture.)
12. The area of a triangle ABC is given by the formula

    $$K = \tfrac{1}{2}ch$$

    where h is the altitude from C to line AB. Show that $K = \frac{1}{2}bc \sin A$. (First draw appropriate figures.)

3. General Triangles

In Section 2, we saw that we could use trigonometry to determine the unknown parts of a right triangle if we knew two sides or one side and an acute angle. We now develop formulas which enable us to determine unknown parts of general triangles given three sides, two sides and the angle between them, or two angles and one side. We will also see that although two sides and one of the angles other than the one between them do not, in general, determine a unique triangle, there are, at most, only two possibilities in such a situation and we can determine both of these.

THE LAW OF COSINES

Consider a triangle ABC with sides a, b, and c opposite angles A, B, and C, respectively. First we consider the cases where A is acute as in Figure 6–8. Let h be the altitude of the triangle drawn from C to point P on line AB.

Figure 6–8

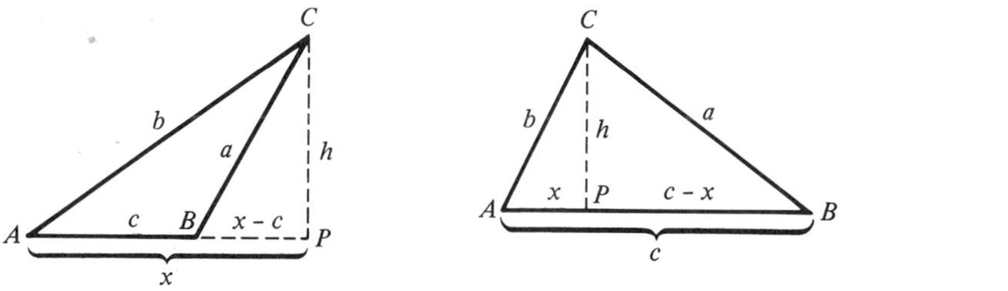

Furthermore, if we let x be the distance from A to P, then either $x - c$ or $c - x$ is the distance from B to P, that is, the distance from B to P is $|x - c|$. Applying the Pythagorean theorem to $\triangle CPB$ we have

$$a^2 = h^2 + |c - x|^2$$

that is,

$$a^2 = h^2 + (c - x)^2 \qquad \text{because } |c - x|^2 = (c - x)^2$$

Thus

$$a^2 = h^2 + c^2 - 2cx + x^2$$

and

$$a^2 = (h^2 + x^2) + c^2 - 2cx$$

Applying the Pythagorean theorem to $\triangle CPA$ we have

$$b^2 = h^2 + x^2$$

Thus, by substitution,

$$a^2 = b^2 + c^2 - 2cx$$

But

$$\frac{x}{b} = \cos A$$

and, therefore,

$$x = b \cos A$$

Substituting yields

$$a^2 = b^2 + c^2 - 2c(b \cos A)$$

or

$$a^2 = b^2 + c^2 - 2bc \cos A$$

If A is obtuse, the triangle is illustrated by the drawing in Figure 6–9. In this case

$$\begin{aligned} a^2 &= h^2 + (x + c)^2 \\ &= h^2 + 2cx + c^2 + x^2 \\ &= h^2 + x^2 + c^2 + 2cx \end{aligned}$$

As before,

$$b^2 = h^2 + x^2$$

and, therefore,

$$a^2 = b^2 + c^2 + 2cx$$

Figure 6–9

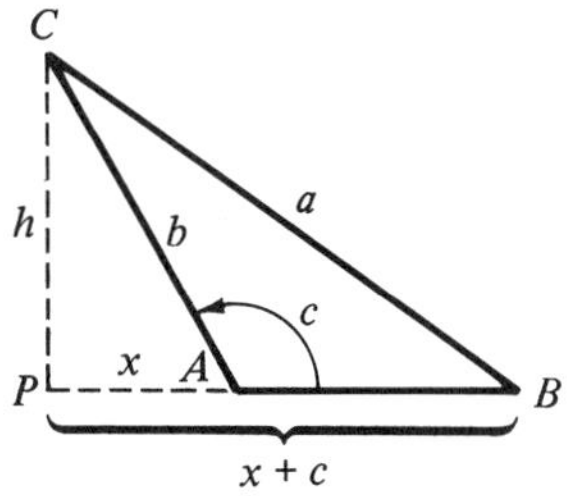

But

$$\frac{x}{b} = \cos(180^\circ - A) = -\cos A$$

Thus

$$x = -b \cos A$$

Substituting yields

$$a^2 = b^2 + c^2 + 2c(-b \cos A)$$

or

$$a^2 = b^2 + c^2 - 2bc \cos A$$

Thus for any triangle A, B, C with corresponding sides a, b, and c we have

$$a^2 = b^2 + c^2 - 2bc \cos A$$

This is known as the *law of cosines.*

THE LAW OF SINES

Now suppose we have a triangle ABC with corresponding sides a, b, and c and altitude h. We consider three possibilities at once but assume that angles A and B are not right angles. (See Figure 6–10.)

In all three cases we have

$$\frac{h}{b} = \sin A \quad \text{and} \quad \frac{h}{a} = \sin B$$

[In the second case $\sin A = \sin(180^\circ - A)$, in the third case $\sin B = \sin(180^\circ - B)$.]

Thus

$$h = b \sin A \quad \text{and} \quad h = a \sin B$$

Figure 6–10

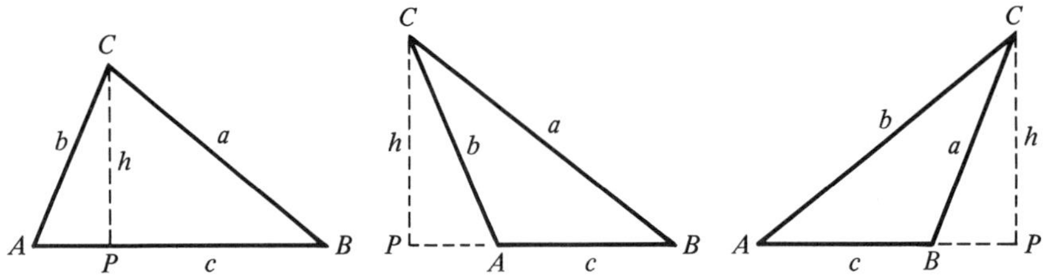

Substituting yields the equation

$$a \sin B = b \sin A$$

Since

$$\sin B \neq 0 \quad \text{and} \quad \sin A \neq 0$$
$$\sin A \sin B \neq 0$$

Dividing both sides by $\sin A \sin B$ yields

$$\frac{a}{\sin A} = \frac{b}{\sin B}$$

If we had drawn the altitude from A to a, we would have obtained

$$\frac{b}{\sin B} = \frac{c}{\sin C}$$

Thus we have

$$\frac{a}{\sin A} = \frac{b}{\sin B} = \frac{c}{\sin C}$$

This equation is known as the *law of sines.* If one angle is a right angle, the equation will still hold and the justification is, in fact, a trivial algebraic verification. We leave it as an exercise.

From geometry recall that two triangles are congruent if three sides of one are congruent to three sides of the other. (SSS) Thus a triangle is completely determined by its sides. If we are given the three sides of a triangle, we can use the law of cosines to determine the angles.

Example 1

Determine angle C in triangle ABC if $a = 12$, $b = 14$, and $c = 9$. Using the law of cosines in the form

$$c^2 = a^2 + b^2 - 2ab \cos C$$

we have

$$
\begin{aligned}
9^2 &= 12^2 + 14^2 - 2(12)(14)\cos C \\
81 &= 144 + 196 - 336\cos C \\
&= 340 - 336\cos C \\
-259 &= -336\cos C \\
\cos C &= \frac{259}{336} \\
\log\cos C &= \log 259 - \log 336 \\
&= 2.4133 - 2.5263 \\
&= (12.4133 - 10) - (2.5263)
\end{aligned}
$$

Thus

$$
\begin{aligned}
\log\cos C &= 9.8870 - 10 \\
C &\doteq 39°26'
\end{aligned}
$$

We could find the other angles by using the law of cosines in the form

$$b^2 = a^2 + c^2 - 2ac\cos B$$
$$a^2 = b^2 + c^2 - 2bc\cos A$$

Of course, once A or B is found, the third can be computed easily using the fact that the sum of the measurements of the three angles is 180°. □

From geometry we also know that the two triangles are congruent if two sides and the included angle of one are congruent to two sides and the included angle of the other (SAS). Thus a triangle is completely determined by two sides and their included angle. The law of cosines can be used to determine the third side of a triangle when two sides and the included angle are known.

Example 2

Suppose we have a triangle with sides 4 and 9 and the included angle is 34°. We draw the triangle and label it as in Figure 6–11. By the law of

Figure 6–11

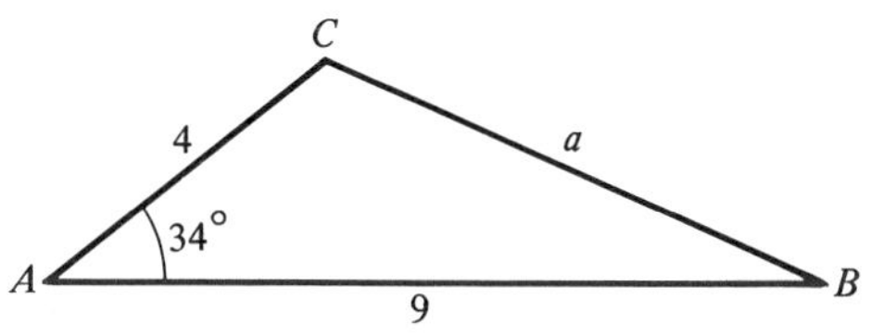

cosines

$$a^2 = 4^2 + 9^2 - 2(4)(9)\cos 34°$$
$$a^2 = 16 + 81 - 72(.8290)$$
$$= 97 - 59.6880$$
$$= 37.3120$$
$$a = \sqrt{37.3120}$$

Using logarithms

$$\log a = \tfrac{1}{2}\log 37.312$$
$$= \tfrac{1}{2}(1.5718)$$
$$= .7859$$
$$a \doteq 6.109$$

We now can use the law of sines to determine the other angles. Since

$$\frac{4}{\sin B} = \frac{a}{\sin 34°}$$
$$a \sin B = 4 \sin 34°$$
$$\sin B = \frac{4 \sin 34°}{a}$$

$$\log \sin B = \log 4 + \log \sin 34° - \log a$$
$$= .6021 + (9.7476 - 10) - .7859 \quad \text{(We computed } \log a \text{ above.)}$$

Thus

$$\log \sin B = 9.5638 - 10$$
$$B \doteq 21°29' \quad \text{or else} \quad B \doteq 158°31'$$

But if

$$B \doteq 158°31'$$
$$A + B > 180°$$

Thus

$$B \doteq 21°29'$$

Then

$$C \doteq 180° - (34° + 29°29') = 124°31' \;\square$$

Recall also from geometry that two angles and the included side completely determine a triangle. (ASA) We can use the law of sines to determine the other parts.

Example 3

Suppose $A = 42°$, $B = 55°$, and $c = 10$. (See Figure 6–12.) Find a, b, and C.

Figure 6–12

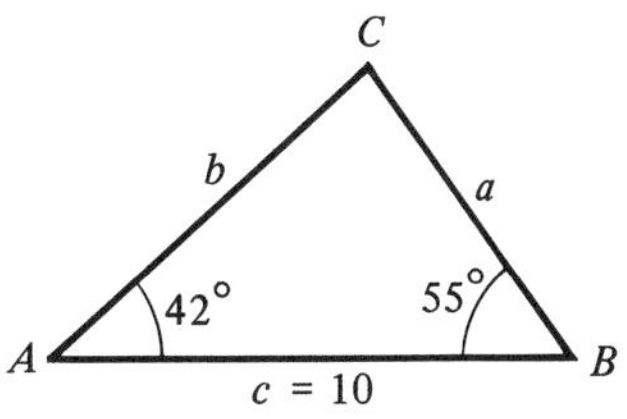

We can determine C immediately:

$$C = 180° - (42° + 55°) = 83°$$

Now using the law of sines

$$\frac{a}{\sin 42°} = \frac{10}{\sin 83°}$$

$$a = \frac{10 \sin 42°}{\sin 83°}$$

$$\begin{aligned}\log a &= \log 10 + \log \sin 42° - \log \sin 83° \\ &= 1 + (9.8255 - 10) - (9.9968 - 10)\end{aligned}$$

Thus

$$\log a = .8287$$

and

$$a \doteq 6.74$$

Also

$$\frac{b}{\sin 55°} = \frac{10}{\sin 83°}$$

$$b = \frac{10 \sin 55°}{\sin 83°}$$

$$\begin{aligned}\log b &= \log 10 + \log \sin 55° - \log \sin 83° \\ b &= 1 + (9.9134 - 10) - (9.9968 - 10)\end{aligned}$$

Thus

$\log b = .9166$

and

$b \doteq 8.25$ □

Two sides and one of the angles not between them do not determine a triangle but they do determine at most two possibilities.

Example 4

Determine the unknown parts of a triangle containing one angle of 30°, in which the side opposite the angle is 8 and one of the other sides is 14. (See Figure 6–13.)

Figure 6–13

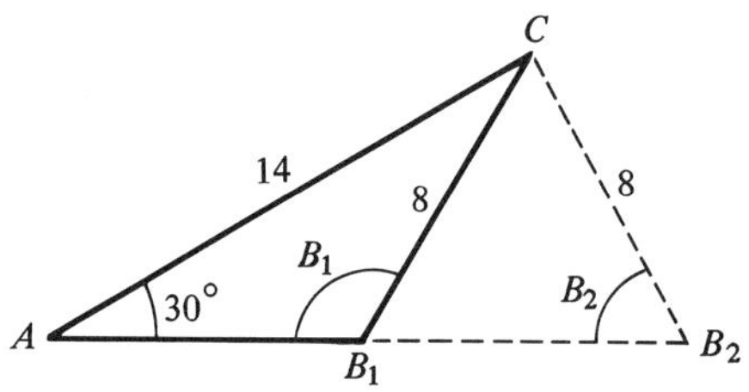

By the law of sines

$$\frac{8}{\sin A} = \frac{14}{\sin B_1} = \frac{14}{\sin B_2}$$

Since triangle B_1CB_2 is isosceles, its base angles are the same and therefore $180° - B_1 = B_2$. From the equation we have

$$\sin B_1 = \frac{14 \sin A}{8} \quad \text{and} \quad \sin B_2 = \frac{14 \sin A}{8}$$

Since $\sin (180° - x) = \sin x$ for $0° \leq x \leq 180°$, there are two possibilities for B_1 and B_2 but $B_2 = 180° - B_1$. □

In the above example, since $A = 30°$, $\sin A = \frac{1}{2}$. Thus we have

$$\sin B_1 = \frac{14(\frac{1}{2})}{8} = \frac{14}{16} = \frac{7}{8} = .875$$

and

$$\sin B_2 = \frac{7}{8} = .875$$

From the table we see that one angle is $61°03'$ and the other is $118°57'$. We see that the information $A = 30°$, $b = 14$, and $a = 8$ does not determine a unique triangle but there are only two possibilities and we have determined both of them. In general, given A, a, and b, the other sides and angles of the triangle are not uniquely determined. There may be two possibilities if, in fact, such a triangle exists; in one case B is acute, in the other it is obtuse. It may happen that no such triangle exists or that $B = 90°$. These cases are illustrated in the exercises.

In summary, we now have formulas and methods which enable us to determine all parts of a triangle as follows:

Given:	*Use*:
SSS	Law of cosines
ASA	Law of sines
SAS	Law of cosines and law of sines

Note that if we have SAA or AAS, we can immediately determine the third angle and this reduces to ASA. Furthermore, if we know two sides and one of the angles other than the included angle, we can determine the possibilities for the other parts.

The methods developed for determining the unknown parts of triangles have many practical applications. Among these are problems requiring us to determine distances which are difficult or impossible to measure directly.

Example 5

From our cottage on the shore of Battersea Bay, we wish to determine the distance from our dock (point D) to Goat Island (point G). We choose a point P 100 ft from the dock such that the point P has a good line of sight to Goat Island. By measurement we find that the resultant data are as follows and are as in Figure 6–14.

(a) $\angle GDP$ measures 145°.
(b) $\angle GPD$ measures 32°.

Figure 6–14

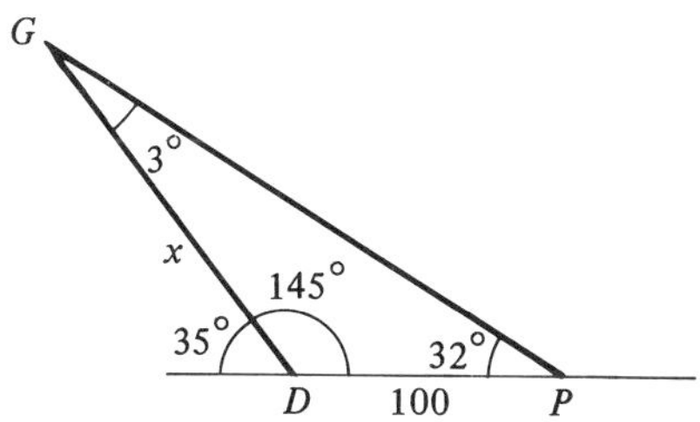

By arithmetic, $\angle DGP$ is 3°. Now, by the law of sines,

$$\frac{x}{\sin \angle GPD} = \frac{100}{\sin \angle DGP}$$

$$x = \frac{100 \sin \angle GPD}{\sin \angle DGP}$$

$$= \frac{100 \sin 32^\circ}{\sin 3^\circ}$$

$$\log x = \log 100 + \log \sin 32^\circ - \log \sin 3^\circ$$
$$= 2 + (9.7242 - 10) - (8.7188 - 10)$$

Thus

$$\log x = 3.0054$$

and

$$x \doteq 1013 \; \square$$

If two forces are applied at the same point, the resultant force is in the direction of the diagonal of the parallelogram so determined and its magnitude is the length of the diagonal. (See Figure 6–15.) This is the vector addition of forces. A special case of this principle was discussed in Example 5 of Section 2. The resultant force can be determined by the techniques developed in this section provided we know the magnitudes and directions of the original forces.

Figure 6–15

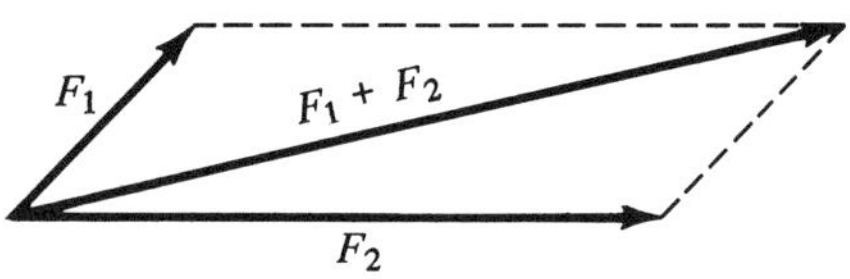

Example 6

Suppose a plane is heading east at 200 mph and a wind of 32 mph is blowing in a northeasterly direction in a direction 60° east of due north. What course is the plane actually flying and at what speed?

First we draw a diagram as in Figure 6–16. Since the wind is blowing 60° east of north, it makes an angle of 30° with the force of the thrust of the plane's engines. From geometry we know that the four angles of the

Figure 6-16

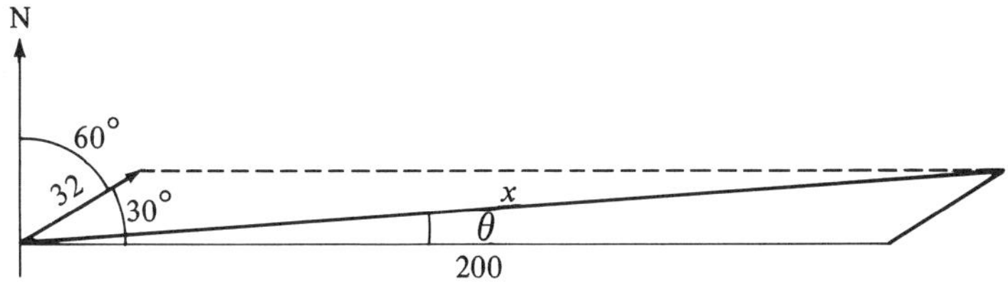

parallelogram are 30°, 150°, 30°, and 150°. By the law of cosines,

$$\begin{aligned} x^2 &= (200)^2 + (32)^2 - 2(200)(32)\cos 150^\circ \\ &= 40{,}000 + 1024 - (12{,}800)(-.8660) \\ &= 52{,}109 \end{aligned}$$

Using logarithms we now have

$$\begin{aligned} \log x^2 &= \log 52{,}109 \\ 2\log x &= 4.7169 \\ \log x &= 2.358 \\ x &= 228.3 \end{aligned}$$

The actual speed of the plane is about 228 mph.

To find the direction of the force, we first find θ. By the law of sines,

$$\frac{x}{\sin 150^\circ} = \frac{32}{\sin \theta}$$

or

$$\frac{\sin \theta}{32} = \frac{\sin 150^\circ}{x}$$

Thus

$$\sin \theta = \frac{32 \sin 150^\circ}{x}$$

$$\begin{aligned} \log \sin \theta &= \log 32 + \log \sin 150^\circ - \log x \\ &= \log 32 + \log \sin 30^\circ - 2.3584 \qquad \text{(Using our previous computation of } \log x\text{.)} \\ &= 1.5051 + 9.6990 - 10 - 2.3584 \end{aligned}$$

Thus

$$\log \sin \theta = 8.8457 - 10$$

and

$\theta = 4°1'$

The true course is $85°59'$ east of north. □

EXERCISES

In Exercises 1–12, solve the triangles for the parts not given.

1. $a = 12$, $b = 34$, and $C = 54°$
2. $a = 40$, $b = 10$, and $C = 151°$
3. $a = 35$, $b = 25$, and $C = 53°20'$
4. $b = 500$, $C = 32°$, and $A = 72°$
5. $c = 15$, $A = 42°$, and $B = 63°$
6. $a = 17.31$, $B = 61°17'$, and $C = 52°9'$
7. $a = 11$, $b = 15$, and $c = 9$
8. $a = 12.3$, $b = 19.4$, and $c = 10.7$
9. $a = 45$, $A = 35°$, and $b = 47.5$
10. $b = 17.3$, $B = 120°$, and $c = 5.23$
11. $a = 6$, $b = 14$, and $B = 31°$
12. $c = 10$, $b = 20$, and $C = 30°$
13. Draw a diagram in which you give SSA and no such triangle exists. Explain why.
14. Draw a diagram in which you give SSA with an acute angle and exactly one such triangle exists. Explain why.
15. Verify that the law of sines is valid even if one of the angles is a right angle.
16. Express $\cos A$ in terms of a, b, and c. [*Hint*: Use the law of cosines.]

*17. Show that if we are given a, b, and A, there are two triangles satisfying the conditions if and only if $b \sin A < a < b$.

18. If you are flying a plane heading 10° east of north at 160 mph and the wind is blowing in a southeasterly direction 130° east of north, what is your actual direction and speed?
19. From one point, the angle of elevation of a church spire is 38°. From a point 50 ft further away it is 29°. How high is the spire?
20. You wish your boat to travel due north. Suppose you have a cruising speed of 35 mph and the wind is blowing southeast at a velocity of 10 mph. What direction should your boat take? What actual speed will it make?
21. As your boat travels toward shore, the angle of elevation of the top of the lighthouse is 15° at 1:30 and 20° at 1:40. At this rate, at what time will you reach shore?
22. We wish to determine the distance from Garden Island (point G) to Noble's Point (point N) on Battersea Bay. Our observation points are points P and Q on the west shore and they are chosen to be 100 ft apart. By measurement, we find that $\angle GPQ$ measures 114°, $\angle NPQ$ measures 99°, $\angle GQP$ measures 63°, and $\angle NQP$ measures 78°. Find the distance from B to N. (This is a two-step problem. First find the distance from G to P or at least the logarithm of this distance.)
23. A quadrilateral has sides $a = 72$ in., $b = 58$ in., $c = 69$ in., and $d = 82$ in. Determine the diagonals if the angle between a and b is 68°. Determine the other angles.

*24. Let a parallelogram have sides a, b, a, b, and diagonals d_1 and d_2. Prove that $d_1{}^2 + d_2{}^2 = 2(a^2 + b^2)$.

4. The Area of a Triangle

From geometry, we know that the area K of triangle ABC is given by

$$K = \tfrac{1}{2}ch$$

where h is the altitude of the triangle drawn from C to line AB. (See Figure 6–17.) Notice that if A is an acute angle,

$$\frac{h}{b} = \sin A$$

Figure 6–17

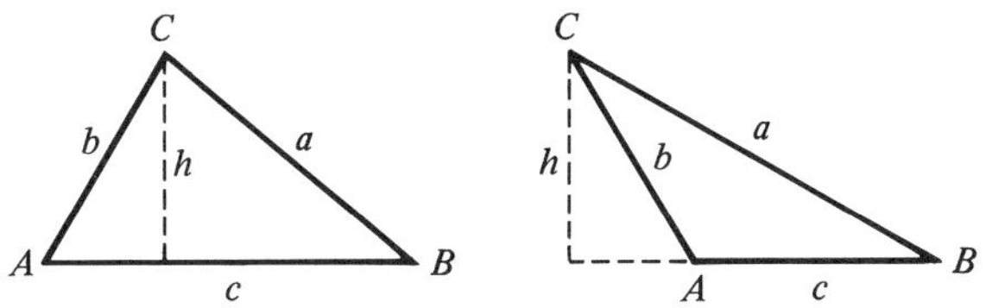

and if A is obtuse

$$\frac{h}{b} = \sin (180° - A)$$

Since

$$\sin (180° - A) = \sin A$$

in either case we have

$$\frac{h}{b} = \sin A$$

and, therefore,

$$h = b \sin A$$

Substituting in the known area formula yields

$$K = \tfrac{1}{2}ch = \tfrac{1}{2}c(b \sin A)$$

Thus the area of the triangle is given by

$$K = \tfrac{1}{2}bc \sin A$$

Example 1

Find the area of a triangle with sides 62.34, 147.5, and included angle $43°37'$.

We use the formula

$$K = \tfrac{1}{2}ab \sin C$$

with $a = 62.34$, $b = 147.5$, and $C = 43°37'$.

$$K = \tfrac{1}{2}(62.34)(147.5)(\sin 43°37')$$

The computation is made easier by using logarithms:

$$\begin{aligned} \log K &= \log \tfrac{1}{2} + \log 62.34 + \log 147.5 + \log \sin 43°37' \\ &= (9.6990 - 10) + 1.7948 + 2.1688 + 9.8387 - 10 \end{aligned}$$

Thus

$$\log K = 3.4713$$

and

$$K \doteq 2960 \quad \square$$

Another useful formula for the area of triangle ABC is given by

$$K = \sqrt{s(s-a)(s-b)(s-c)}$$

where $s = \tfrac{1}{2}(a + b + c)$. (See Figure 6–18.)

Figure 6–18

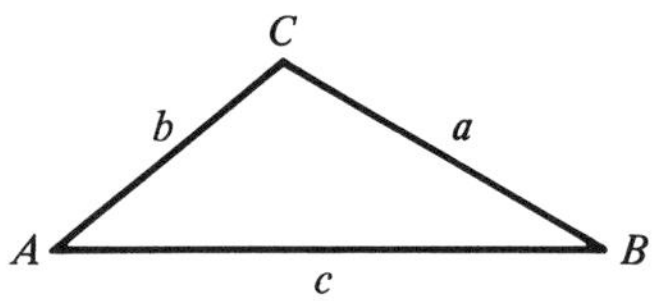

The proof of this formula begins with the area formula previously established and uses the law of cosines.

Since $K = \tfrac{1}{2}bc \sin A$,

$$\begin{aligned} K^2 &= \tfrac{1}{4}b^2c^2 \sin^2 A \\ &= \tfrac{1}{4}b^2c^2(1 - \cos^2 A) \\ &= \frac{bc}{2}(1 + \cos A)\frac{bc}{2}(1 - \cos A) \end{aligned}$$

Now using the law of cosines (solve the formula for cos A) we have

$$K^2 = \frac{bc}{2}\left(1 + \frac{b^2 + c^2 - a^2}{2bc}\right)\frac{bc}{2}\left(1 - \frac{b^2 + c^2 - a^2}{2bc}\right)$$

$$= \frac{2bc + b^2 + c^2 - a^2}{4} \cdot \frac{2bc - b^2 - c^2 + a^2}{4}$$

$$= \frac{(b + c)^2 - a^2}{4} \cdot \frac{a^2 - (b - c)^2}{4}$$

$$= \left(\frac{b + c + a}{2}\right)\left(\frac{b + c - a}{2}\right)\left(\frac{a - b + c}{2}\right)\left(\frac{a + b - c}{2}\right)$$

$$= s(s - a)(s - b)(s - c)$$

Thus

$$K = \sqrt{s(s - a)(s - b)(s - c)}$$

Example 2

Determine the area of a triangle with sides 42.3, 71.4, and 62.7. To use the formula

$$K = \sqrt{s(s - a)(s - b)(s - c)}$$

first we must determine s:

$$s = \tfrac{1}{2}(42.3 + 71.4 + 62.7)$$
$$= 88.2$$

Then

$$s - a = 43.9$$
$$s - b = 16.8$$
$$s - c = 25.5$$

By the formula,

$$K = \sqrt{(88.2)(43.9)(16.8)(25.5)}$$

This computation could be completed very quickly if we had a small electronic calculator. We will compute K using the methods of Chapter 4.

Using logarithms we have

$$\log K = \tfrac{1}{2}(\log 88.2 + \log 43.9 + \log 16.8 + \log 25.5)$$
$$= \tfrac{1}{2}(1.9455 + 1.6425 + 1.2253 + 1.4065)$$

Thus

$$\log K = 3.1099$$

and

$K \doteq 1287$ □

A third formula for the area of a triangle is given by

$$K = \frac{a^2 \sin B \sin C}{2 \sin A}$$

The proof of this formula is left as an exercise. Notice that it can be used when we know two angles and the side between them.

Example 3

Determine the area of a triangle having angles 47°20′ and 39°30′ with included side 92.5.

If we let $a = 92.5$, $B = 47°20'$, and $C = 39°30'$, then

$$A = 180° - (47°20' + 39°30')$$

that is,

$$A = 93°10'$$

Using the formula we have

$$K = \frac{(92.5)^2 \sin 47°20' \sin 39°30'}{2 \sin 93°10'}$$

Then

$$\log K = 2 \log 92.5 + \log \sin 47°20' + \log \sin 39°30' - (\log 2 + \log \sin 93°10')$$

Since

$$\sin 93°10' = \sin 86°50'$$

we have

$$\log K = (2 \log 92.5 + \log \sin 47°20' + \log \sin 39°30') - (\log 2 + \log \sin 86°50')$$

Thus

$$\log K = 3.3027$$

and

$K \doteq 2008$ □

In summary, we now have formulas which enable us to determine the area of a triangle as follows:

Given:	*Use:*
SSS	$K = \sqrt{s(s-a)(s-b)(s-c)}$
SAS	$K = \frac{1}{2}ab \sin C$
ASA	$K = \dfrac{a^2 \sin B \sin C}{2 \sin A}$

EXERCISES

In Exercises 1–6 compute the areas of the triangles.

1. $a = 15.2, b = 17.3, c = 16.1$
2. $a = 19.2, b = 24.7, C = 42°20'$
3. $a = 73.6, B = 73°30', C = 51°40'$
4. $a = 112.9, b = 98.76, c = 137.6$
5. $a = 423.6, b = 710.2, C = 39°43'$
6. $a = 3217, B = 62°17', C = 34°47'$
7. Solve the law of sines equation

$$\frac{a}{\sin A} = \frac{b}{\sin B}$$

for b and substitute this expression for b in the formula $K = \frac{1}{2}ab \sin C$. The result should be the third area formula.

8. Prove: In triangle ABC,

$$\sin A = \frac{2}{bc}\sqrt{s(s-a)(s-b)(s-c)}$$

Notice that this formula provides another method of finding angles given SSS.

9. A triangular lot has sides 145 ft, 105 ft, and 190 ft. Determine its area.
10. A lot in the form of a parallelogram has sides 192 ft, 247 ft, 192 ft, and 247 ft. Determine the area of the lot if one of the angles of the parallelogram is 112°.
11. A quadrilateral lot $ABCD$ has sides $a = 113$ ft, $b = 147$ ft, $c = 154$ ft, and $d = 131$ ft. Determine the area of the lot if the angle between a and b is 76°.
12. What should the angles of a parallelogram be in order for the area to be a maximum? Justify your answer.

*13. A quadrilateral has sides 52 ft, 70 ft, 63 ft, and 73 ft. What should the angles be in order for the area to be a maximum?

14. Determine the area of a regular pentagon with side 15 in.

*15. Generalize Exercise 14 to obtain a formula for the area of a regular pentagon.

5. Trigonometric Equations

Frequently in applications and in higher mathematics courses, we are confronted with conditional equations involving trigonometric functions.

Example 1

$2 \sin x = 1$ □

Example 2

$\sin^2 x = \frac{1}{2}$ □

Example 3

$\sin x \tan x + \sin x + \tan x + 1 = 0$ □

The problem in examples such as these is, as usual, to determine the values of x which make the equation true. Since the trigonometric functions are periodic, we can expect more than one correct answer. The most general process is to reduce the given equation to simpler equations such as $\sin x = .8$, $\tan x = 4$, and so on, and then use our knowledge of values of functions or tables of values in order to determine x.

To illustrate, let us deal first with the examples above.

Example 1

Solve

$$2 \sin x = 1$$

Dividing both sides by 2 yields

$$\sin x = \frac{1}{2}$$

We know that

$$\sin \frac{\pi}{6} = \frac{1}{2}$$

and that

$$\sin \frac{5\pi}{6} = \frac{1}{2}$$

Thus

$$\frac{\pi}{6} \quad \text{and} \quad \frac{5\pi}{6}$$

are solutions. Since $\sin x$ is periodic with period 2π, the solutions are all numbers of the forms

$$\frac{\pi}{6} + 2n\pi \qquad \frac{5\pi}{6} + 2n\pi$$

for any integer n. □

Example 2

Solve

$$\sin^2 x = \frac{1}{2}$$

First of all,

$$\sin^2 x = \frac{1}{2}$$

if and only if

$$\sin x = \pm\frac{1}{\sqrt{2}}$$

that is, the equation

$$\sin^2 x = \frac{1}{2}$$

is equivalent to the system

$$\sin x = \frac{1}{\sqrt{2}} \quad \text{or} \quad \sin x = -\frac{1}{\sqrt{2}}$$

We know that

$$\sin\frac{\pi}{4} = \frac{1}{\sqrt{2}} \quad \text{and} \quad \sin\frac{3\pi}{4} = \frac{1}{\sqrt{2}}$$

Also

$$\sin -\frac{\pi}{4} = -\frac{1}{\sqrt{2}} \quad \text{and} \quad \sin -\frac{3\pi}{4} = -\frac{1}{\sqrt{2}}$$

Thus

$$\frac{\pi}{4}, \qquad -\frac{\pi}{4}, \qquad \frac{3\pi}{4}, \quad \text{and} \quad -\frac{3\pi}{4}$$

all are solutions.

Since $\sin x$ is periodic with period 2π,

$$\frac{\pi}{4} + 2n\pi, \qquad -\frac{\pi}{4} + 2n\pi, \qquad \frac{3\pi}{4} + 2n\pi, \qquad -\frac{3\pi}{4} + 2n\pi$$

are also solutions for any integer n. □

Example 3

Solve

$$\sin x \tan x + \sin x + \tan x + 1 = 0$$

If we factor the left side we have

$$(\sin x + 1)(\tan x + 1) = 0$$

which implies that

$$\sin x + 1 = 0 \quad \text{or} \quad \tan x + 1 = 0$$

Hence

$$\sin x = -1 \quad \text{or} \quad \tan x = -1$$

Now

$$\sin x = -1 \quad \text{if} \quad x = -\tfrac{1}{2}\pi$$

In fact,

$$\sin x = -1 \quad \text{if and only if} \quad x = (2n - \tfrac{1}{2})\pi$$

for some integer n. Also

$$\tan x = -1 \quad \text{if} \quad x = \frac{\pi}{4} \quad \text{or} \quad x = -\frac{\pi}{4}$$

Since the tangent function has period π,

$$\tan x = -1 \quad \text{if and only if} \quad x = \pm\frac{\pi}{4} + n\pi$$

for some integer n. Thus the solutions of the equation are

$$(2n - \tfrac{1}{2})\pi, \qquad \frac{\pi}{4} + n\pi, \qquad -\frac{\pi}{4} + n\pi$$

for any integer n. □

We can use any of our usual equation-solving techniques to obtain these simpler equations.

Example 4

Solve

$$\tan^2 x - 3 \tan x - 5 = 0$$

If we view $\tan x$ as an unknown, this is just a quadratic equation in the variable $\tan x$.

By the quadratic formula,

$$\tan x = \frac{3 \pm \sqrt{9 + 20}}{2}$$
$$= \frac{3 \pm \sqrt{29}}{2}$$

If we approximate $\sqrt{29}$ by 5.386, then

$$\tan x \doteq \frac{3 \pm 5.386}{2}$$
$$\doteq 4.193 \quad \text{or} \quad -1.193$$

From the table we see that

$$x = 76°35' \quad \text{or} \quad x = -50°2'$$

Since the tangent function is periodic with period 180°, the solution set is all angles

$$76°35' + n180° \qquad -50°2' + n180°$$

for integers n. □

In the first three examples we gave the answers in radians and in the last case we gave them in degrees. The form of the answer depends on the use to which it will be put. In this text, the tables are in both degrees and radians and you should be able to use both.

Sometimes we need to apply identities in the solution process.

Example 5

Solve

$$\cos^2 x + 2 \sin x = 1$$

We can reduce the given equation to an equation involving only $\sin x$ by using an identity. Since $\cos^2 x = 1 - \sin^2 x$, the original equation is equivalent to the equation

$$1 - \sin^2 x + 2 \sin x = 1$$

Thus

$$-\sin^2 x + 2\sin x = 0$$
$$\sin^2 x - 2\sin x = 0$$
$$\sin x(\sin x - 2) = 0$$

Hence

$$\sin x = 0 \quad \text{or} \quad \sin x - 2 = 0$$

that is,

$$\sin x = 0 \quad \text{or} \quad \sin x = 2$$

Since $-1 \leq \sin x \leq 1$, $\sin x = 2$ has no solutions. Thus the only solutions are solutions of

$$\sin x = 0$$

and these are all numbers of the form

$$n\pi \qquad \text{for an integer } n \ \square$$

Example 6

Solve

$$\sin 2x - \cos x = 0$$

Using the identity

$$\sin 2x = 2\sin x \cos x$$

we have

$$2\sin x \cos x - \cos x = 0$$
$$(2\sin x - 1)\cos x = 0$$

Hence

$$2\sin x - 1 = 0 \quad \text{or} \quad \cos x = 0$$
$$\sin x = \tfrac{1}{2} \quad \text{or} \quad \cos x = 0$$

The solutions of $\sin x = \frac{1}{2}$ are numbers of the form

$$\frac{\pi}{6} + 2n\pi \quad \text{and} \quad \frac{5\pi}{6} + 2n\pi \qquad \text{for an integer } n$$

The solutions of $\cos x = 0$ are numbers of the form

$$\frac{\pi}{2} + n\pi \qquad \text{for an integer } n \ \square$$

Not all trigonometric equations contain only functions such as $\sin x$, $\cos x$, and so on. When multiples of x are involved we proceed as before to determine what the multiple of x must be and then find x.

Example 7

Solve

$$\sin^2 3x + \sin 3x = 0$$

Factoring yields

$$\sin 3x(\sin 3x + 1) = 0$$

Hence

$$\sin 3x = 0 \quad \text{or} \quad \sin 3x = -1$$

Hence

$$3x = n\pi$$

or

$$3x = -\frac{\pi}{2} + 2n\pi \qquad \text{for some integer } n$$

Thus

$$x = \frac{n\pi}{3}$$

or

$$x = -\frac{\pi}{6} + \frac{2n\pi}{3} \qquad \text{for some integer } n$$

EXERCISES

Solve the following equations; give the solutions in either radians or degrees.

1. $\tan^2 x = 3$
2. $\cos^2 x = \frac{1}{2}$
3. $\sin^2 x = \frac{3}{4}$
4. $8 \sin x + 1 = 5 \sin x + 2$
5. $4 \tan 5x + 3 = 2 \tan 5x + 8$
6. $\cos 2x + \cos x = 0$
7. $\sin 2x + \sin x = 0$
8. $\sin^2 x + 2 \sin x = 1$
9. $2 \sin^2 x + \sin x = 1$
10. $\sin x + \cos x = 0$
11. $\cos \frac{x}{2} - \sin x = 1$
12. $4 \sin^2 x \tan x = \tan x$
13. $\sec^2 5x = 4$
14. $\cos^2 x + \cos x = 6$
15. $2 \tan 3x = \sec^2 3x$
16. $\sin x = \sec x$

7

The Complex Numbers

The complex number system is an enlargement of the real number system and has many uses in both pure and applied mathematics. In this chapter we develop the fundamental properties of complex numbers and some of the more important computational techniques.

1. The System of Complex Numbers

Historically, number systems evolved from necessity as did most of science and technology. The counting numbers or natural numbers arose first and positive rational numbers were next. The concept of numbers representing distances forced the development of irrational numbers such as $\sqrt{2}$ and π and negative numbers were needed to consider directed distances. In order to coordinatize the line we need the entire real number system R. This system, however, is not sufficient for solving certain algebraic and technical problems and, therefore, the complex number system was developed.

Notice that the equation

$$4x - 12 = 0$$

has a solution in N whereas the equation

$$4x + 12 = 0$$

has no solution in N but does have a solution in Z. Similarly, the equation

$$3x - 2 = 0$$

has no solution in Z but does have a solution in Q and the equation

$$x^2 - 2 = 0$$

has no solution in Q but does have a solution in R.

We can view each of these systems as an extension or enlargement of the previously mentioned one, the purpose being, to some extent, to construct a system in which certain equations are solvable.

The equation

$$x^2 + 1 = 0$$

has no solution in R and in order to provide solutions to this equation and many others, we need the system of complex numbers.

AXIOMS FOR THE COMPLEX NUMBERS

The complex number system, denoted by C, is a set with two operations, called addition and multiplication, satisfying the following conditions.

1. $R \subseteq C$; that is, R is a subsystem of C.
2. The sum and product of two complex numbers is a complex number.
3. Addition and multiplication are commutative.
4. Addition and multiplication are associative.
5. The distributive law holds; that is,

   $$x(y + z) = xy + xz$$

   for any complex numbers x, y, and z.
6. There is a complex number, denoted by i, such that $i^2 = -1$.*
7. Every complex number can be expressed in the form $a + bi$ with a and b real numbers.

Numbers of the form bi with $b \in R$ are called *imaginary* numbers. (This is an unfortunate tradition. The number $4i$ is no more imaginary than is $\frac{2}{3}$ or $\sqrt{2}$.) In the representation of a complex number by $a + bi$, a is called the real part and bi the imaginary part. Axiom 7 assures us that every complex number can be expressed as the sum of a real number and an imaginary number. In fact, this representation is unique as we establish in the following assertion.

* In physics, i is used to denote electric current so scientists and engineers frequently use j to denote $\sqrt{-1}$.

PROPOSITION 1

If $a + bi$ and $c + di$ are complex numbers such that $a + bi = c + di$, then $a = c$ and $b = d$.

Proof:

If

$$a + bi = c + di$$

then

$$a - c = di - bi$$
$$a - c = (d - b)i$$

Thus

$$(a - c)^2 = [(d - b)i]^2$$
$$(a - c)^2 = (d - b)^2 i^2$$

and

$$(a - c)^2 = -(d - b)^2$$

Since $(a - c)$ is a real number, $(a - c)^2 \geq 0$. Since $(d - b)$ is a real number, $-(d - b)^2 \leq 0$. Since $(a - c)^2 = -(d - b)^2$, both are zero.

Thus $(a - c)^2 = 0$, $a - c = 0$, and $a = c$. Also $-(d - b)^2 = 0$, $d - b = 0$, and $b = d$.

Certain other expected properties of C are established by the following sequence of assertions. The proofs are rather straightforward and we leave them as exercises.

PROPOSITION 2

If $z \in C$, then $z + 0 = z$ and $z \cdot 0 = 0$. Also $z \cdot 1 = z$.
[*Hint*: Let z be expressed in the form $a + bi$.]

PROPOSITION 3

If $z \in C$, then there exists a unique complex number $-z$, called the negative of z, such that

$$z + (-z) = 0$$

We define subtraction as we did for R:

$$x - y = x + (-y) \qquad \text{for any complex numbers } x \text{ and } y$$

PROPOSITION 4

If y and z are nonzero complex numbers, then $yz \neq 0$.

Since C satisfies the same basic properties of addition and multiplication that R does, all techniques using these principles extend to C. Thus we can use these techniques to compute in C just as we did in R.

Example 1

$$\begin{aligned}(2 + 5i) + (7 + 3i) &= (2 + 7) + (5 + 3)i \\ &= 9 + 8i \ \square\end{aligned}$$

Example 2

$$\begin{aligned}(2 + 5i)(7 + 3i) &= 2 \cdot 7 + 2 \cdot 3i + 5i \cdot 7 + (5i)(3i) \\ &= 14 + 6i + 35i + 15i^2 \\ &= 14 + 41i - 15 \\ &= -1 + 41i \ \square\end{aligned}$$

We add and multiply complex numbers in the form $a + bi$ just as we do polynomials. Then we replace i^2 by -1 and simplify. With some practice, you will do this mentally and save some writing.

Example 3

$$\begin{aligned}(5 + 6i)(2 + 4i) &= 10 - 24 + (20 + 12)i \\ &= -14 + 32i \ \square\end{aligned}$$

Example 4

$$(2 - i)(11 + 2i) = 24 - 7i \ \square$$

The question of division of complex numbers arises quite naturally. In Q and R, we have all the division we could hope for in that a/b exists if $b \neq 0$.

An analogous assertion holds for C but since we did not assume it in the axioms, we need to justify such a claim. As before, we denote x divided by y by $x \div y$ or x/y and define x/y to be the unique complex number whose product with y is x if such a number exists. Division by 0 is impossible as it is in R for exactly the same reasons.

PROPOSITION 5

If x and y are complex numbers and $y \neq 0$, then x/y exists in C.

Proof:

All we need do is exhibit a number whose product with y is x.

Let $x = a + bi$ and $y = c + di$ with a, b, c, and d real numbers.

Since $c \neq 0$ or $d \neq 0$, $c^2 + d^2 \neq 0$. Thus

$$\frac{ac + bd}{c^2 + d^2} - \frac{ad - bc}{c^2 + d^2} i$$

is a complex number. Straightforward calculations show that

$$\left(\frac{ac + bd}{c^2 + d^2} - \frac{ad - bc}{c^2 + d^2} i\right)(c + di) = a + bi$$

We leave the computation to the reader. Thus we have a complex number whose product with $c + di$ is $a + bi$, that is, whose product with y is x. Proposition 4 assures uniqueness since if

$$zy = x \quad \text{and} \quad wy = z$$

then

$$zy - wy = 0$$

$$(z - w)y = 0$$

Since $y \neq 0$, $z - w = 0$ and $z = w$. Hence

$$\frac{ac + bd}{c^2 + d^2} - \frac{ad - bc}{c^2 + d^2} i$$

is the unique complex number whose product with y is x; that is, it is x/y.

Our rules for division established in Chapter 1 were dependent only on Axioms 0–7 and 11 for R and since we have established analogous assertions for C, all the properties of division not dependent upon the ordering will extend to C. Thus all our rules for handling fractions hold for complex numbers.

The proof of Proposition 5 may seem a bit artificial or like a "put up job." Where did we get the number

$$\frac{ac + bd}{c^2 + d^2} - \frac{ad - bc}{c^2 + d^2} i?$$

It comes from looking at what the fraction must be and rewriting it with a real denominator by multiplying the numerator and denominator by a suitable complex number. We multiply the numerator and denominator of

$$\frac{a + bi}{c + di} \quad \text{by} \quad c - di$$

and obtain the desired result:

$$\frac{a+bi}{c+di} = \frac{a+bi}{c+di} \cdot \frac{c-di}{c-di} = \frac{ac - bdi^2 + bci - adi}{c^2 - d^2i^2}$$
$$= \frac{ac + bd + (bc - ad)i}{c^2 + d^2}$$
$$= \frac{ac+bd}{c^2+d^2} + \frac{bc-ad}{c^2+d^2}\, i$$
$$= \frac{ac+bd}{c^2+d^2} - \frac{ad-bc}{c^2+d^2}\, i$$

The *conjugate* of the complex number $a + bi$ is the complex number $a - bi$. If z denotes a complex number, then $\bar{z}$ denotes the conjugate of z. ("$\bar{z}$" is usually read "z bar".)

Example 5

The conjugate of $3 + 4i$ is $3 - 4i$.
The conjugate of $-5 - 6i$ is $-5 + 6i$.
The conjugate of $\sqrt{3} + \sqrt{5}i$ is $\sqrt{3} - \sqrt{5}i$.
The conjugate of $7i$ is $-7i$.
The conjugate of 19 is 19.

Notice that for any complex number z, $z\bar{z}$ is a nonnegative real number since

$$(a + bi)(a - bi) = a^2 - b^2i^2 = a^2 + b^2$$

Furthermore, if $z \neq 0$, then $z\bar{z} > 0$. □

The computational technique we use for division of complex numbers is to multiply the numerator and denominator of the fraction by the conjugate of the denominator.

Example 6

$$\frac{2+3i}{7+2i} = \frac{(2+3i)(7-2i)}{(7+2i)(7-2i)}$$
$$= \frac{14 + 6 + 21i - 4i}{49 + 4}$$
$$= \frac{20}{53} + \frac{17}{53}\, i \quad \square$$

The following proposition establishes certain fundamental relationships between conjugation or "taking the conjugate of a number" and the arithmetic operations of C.

PROPOSITION 6

If z and w are complex numbers, then

(a) $\overline{z + w} = \bar{z} + \bar{w}$
(b) $\overline{zw} = \bar{z}\,\bar{w}$
(c) $\overline{z - w} = \bar{z} - \bar{w}$
(d) $\overline{(z/w)} = \bar{z}/\bar{w}$, if $w \neq 0$

Proof:

We give a proof of part (a) and leave the rest as an exercise.
Let $z = a + bi$ and $w = c + di$; then

$$\overline{z + w} = \overline{(a + bi) + (c + di)} = \overline{(a + c) + (b + d)i} = (a + c) - (b + d)i$$

and

$$\bar{z} + \bar{w} = \overline{a + bi} + \overline{c + di} = (a - bi) + (c - di) = (a + c) - (b + d)i$$

PROPOSITION 7

If z is a complex number such that $z = \bar{z}$, then $z \in R$.

Proof: Exercise.

Thus far we have not mentioned any ordering or comparability of complex numbers. If we examine the list of assertions that can be proved about R, we see that $0 < a^2$ for any nonzero real number a. If we put an ordering on C satisfying properties analogous to Axioms 8, 9, and 10 for R, we would have

$$0 < 1^2 \quad \text{and} \quad 0 < i^2$$

that is,

$$0 < 1 \quad \text{and} \quad 0 < -1$$

This would imply

$$0 < 1 \quad \text{and} \quad 1 < 0$$

which contradicts Axiom 8.

Thus C cannot be ordered in a way that is compatible with both addition and multiplication.

As we noted, the complex number system does provide a solution to the equation

$$x^2 = -1$$

In fact, it provides two of them, i and $-i$ since $(-i)^2 = i^2 = -1$. In general, the equation

$$x^2 = -c$$

has two complex solutions if c is a positive real number. This equation is equivalent to

$$x^2 + c = 0$$

which can be rewritten

$$(x + i\sqrt{c})(x - i\sqrt{c}) = 0$$

and

$$x = -i\sqrt{c} \quad \text{or} \quad x = i\sqrt{c}$$

Thus in general, a negative real number has two possible square roots; that is, if $c > 0$,

$$(i\sqrt{c})^2 = -c \quad \text{and} \quad (-i\sqrt{c})^2 = -c$$

The number $i\sqrt{c}$ is called the *principal square root* of $-c$ and when we write $\sqrt{-c}$ we mean $i\sqrt{c}$.

Example 7

$$\sqrt{-5} = i\sqrt{5}$$
$$\sqrt{-4} = 2i$$
$$\sqrt{-32} = 4i\sqrt{2} \ \square$$

Since all real numbers have square roots in C, all quadratic equations

$$Ax^2 + Bx + C = 0 \qquad \text{with } A, B, \text{ and } C \text{ in } R \text{ and } A \neq 0$$

have solutions in C, namely

$$\frac{-B + \sqrt{B^2 - 4AC}}{2A} \quad \text{and} \quad \frac{-B - \sqrt{B^2 - 4AC}}{2A}$$

using the formula previously established. If $B^2 - 4AC \geq 0$, then the roots are real as before. If $B^2 - 4AC < 0$, then they are complex. The formula is also applicable if A, B, and C are complex numbers.

Example 8

The solutions of

$$x^2 + x + 4 = 0$$

are given by

$$x = \frac{-1 \pm \sqrt{-15}}{2} \quad \text{and} \quad x = \frac{-1 \pm i\sqrt{15}}{2} \ \square$$

From the uniqueness of representation established by Proposition 1, it follows that an equation involving complex numbers is equivalent to a system of two equations involving real numbers.

Example 9

Solve the equation

$$z^2 = 3 + 4i$$

Let $z = x + yi$ with x and y real variables. Then

$$\begin{aligned}(x + yi)^2 &= 3 + 4i\\ (x^2 - y^2) + 2xyi &= 3 + 4i\end{aligned}$$

Thus

$$x^2 - y^2 = 3 \quad \text{and} \quad 2xy = 4$$

Solving the second equation yields $y = 2/x$. Then substituting in the first yields

$$\begin{aligned}x^2 - \left(\frac{2}{x^2}\right) &= 3\\ x^2 - \left(\frac{4}{x^2}\right) &= 3\\ x^4 - 4 &= 3x^2\\ x^4 - 3x^2 - 4 &= 0\\ (x^2 - 4)(x^2 + 1) &= 0\\ x = 2 \quad \text{or} \quad x &= -2\end{aligned}$$

(We want only real solutions for x.) If $x = 2$, then $y = 1$. If $x = -2$, $y = -1$. Thus $2 + i$ and $-2 - i$ are the possible solutions.

Checking shows that they both are, in fact, solutions.

$$\begin{aligned}(2 + i)^2 &= 3 + 4i\\ (-2 - i)^2 &= 3 + 4i \ \square\end{aligned}$$

EXERCISES

1. Compute the following.
   (a) $(2 + 5i) + (8 + 11i)$ (b) $(2 + 5i)(8 + 11i)$
   (c) $(\frac{1}{3} + \frac{4}{3}i)(\frac{1}{2} - \frac{6}{5}i)$ (d) $(\frac{1}{3} + \frac{4}{3}i) - (\frac{1}{2} - \frac{6}{5}i)$
   (e) $(\sqrt{2} + \sqrt{3}i)(\sqrt{6} - \sqrt{2}i)$ (f) $(\sqrt{10} + \sqrt{6}i)(-\sqrt{5} - \sqrt{3}i)$
2. Express each of the following in the form $p + qi$ where p and q are real numbers.
   (a) $\dfrac{4 + 5i}{7 + 2i}$ (b) $\dfrac{11 - 6i}{2 - 5i}$ (c) $\dfrac{-.4 + .9i}{.05 + .3i}$ (d) $\dfrac{\sqrt{2} + \sqrt{3}i}{\sqrt{5} + \sqrt{7}i}$

(e) $\dfrac{\sqrt[3]{7} + \sqrt{5}i}{\sqrt[3]{6} + \sqrt{11}i}$ (f) $\dfrac{\pi + ei}{\sqrt{17} + \sqrt{19}i}$ (g) $\dfrac{2 + 3i}{5i}$ (h) $\dfrac{a + bi}{a - bi}$

3. Simplify the following.
 (a) $(2i)^3$ (b) $(4.1i)^2$ (c) $(\sqrt{7}i)^3$ (d) $\sqrt{-16}$
 (e) $\sqrt{-12}$ (f) $\sqrt{-192}$ (g) $\sqrt{-\sqrt{2}}$ (h) $\sqrt{-\pi^3}$
4. (a) Compute i^3, i^4, i^5, i^6, i^7, i^8, and i^9.
 (b) From this pattern, can you develop a method for computing i^n for any positive integer n?
5. (a) Compute i^{-1}, i^{-2}, i^{-3}, i^{-4}, i^{-5}, i^{-6}, i^{-7}, i^{-8}, and i^{-9}.
 (b) From this pattern, can you develop a method for computing i^n for any negative integer n?
6. Prove Proposition 2.
7. Prove Proposition 3.
8. Prove Proposition 4. [*Hint*: If $xy = 0$, then $xy\,\bar{x}\,\bar{y} = 0$.]
9. Complete the proof of Proposition 5 by verifying the computation directly.
10. Complete the proof of Proposition 6.
11. Solve the following equations.
 (a) $z^2 + 4z + 8 = 0$ (b) $z^2 + z + 1 = 0$
 (c) $z^2 + 3z + 9 = 0$ (d) $z^2 - z + 1 = 0$
 (e) $z^3 - 1 = 0$ (f) $z^3 - b^3 = 0$ for b a real number
 (g) $z^4 - 1 = 0$ (h) $z^4 - b^4 = 0$ for b a real number
12. Solve the following equations.
 (a) $z^2 = i$ (b) $z^2 = 5i$
 (c) $z^2 = bi$ for b any real number (d) $z^2 = a + bi$ for a and b real numbers
13. Solve the following equations.
 (a) $z^2 + 4iz - (2 + 3i) = 0$
 (b) $iz^2 - (2 + 5i)z - (6 + 4i) = 0$
 (c) $(5 - 4i)z^2 + (6 + 5i)z + (7 - 3i) = 0$
 (d) $(\frac{1}{3} - \frac{2}{3}i)z^2 + \sqrt{2}iz + (\frac{4}{5} - 3i) = 0$
14. If A, B, and C are real numbers with $B^2 - 4AC < 0$, what is the relationship between the two roots of the equation

 $Az^2 + Bz + C = 0$?

2. Graphic Representation of Complex Numbers

One of the nicer properties of the real numbers is that we can get an intuitive grasp of the system by thinking of real numbers as points on a line. In order to grasp C in a similar manner, we can think of complex numbers as points in the plane. After all, each complex number is represented uniquely in the form $a + bi$ and it is rather natural to associate the complex number $a + bi$ with the point (a, b) in the plane and vice versa. Thus to view complex numbers as points in the plane we need only write $a + bi$ instead of (a, b). (See Figure 7–1.)

Figure 7-1

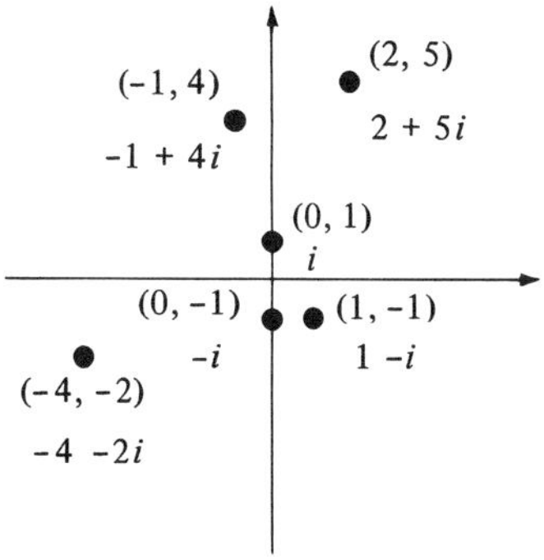

When using a coordinate system to represent complex numbers, we can drop the X and Y designations and call the horizontal axis the *real axis* R and the vertical axis the *imaginary axis* I. (See Figure 7–2.)

Figure 7-2

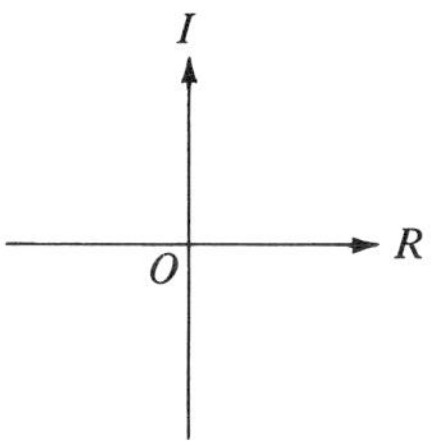

Adding complex numbers graphically turns out to follow the "parallelogram law" of vector addition; that is, if we draw the line segments from 0 to $a + bi$ and $c + di$ and complete the parallelogram which has these two segments as sides, the sum of the two numbers will be the vertex opposite 0.

Example 1

The sum of $3 + 5i$ and $6 + 2i$ is illustrated in Figure 7–3. □

Although we cannot order C as we can R, we can assign a "size" to each complex number by defining the *absolute value* of a complex number to be its distance from the origin 0. For the complex number $a + bi$, the distance formula requires this to be $\sqrt{a^2 + b^2}$. We denote the absolute value of the

Figure 7-3

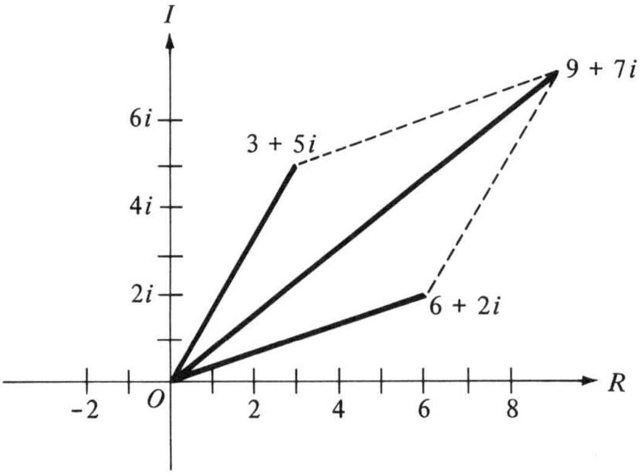

complex number z by $|z|$. Thus

$$|a + bi| = \sqrt{a^2 + b^2}$$

See Figure 7–4.

Figure 7-4

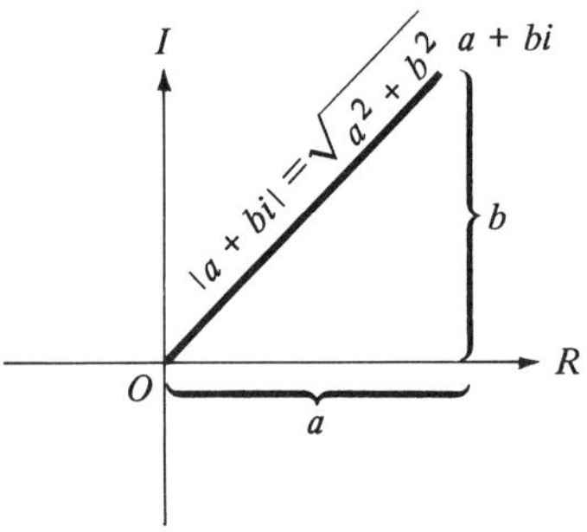

Notice that if z is a complex number, then

$$|z| = \sqrt{z\bar{z}}$$

If $z \in R$, then

$$|z| = \sqrt{z^2}$$

as before.

Figure 7-5

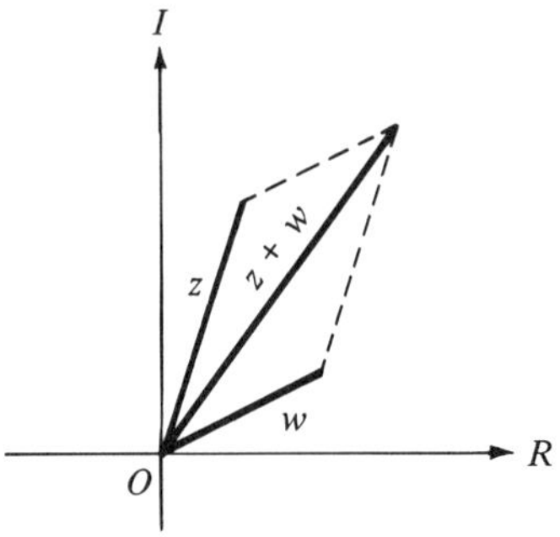

Since addition of complex numbers follows the parallelogram law, the fact that the sum of the lengths of two sides of a triangle is not less than the length of the third side yields the following assertion:

PROPOSITION 1 (the triangle inequality)

If z and w are complex numbers (see Figure 7–5), then

$$|z + w| \leq |z| + |w|$$

(This can be proved formally using only algebra and we leave the proof as an exercise for the interested student.)

The following assertion is a generalization of a fact that we know holds for real numbers. The proof is quite easy using conjugates.

PROPOSITION 2

If z and w are complex numbers, then $|zw| = |z|\,|w|$.

Proof:

$$\begin{aligned}|zw| &= \sqrt{(zw)(\overline{zw})}\\ &= \sqrt{zw\bar{z}\bar{w}}\\ &= \sqrt{z\bar{z}w\bar{w}}\\ &= \sqrt{z\bar{z}}\,\sqrt{w\bar{w}}\\ &= |z|\,|w|\end{aligned}$$

A point in the plane, say (a, b), is completely determined if we know

(i) the distance r from the origin to the point, and
(ii) the angle θ that the point makes with the positive X-axis.

See Figure 7–6.

Figure 7-6

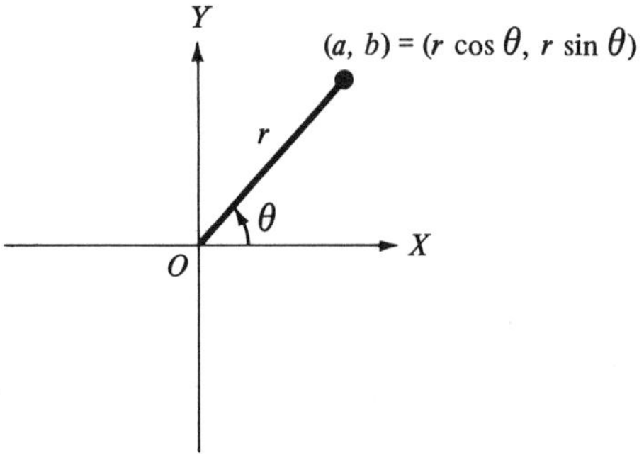

The numbers r and θ are called the *polar coordinates* of the point. Notice that the usual rectangular coordinates of a point are just $(r \cos \theta, r \sin \theta)$.

A complex number $a + bi$, then, is completely determined by:

(i) its absolute value, and
(ii) the angle that the segment from 0 to $a + bi$ makes with the positive real axis.

See Figure 7–7.

Figure 7-7

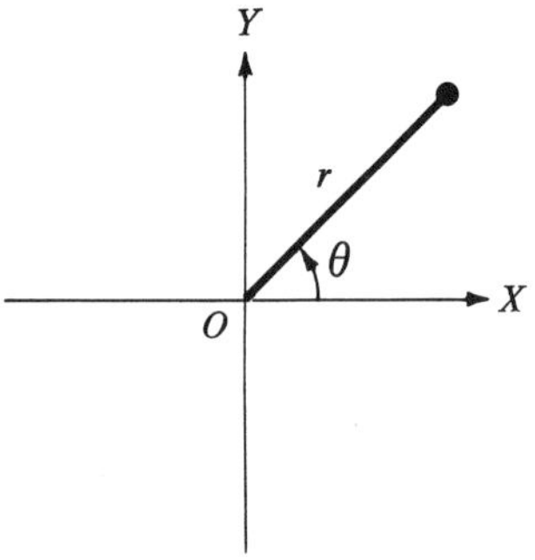

Thus every complex number can be expressed in the form

$$r(\cos \theta + i \sin \theta)$$

where θ is the angle between the segment from 0 to the number and the positive real axis and r is the length of this segment. This form is called the *trigonometric form* of a complex number. When using this form, the absolute value is sometimes called the *modulus* and the angle θ is called the *argument* or *amplitude*. Notice that the trigonometric form is not a unique representation because the argument θ is not unique.

Example 2

$$\begin{aligned} 5(\cos 30^\circ + i \sin 30^\circ) &= 5(\cos 390^\circ + i \sin 390^\circ) \\ &= 5(\cos -330^\circ + i \sin -330^\circ) \ \square \end{aligned}$$

In general,

$$a + bi = \sqrt{a^2 + b^2}\left(\frac{a}{\sqrt{a^2 + b^2}} + \frac{ib}{\sqrt{a^2 + b^2}}\right)$$

Thus

$$\cos \theta = \frac{a}{\sqrt{a^2 + b^2}} \quad \text{and} \quad \sin \theta = \frac{b}{\sqrt{a^2 + b^2}}$$

Example 3

$$|3 + 3i| = \sqrt{18} = 3\sqrt{2}$$

Thus

$$\begin{aligned} 3 + 3i &= 3\sqrt{2}\left(\frac{1}{\sqrt{2}} + \frac{i1}{\sqrt{2}}\right) \\ &= 3\sqrt{2}\,(\cos 45^\circ + i \sin 45^\circ) \ \square \end{aligned}$$

Example 4

$$|\sqrt{3} - i| = \sqrt{3 + 1} = 2$$

Thus

$$\sqrt{3} - i = 2\left(\frac{\sqrt{3}}{2} - \frac{1}{2}i\right)$$

Hence

$$\cos \theta = \frac{\sqrt{3}}{2} \text{ and } \sin \theta = \frac{-1}{2}$$

The angle -30° satisfies these equations and

$$\sqrt{3} - i = 2[\cos(-30^\circ) + i(\sin - 30^\circ)] \ \square$$

In general, the reference angle for θ will be $|\text{Arcsin}\,(b/\sqrt{a^2 + b^2})|$; we decide the quadrant according to the signs of $\cos\theta$ and $\sin\theta$ or just by examining the number.

Example 5

$$-5 - 5i = 5\sqrt{2}\left(\frac{-1}{\sqrt{2}} - \frac{i}{\sqrt{2}}\right)$$

$$\text{Arcsin}\,\frac{1}{\sqrt{2}} = 45^\circ$$

The reference angle is 45°. Since the point is in quadrant III,

$$\theta = 225^\circ$$

satisfies the conditions as do

$$\theta = -135^\circ$$

and

$$\theta = 585^\circ \quad \square$$

The trigonometric form turns out to be very useful in the multiplication of complex numbers because of the following fact.

PROPOSITION 3

If z and w are complex numbers represented in trigonometric form by

$$z = r(\cos\alpha + i\sin\alpha) \quad \text{and} \quad w = s(\cos\beta + i\sin\beta)$$

then

$$zw = (rs)[\cos(\alpha + \beta) + i\sin(\alpha + \beta)]$$

Proof:

$$\begin{aligned} zw &= r(\cos\alpha + i\sin\alpha)s(\cos\beta + i\sin\beta) \\ &= (rs)[(\cos\alpha\cos\beta - \sin\alpha\sin\beta) + i(\sin\alpha\cos\beta + \cos\alpha\sin\beta)] \end{aligned}$$

From trigonometry we know

$$\cos\alpha\cos\beta - \sin\alpha\sin\beta = \cos(\alpha + \beta)$$
$$\sin\alpha\cos\beta + \cos\alpha\cos\beta = \sin(\alpha + \beta)$$

Thus

$$zw = (rs)[\cos(\alpha + \beta) + i\sin(\alpha + \beta)]$$

Example 6

$$[2(\cos 18^\circ + i \sin 18^\circ)][3(\cos 43^\circ + i \sin 43^\circ)] = 6(\cos 61^\circ + i \sin 61^\circ)$$

See Figure 7–8. □

Figure 7–8

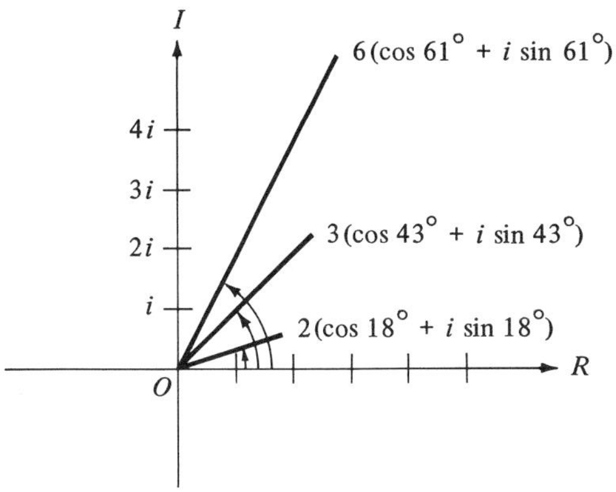

Example 7

$$[\tfrac{1}{2}(\cos 22^\circ + i \sin 22^\circ)][8(\cos 23^\circ + \sin 23^\circ)] = 4(\cos 45^\circ + i \sin 45^\circ)$$ □

Example 8

$$[5(\cos 20^\circ + i \sin 20^\circ)]^2 = 25(\cos 40^\circ + i \sin 40^\circ)$$ □

As we see in Example 8, the formula established in Proposition 3 leads immediately to the following corollary.

COROLLARY 4 (De Moivre's theorem)

If z is a complex number with trigonometric representation

$$z = r(\cos \theta + i \sin \theta)$$

and n is a positive integer, then

$$z^n = r^n(\cos n\theta + i \sin n\theta)$$

Example 9

$$\begin{aligned}[2(\cos 15^\circ + i \sin 15^\circ)]^8 &= 2^8(\cos 120^\circ + i \sin 120^\circ) \\ &= 256(\cos 120^\circ + i \sin 120^\circ) \quad \square\end{aligned}$$

Example 10

$$\begin{aligned}[\sqrt{3} + i]^5 &= \left[2\left(\frac{\sqrt{3}}{2} + i\frac{1}{2}\right)\right]^5 \\ &= [2(\cos 30^\circ + i \sin 30^\circ)]^5 \\ &= 2^5(\cos 150^\circ + i \sin 150^\circ) \\ &= 32\left(\frac{-\sqrt{3}}{2} + \frac{i}{2}\right) = -16\sqrt{3} + 16i \quad \square\end{aligned}$$

Corollary 4 also provides us with a method for determining all solutions to an equation of the form

$$x^n = a + bi$$

that is, for finding all nth roots of a complex number.

If

$$a + bi = r(\cos\theta + i \sin\theta)$$

then

$$\left[\sqrt[n]{r}\left(\cos\frac{\theta}{n} + i \sin\frac{\theta}{n}\right)\right]^n = r(\cos\theta + i \sin\theta) = a + bi$$

and, therefore, we have one nth root of $a + bi$.

Consider also the representation

$$a + bi = r[\cos(360^\circ + \theta) + i \sin(360^\circ + \theta)]$$

Then

$$\begin{aligned}&\left[\sqrt[n]{r}\left(\cos\frac{360^\circ + \theta}{n} + i \sin\frac{360^\circ + \theta}{n}\right)\right]^n \\ &\qquad = r[\cos(360^\circ + \theta) + i \sin(360^\circ + \theta)] \\ &\qquad = r(\cos\theta + i \sin\theta) \\ &\qquad = a + bi\end{aligned}$$

In fact, for any integer k, $0 \le k < n$, we have

$$\begin{aligned}&\left[\sqrt[n]{r}\left(\cos\frac{360^\circ k + \theta}{n} + i \sin\frac{360^\circ k + \theta}{n}\right)\right]^n \\ &\qquad = r[\cos(360^\circ k + \theta) + i \sin(360^\circ k + \theta)] \\ &\qquad = r(\cos\theta + i \sin\theta) \\ &\qquad = a + bi\end{aligned}$$

Furthermore, if $0 \le p < q < n$, then

$$\frac{360^\circ p + \theta}{n} \neq \frac{360^\circ q + \theta}{n}$$

and all of these complex numbers are different. Thus in general, n nth roots of

$$r(\cos \theta + i \sin \theta)$$

are given by

$$\sqrt[n]{r}\left[\cos\left(360^\circ \frac{k}{n} + \frac{\theta}{n}\right) + i \sin\left(360^\circ \frac{k}{n} + \frac{\theta}{n}\right)\right] \qquad \text{for } k = 0, 1, 2, \ldots, n-1$$

If

$$\begin{aligned} r(\cos \theta + i \sin \theta) &= [s(\cos \alpha + i \sin \alpha)]^n \\ &= s^n(\cos n\alpha + i \sin n\alpha) \end{aligned}$$

then $s = \sqrt[n]{r}$ and θ and $n\alpha$ must have the same terminal side. Thus $n\alpha - \theta$ must be a multiple of 360°.

Since

$$n\alpha - \theta = 360^\circ p$$

for some integer p,

$$n\alpha = 360^\circ p + \theta$$

and

$$\alpha = 360^\circ \frac{p}{n} + \frac{\theta}{n}$$

Furthermore, if $p \ge n$ or $p < 0$, then the terminal side of $360^\circ p/n$ is the same as the terminal side of $360^\circ k/n$ for some integer k, $0 \le k < n$. Thus all nth roots of $a + bi$ are given by the formula.

The above discussion establishes the following assertion.

PROPOSITION 5

For each positive integer n, the complex number $r(\cos \theta + i \sin \theta)$, with $r > 0$, has exactly n nth roots; that is, the equation

$$z^n = r(\cos \theta + i \sin \theta)$$

has exactly n solutions. These are given by

$$\sqrt[n]{r}\left[\cos\left(360^\circ \frac{k}{n} + \frac{\theta}{n}\right) + i \sin\left(360^\circ \frac{k}{n} + \frac{\theta}{n}\right)\right] \qquad \text{for } k = 0, 1, \ldots, n-1$$

Example 11

Since $360°/5 = 72°$, $40°/5 = 8°$, and $2^5 = 32$, the five 5th roots of $32(\cos 40° + i \sin 40°)$ are given by

$$2[\cos 8° + i \sin 8°]$$
$$2[\cos (72° + 8°) + i \sin (72° + 8°)] = 2(\cos 80° + i \sin 80°)$$
$$2[\cos (144° + 8°) + i \sin (144° + 8°)] = 2(\cos 152° + i \sin 152°)$$
$$2[\cos (216° + 8°) + i \sin (216° + 8°)] = 2(\cos 224° + i \sin 224°)$$
$$2[\cos (288° + 8°) + i \sin (288° + 8°)] = 2(\cos 296° + i \sin 296°)$$ □

Example 12

Three cube roots of 1 are given by

$$\cos 0° + i \sin 0° = 1$$

$$\cos \frac{360°}{3} + i \sin \frac{360°}{3} = \cos 120° + i \sin 120° = \frac{1}{2} + i\frac{\sqrt{3}}{2}$$

$$\cos 240° + i \sin 240° = \frac{1}{2} - i\frac{\sqrt{3}}{2}$$ □

Example 13

Find all solutions of the equation $z^4 = -1$.

First

$$-1 = 1(\cos 180° + i \sin 180°)$$

Then

$$\frac{360°}{4} = 90° \quad \text{and} \quad \frac{180°}{4} = 45°$$

Thus the four roots are

$$\cos 45° + i \sin 45° = \frac{1}{\sqrt{2}} + i\frac{1}{\sqrt{2}}$$

$$\cos (90° + 45°) + i \sin (90° + 45°) = \cos 135° + i \sin 135° = \frac{-1}{\sqrt{2}} + i\frac{1}{\sqrt{2}}$$

$$\cos (180° + 45°) + i \sin (180° + 45°) = \cos 225° + i \sin 225° = \frac{-1}{\sqrt{2}} - i\frac{1}{\sqrt{2}}$$

$$\cos(270° + 45°) + i \sin(270° + 45°) = \cos 315° + i \sin 315°$$
$$= \frac{1}{\sqrt{2}} - i\frac{1}{\sqrt{2}} \quad \square$$

If b is a positive real number, then the n nth roots of b are given by

$$\sqrt[n]{b}\left(\cos 360°\frac{k}{n} + i \sin 360°\frac{k}{n}\right) \qquad \text{for } k = 0, 1, 2, \ldots, n - 1$$

If b is a negative real number, then the n nth roots of b are given by

$$\sqrt[n]{|b|}\left[\cos\left(360°\frac{k}{n} + \frac{180°}{n}\right) + i \sin\left(360°\frac{k}{n} + \frac{180°}{n}\right)\right]$$
$$\text{for } k = 0, 1, 2, \ldots, n - 1$$

Graphically they are just the nth roots of b, rotated through an angle of $180°/n$.

In general, the nth roots of $a + bi$ are just the nth roots of $|a + bi|$ rotated through an angle of θ/n where $a + bi = r(\cos\theta + i \sin\theta)$.

Example 14

The four 4th roots of 16 are 2, $2i$, -2, and $-2i$. (See Figure 7–9.) □

Figure 7–9

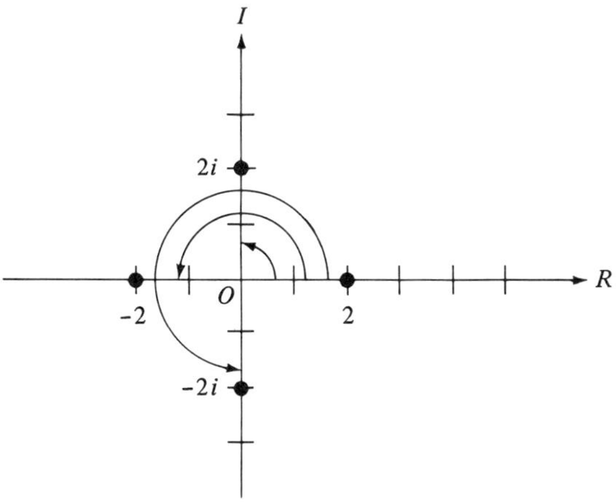

Example 15

The fifth roots of 13 all have absolute value $\sqrt[5]{13}$ and are equally spaced around the circle with center at 0 and radius $\sqrt[5]{13}$. One of them, of course, is $\sqrt[5]{13}$ and the others can be drawn by using arguments that are multiples of 72°. (See Figure 7–10.) □

Figure 7–10

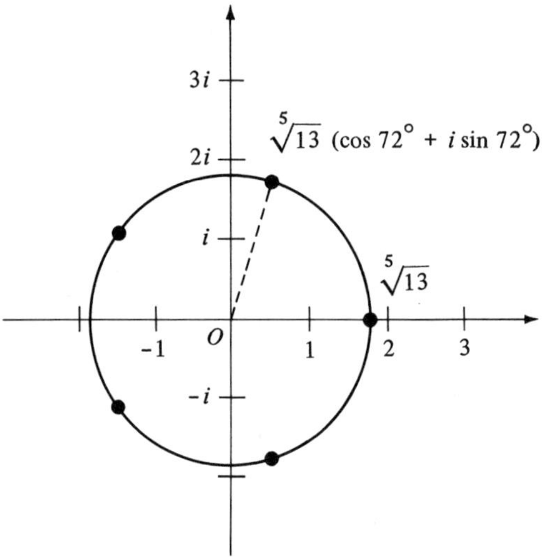

Example 16

The fifth roots of $5 + 12i$ all have absolute value $\sqrt[5]{|5 + 12i|} = \sqrt[5]{13}$ and, therefore, are on the same circle as the fifth roots of 13. They can be plotted by rotating the fifth roots of 13 through an angle of $\frac{1}{5}$ arcsin $\frac{12}{13}$. (See Figure 7–11.) □

Figure 7-11

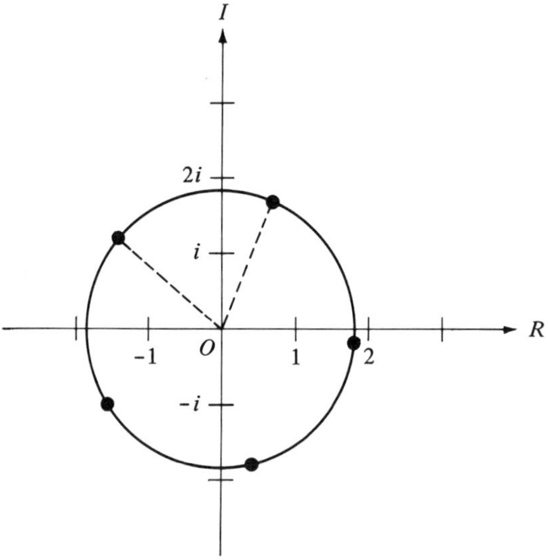

When we began this chapter, we said that C was needed because certain equations were not solvable in R. The natural question now is: Is C large enough? In the sense of classical algebra, the answer is yes and is given by the following very important assertion.

THE FUNDAMENTAL THEOREM OF ALGEBRA

If $p(z)$ is a polynomial with complex coefficients, then the equation $p(z) = 0$ has a solution in C.

The proof of this theorem is long and difficult. It was, in fact, the doctoral dissertation of one of the greatest mathematicians in history and is far beyond the scope of this text.

EXERCISES

1. Plot each of the following on a coordinate system and determine its absolute value.
   (a) $2 + 5i$ (b) $7 - 3i$ (c) $11 - 14i$
   (d) $\frac{2}{3} + \frac{3}{4}i$ (e) $\frac{7}{2} - \frac{6}{7}i$ (f) $\sqrt{2} + \sqrt{3}i$
   (g) $.4 - .7i$ (h) $-1.3 + 2.2i$ (i) $\sqrt{-5}$
   (j) $\sqrt{-a}$, for a positive
2. Write each of the complex numbers in Exercise 1 in trigonometric form. (You need not determine the argument, only the sine and cosine of the argument.)
3. Use your drawings in Exercise 1 and a protractor to estimate the arguments of each of the numbers.
4. Write each of the following complex numbers in the form $a + bi$.
   (a) $5(\cos 30^\circ + i \sin 30^\circ)$ (b) $6(\cos 45^\circ + i \sin 45^\circ)$
   (c) $11(\cos 120^\circ + i \sin 120^\circ)$ (d) $.4(\cos 210^\circ + i \sin 210^\circ)$
   (e) $.6(\cos 315^\circ + i \sin 315^\circ)$ (f) $5(\cos 270^\circ + i \sin 225^\circ)$
   (g) $3(\cos -225^\circ + i \sin -225^\circ)$ (h) $14(\cos 3\pi/4 + i \sin 3\pi/4)$
   (i) $.7(\cos 75^\circ + i \sin 75^\circ)$ (j) $1.3(\cos 5\pi + i \sin 5\pi)$
5. Determine the arguments and moduli of the following complex numbers and write the numbers in trigonometric form.
   (a) $\sqrt{3} + i$ (b) $5 + 5i$ (c) $1 - \sqrt{3}i$
   (d) $-7i$ (e) $\sqrt{-18}$ (f) $3 - 4i$
6. Draw diagrams to illustrate the following two sums and demonstrate the associative law of addition.
   (a) $[(2 + 3i) + (1 + 4i)] + (7 + 5i)$
   (b) $(2 + 3i) + [(1 + 4i) + (7 + 5i)]$
7. Compute the following.
   (a) $[4(\cos 30^\circ + i \sin 30^\circ)][5(\cos 15^\circ + i \sin 15^\circ)]$
   (b) $[.3(\cos 22^\circ + i \sin 22^\circ)][1.6(\cos 76^\circ + i \sin 76^\circ)]$
   (c) $[11(\cos 211^\circ + i \sin 211^\circ)][1.4(\cos 312^\circ + i \sin 312^\circ)]$
   (d) $[\frac{2}{3}(\cos -43^\circ + i \sin -43^\circ)][\frac{4}{7}(\cos 92^\circ + i \sin 92^\circ)]$
8. Use trigonometric tables to express your answers to Exercise 7 in the form $a + bi$.
9. Compute the following and express in the form $a + bi$.
   (a) $[3(\cos 10^\circ + i \sin 10^\circ)]^3$ (b) $[2(\cos 15^\circ + i \sin 15^\circ)]^4$
   (c) $[1.2(\cos 210^\circ + i \sin 210^\circ)]^2$ (d) $[\frac{2}{3}(\cos 45^\circ + i \sin 45^\circ)]^5$
   (e) $[\cos -20^\circ + i \sin -20^\circ]^9$ (f) $[\cos 195^\circ + i \sin 195^\circ]^6$

10. Compute all the cube roots of the following complex numbers.
    (a) -1 (b) i (c) $-i$
    (d) 8 (e) $\frac{1}{2}(\sqrt{3} + i)$ (f) $\frac{2}{3} + \frac{4}{5}i$
11. Compute all the fourth roots of the following complex numbers.
    (a) -1 (b) i (c) 9
    (d) $\frac{1}{2}(\sqrt{3} + i)$ (e) $\frac{1}{32} - \sqrt{3}i/32$ (f) $\frac{4}{5} - 4i$
12. Solve the following equations.
    (a) $x^2 + 3x + 5 = 0$ (b) $2x^2 + 3ix + 7 = 0$
    (c) $ix^2 + ix + 1 = 0$ (d) $3x^2 + 5i = 0$
    (e) $x^2 + 4x - 4i = 0$ (f) $5x^2 - 3ix + 8 = 0$
    (g) $2ix^2 - 3x + 7i = 0$ (h) $\sqrt{3}x^2 - \pi x - 5i = 0$
13. Prove that the set of all complex numbers with absolute value 1 is closed under multiplication and division, that is, if z and w are two such numbers, so are zw and z/w.
14. If z is a complex number with absolute value 1, what can be said about $\bar{z}$?

8 Sequences

The intuitive notion of a sequence is a "string of objects" or a "list." In this chapter we will make this notion precise and will investigate certain types of sequences of numbers and properties of these sequences.

1. The Concept of a Sequence

Example 1

Consider the following list of rational numbers:

$$\left\{\frac{1}{2}, \frac{2}{3}, \frac{3}{4}, \frac{4}{5}, \frac{5}{6}, \frac{6}{7}, \frac{7}{8}, \frac{8}{9}, \frac{9}{10}, \frac{10}{11}, \ldots, \frac{n}{n+1}, \ldots\right\}$$

For any positive integer n, the nth number in this list is $n/(n + 1)$. Thus the nineteenth number is 19/20, the four hundredth number is 400/401, and the millionth number in the sequence is 1,000,000/1,000,001. □

Our intuitive idea of a sequence is

"a string of things, one after another"

and we can imagine sequences which are infinitely long, that is, infinite lists. In order to make this notion precise, we define a *sequence* to be *a function with domain N*, the set of natural numbers. Another way of viewing a sequence is as a rule which associates each natural number with a number. The *terms* of the sequence are the numbers in the range, the nth term being the image of n (intuitively it is the nth number in the list). Instead of using functional notation such as $f(n)$, we usually denote a sequence by an expression such as $\{a_n\}$ where the a_i's are the terms of the sequence. As with other functions, a sequence frequently is described by a formula.

Example 2

The sequence

$$\left\{\frac{1}{2}, \frac{2}{3}, \frac{3}{4}, \frac{4}{5}, \frac{5}{6}, \ldots\right\}$$

can be described by

$$\left\{\frac{n}{(n+1)}\right\}$$

and we obtain the 25th term by substituting 25 for n. Thus the 25th term of $\{n/(n+1)\}$ is $\frac{25}{26}$. □

A sequence of rational numbers is a sequence in which each term is a rational number; a sequence of integers is a sequence in which each term is an integer; and so on.

A sequence in which each term except the first is obtained by adding a fixed number to the previous term is called an *arithmetic* sequence.

Example 3

The sequence

$$\{5, 7, 9, 11, 13, 15, 17, \ldots\}$$

is an arithmetic sequence in which the first term is 5 and each succeeding term is obtained by adding 2 to the previous term. □

Example 4

The sequence

$$\{8, 7.5, 7, 6.5, 6, 5.5, 5, \ldots\}$$

is an arithmetic sequence in which the first term is 8 and each succeeding term is obtained by adding $-.5$ to the previous term. □

Example 5

The sequence

$\{4.3, 4.9, 5.5, 6.1, 6.7, \ldots\}$

is an arithmetic sequence in which the first term is 4.3 and each succeeding term is obtained by adding .6 to the previous term. □

A sequence in which each term, except the first, is obtained by multiplying the previous term by a fixed number is called a *geometric* sequence.

Example 6

The sequence

$\{5, 10, 20, 40, 80, 160, \ldots\}$

is a geometric sequence in which the first term is 5 and each succeeding term is obtained by multiplying the preceding term by 2. □

Example 7

The sequence

$\{2, \frac{2}{3}, \frac{2}{9}, \frac{2}{27}, \frac{2}{81}, \ldots\}$

is a geometric sequence in which the first term is 2 and each succeeding term is obtained by multiplying the preceding term by $\frac{1}{3}$. □

Example 8

The sequence

$\{9, -.9, .09, -.009, .0009, \ldots\}$

is a geometric sequence in which the first term is 9 and each succeeding term is obtained by multiplying the preceding term by $-.1$. □

The number that we add to obtain the next term in an arithmetic sequence is called the *difference* since it is the difference between any two successive terms. The number that we multiply by to obtain the next term in a geometric sequence is called the *ratio* since it is the ratio between any two successive terms.

EXERCISES

1. Compute the fifth, tenth, and one hundredth terms in each of the following sequences.
   (a) $\left\{\dfrac{n+3}{n+7}\right\}$ (b) $\left\{\dfrac{n}{n^2+1}\right\}$ (c) $\left\{\dfrac{n-2}{n+2}\right\}$ (d) $\left\{\dfrac{n^2+2n-3}{n^2}\right\}$
2. Determine a formula or rule which describes each of the following sequences.
   (a) $\{\frac{1}{2}, \frac{1}{3}, \frac{1}{4}, \frac{1}{5}, \ldots\}$
   (b) $\{\frac{1}{3}, \frac{1}{5}, \frac{1}{7}, \frac{1}{9}, \ldots\}$
   *(c) $\{5, 7, 11, 17, 25, 35, 47, \ldots\}$
   (d) $\{\frac{1}{2}, \frac{1}{6}, \frac{1}{18}, \frac{1}{54}, \frac{1}{162}, \ldots\}$
   (e) $\{\frac{3}{2}, \frac{4}{3}, \frac{5}{4}, \frac{6}{5}, \ldots\}$
   (f) $\{1, -1, 1, -1, 1, -1, \ldots\}$
   (g) $\{1, 2, 4, 8, 16, 32, 64, 128, \ldots\}$
   (h) $\{5, 10, 15, 20, 25, \ldots\}$
   (i) $\{\frac{6}{5}, \frac{12}{11}, \frac{18}{17}, \frac{24}{23}, \frac{30}{29}, \ldots\}$
   *(j) $\{0, \frac{3}{2}, \frac{2}{3}, \frac{5}{4}, \frac{4}{5}, \frac{7}{6}, \frac{6}{7}, \ldots\}$
3. Which of the sequences in Exercise 2 are arithmetic or geometric?
4. Let $\{a_n\}$ be an arithmetic sequence in which the first term is a_1, and each succeeding term is obtained by adding the number d to the previous term, that is, the difference is d.
   (a) Write the first six terms of the sequence.
   (b) Give a general formula for the sequence.
5. Let $\{b_n\}$ be a geometric sequence in which the first term is b_1 and each succeeding term is obtained by multiplying the preceding term by the number r, that is, the ratio is r.
   (a) Write the first six terms of the sequence.
   (b) Give a general formula for the sequence.
6. A sequence $\{a_n\}$ is said to be *increasing* if $a_p \geq a_q$ whenever $p \geq q$ and is said to be *decreasing* if $a_p \leq a_q$ whenever $p \geq q$.
   (a) In Exercises 1 and 2, which sequences are increasing?
   (b) In Exercises 1 and 2, which sequences are decreasing?
7. If $\{a_n\}$ is an arithmetic sequence, what conditions on the difference d will assure that $\{a_n\}$ is increasing? Decreasing?
8. If $\{b_n\}$ is a geometric sequence, what conditions on the ratio r will assure that $\{b_n\}$ is increasing? Decreasing?

2. Limits of Sequences

Some sequences have the property that as n gets large, the terms of the sequence get close to some number.

Example 1

In the sequence

$$\left\{\frac{1}{n}\right\} = \left\{1, \frac{1}{2}, \frac{1}{3}, \frac{1}{4}, \frac{1}{5}, \ldots\right\}$$

as n gets larger, $1/n$ gets close to 0. □

Example 2

In

$$\left\{\frac{(n+1)}{n}\right\} = \left\{\frac{2}{1}, \frac{3}{2}, \frac{4}{3}, \frac{5}{4}, \ldots\right\}$$

the terms get close to 1 as n gets large. □

In these examples, we call 0 and 1 the "limits" of the sequences, that is, the number that the sequence "gets close to" is the "limit" of the sequence. Our intuitive idea of what we mean by "gets close" is not, however, sufficient for a mathematical development. We need to give a precise meaning to this idea and to the concept of limit. The reader should remember, however, that we are only attempting to make precise an idea which is intuitively rather easy to grasp. The formal definition may be a little harder.

Let $\{a_n\}$ be a sequence and L a number. We say that L is the *limit* of the sequence $\{a_n\}$ if for every positive integer m, there is some positive integer w (perhaps dependent on m) such that whenever $n > w$, a_n differs from L by less than $1/m$. Thus if $n > w$, $L - 1/m < a_n < L + 1/m$.

If the sequence $\{a_n\}$ has the limit L, we say that $\{a_n\}$ *converges* to L and may write $\{a_n\} \to L$. A sequence is said to be *convergent* if it has a limit.

Example 3

The number 0 is the limit of $\{1/n\}$ because for every positive integer m, if we let $w = m$, then when $n > w = m$, $1/n < 1/m$, and therefore $0 - 1/m < 1/n < 0 + 1/m$. □

Example 4

The number 1 is the limit of $\{(n + 1)/n\}$ because for any positive integer m, if we again let $w = m$ it follows that when $n > w = m$ we have

$$1 - \frac{1}{m} < \frac{(n+1)}{n} < 1 + \frac{1}{m}$$

because

$$1 - \frac{1}{m} < 1 < \frac{(n+1)}{n} = 1 + \frac{1}{n} < 1 + \frac{1}{m} \quad □$$

The task of proving that a sequence $\{a_n\}$ has a limit or that the number L is the limit of $\{a_n\}$ is, in general, quite difficult. Often, however, by just examining a sequence, we can decide intuitively whether or not it has a limit and what the limit is. Such decisions, although not proofs, are useful in helping us to understand the concepts. We will make many of these intuitive decisions in the next few sections.

EXERCISES

1. Consider the sequences listed in Exercises 1 and 2 in Section 1. By examination try to decide which of these are convergent, and find the limits of the convergent ones.
2. Consider the following convergent sequences and try to determine the limits of each sequence.
   (a) $\{.3, .33, .333, .3333, .33333, \ldots\}$ (b) $\{.6, .66, .666, .6666, .66666, \ldots\}$
   (c) $\{.9, .99, .999, .9999, .99999, \ldots\}$ (d) $\{.1, .16, .166, .1666, \ldots\}$
   (e) $\{.8, .83, .833, .8333, \ldots\}$ (f) $\{.4, .41, .416, .4166, .41666, \ldots\}$
   (g) $\{6, 3, \frac{3}{2}, \frac{3}{4}, \frac{3}{8}, \frac{3}{16}, \ldots\}$ (h) $\{8, .8, .08, .008, \ldots\}$
3. Let $\{b_n\}$ be a geometric sequence with ratio r, $0 < r < 1$. Explain why $\{b_n\} \to 0$.
4. Let $\{a_n\}$ be an arithmetic sequence with difference $d > 0$. Explain why $\{a_n\}$ is not convergent.
5. Let $\{b_n\}$ be a geometric sequence with ratio $r > 1$ and $b_1 \neq 0$. Explain why $\{b_n\}$ is not convergent.
6. Suppose $\{a_n\}$, $\{b_n\}$, and $\{c_n\}$ are sequences such that $a_n \leq b_n \leq c_n$ for each positive integer n. If L is a number such that $\{a_n\} \to L$ and $\{c_n\} \to L$, what can be said about $\{b_n\}$? Can you justify your assertion, that is, explain why?

3. Infinite Decimals Representing Rational Numbers

Example 1

Our intuition tells us that the limit of

$$\{.3, .33, .333, .3333, \ldots\}$$

is $\frac{1}{3}$. We now prove it using the definition of limit.

Let m be a positive integer and let w be a positive integer such that $10^w > m$. Then $1/10^w < 1/m$. For $n > w$, we know that the nth term in the sequence differs from $\frac{1}{3}$ by less than $1/10^w$ which is less than $1/m$. Therefore, the limit of this sequence is $\frac{1}{3}$. □

The ideas used in the argument above do not depend on this particular sequence.

Example 2

The sequence of rational numbers

$$\{.3, .36, .363, .3636, \ldots\}$$

is obtained by the usual division process used to obtain finite decimal approximations for $\frac{4}{11}$. The limit of this sequence is $\frac{4}{11}$ and the argument is much the same as in Example 1. □

The usual process of computing finite decimal approximations for rational numbers produces a sequence of finite decimals in which the mth term and all succeeding terms differ from the original number by less than $1/10^m$. If w is any positive number and $m > w$, then $1/10^m < 1/w$ and all terms from the mth term on differ from the original rational number by less than $1/w$, and therefore this sequence of finite decimals converges to the original rational number.

Example 3

The usual division process applied to $\frac{2}{7}$ and carried out to 8 places yields .28571428 and the pattern starts to repeat.

```
   .28571428
 7)2.00000000
   1 4
   ---
     60
     56
     --
      40
      35
      --
       50
       49
       --
        10
         7
        --
         30
         28
         --
          20
          14
          --
           60
           56
           --
```

Since the remainders have started to repeat, the digits in the quotient must start to repeat and the sequence is completely determined.

$\{.2, .28, .285, .2857, .28571, .285714, .2857142, .28571428, .28571428, \ldots\}$

The nth term and all succeeding terms differ from $\frac{2}{7}$ by less than $1/10^n$ and therefore the sequence converges to $\frac{2}{7}$. □

A sequence of finite decimals such as

$\{.2, .24, .242, .2424, .24242, \ldots\}$

can be abbreviated by

$.242424\ldots$

where it is understood that this is really just an abbreviation for the infinite sequence. *The number represented by this infinite decimal is the limit of the sequence.*

Example 4

$\frac{1}{3} = .333\ldots$

because $\frac{1}{3}$ is the limit of

$\{.3, .33, .333, .3333, \ldots\}$ □

Example 5

$\frac{2}{7} = .28514285714\ldots$

because $\frac{2}{7}$ is the limit of

$\{.2, .28, .285, .2857, .28571, .285714, .2857142, \ldots\}$ □

Example 6

$1 = .999\ldots$

because 1 is the limit of

$\{.9, .99, .999, .9999, .99999, \ldots\}$ □

In our discussion thus far we have assured ourselves that every rational number can be represented by an infinite decimal because it is the limit of the sequence of the finite decimals obtained in the approximation process. Note also that if we perform the usual computations to obtain the approximations for p/q, either the division comes out even or the digits start to repeat after q steps.

$q\overline{)p.00\cdots0}$

The digits in the quotient must repeat (including repetition of 0) because after q steps, we will have q remainders all smaller than q. If one of the remainders is 0, the division comes out even and we have a finite decimal. If none of them are zero, then at the qth step, we must get one of the remainders obtained before and the pattern of digits repeats.

Example 7

```
      .8333
   6)5.0000
     4 8
     ---
       20
       18
       --
        20
```

After step 2, all remainders are the same. □

Example 8

```
       .636363
   11)7.00000
      6 6
      ---
        40
        33
        --
         70
```

As soon as 7 appears again, the pattern starts to repeat. □

Example 9

```
       .384615384615384615
   13)5.0000000000
       39
       ---
       110
       104
       ---
         60
         52
         --
          80
          78
          --
           20
           13
           --
            70
            65
            --
             50
```

When the 5 appears again as a remainder, we know that the pattern will repeat. □

Infinite decimals in which the pattern of digits repeats can be described by giving the initial digits and the block of repeating digits. We can just overline the repeating block.

Example 10

$.\overline{47}$ means .474747 . . . □

Example 11

$5.8\overline{23}$ means 5.8232323 . . . □

Example 12

$.\overline{46892}$ means .4689268926892 . . . □

We know that all rational numbers can be represented by an infinite repeating decimal which, in fact, represents an infinite sequence of finite decimals that converges to the rational number. Is such a representation unique? Not in all cases.

Example 13

The rational number $\frac{1}{2}$ is represented by

$$.5, \qquad .50, \qquad .500, \quad \text{and} \quad .5\overline{0}$$ □

Consider also the infinite decimal $.4\overline{9}$ or .4999 . . . . The infinite sequence which this represents is

$$\{.4, .49, .499, .4999, .49999, \ldots\}$$

which we can see has $\frac{1}{2}$ as a limit. Thus the infinite decimal $.4\overline{9}$ represents the rational number $\frac{1}{2}$, and so does the infinite decimal $.5\overline{0}$. Should this trouble us? The fractions $\frac{2}{4}$ and $\frac{3}{6}$ also represent this same number and that does not trouble us. The number $\frac{1}{2}$ has many representations, and it should not bother us that it has two representations as an infinite decimal.

In general, it turns out that the numbers that have two representations as infinite decimals are just those which can be represented as finite decimals. We will not prove this assertion, but the reader can see how two such representations are always available for a finite decimal. If we use the finite representation we can obtain an infinite representation by decreasing the last digit by 1 and affixing an infinite sequence of 9's.

Example 14

$.2 = .1\overline{9}$, $.76 = .75\overline{9}$, and $4.837 = 4.836\overline{9}$ □

If we have a rational number in the form a/b where a and b are integers, we know how to find an infinite decimal representation. Can we reverse the process? How can we find the usual fractional representation given the infinite decimal form?

Example 15

Consider the sequence $\{.3, .36, .363, .3636, \ldots\}$ and let $.\overline{36} = x$, that is x is the limit of the sequence.

If we multiply each term in the sequence above by 100 we get a new sequence

$\{30, 36, 36.3, 36.36, 36.363, \ldots\}$

which we would expect to have as its limit 100 times the limit of the first sequence, or $100x$. The sequence

$\{36.3, 36.36, 36.363, 36.3636, \ldots\}$

will also converge to this same limit, and therefore we would expect, at least intuitively, that the sequence of differences we get by subtracting terms of the first sequence from terms of this one has as its limit $100x - x$. Computing these differences we have the sequence

$\{36.3 - .3, 36.36 - .36, 36.363 - .363, \ldots\}$

which should have as its limit $99x$. This sequence is the constant sequence

$\{36, 36, 36, \ldots\}$

which has 36 for its limit. Thus $99x = 36$, and therefore

$$x = \frac{36}{99} = \frac{4}{11} \quad \square$$

The process described in Example 15 is an explanation and not a proof. The computational technique is illustrated by the next two examples.

Example 16

Let $x = .\overline{36}$.

Then $100x = 36.\overline{36}$ and

$$\begin{aligned} 100x - x &= 36.\overline{36} - .\overline{36} \\ &= (36 + .\overline{36}) - .\overline{36} \\ &= 36 + (.\overline{36} - .\overline{36}) \\ &= 36 \end{aligned}$$

Therefore $99x = 36$, and

$$x = \frac{36}{99} = \frac{4}{11} \quad \square$$

Example 17

If $y = .1\overline{67}$, then

$1000y = 167.\overline{67}$

and

$10y = 1.\overline{67}$

Therefore

$990y = 166$

and

$$y = \frac{166}{990} \square$$

The technique illustrated enables us to compute a fractional representation for a rational number from the infinite repeating decimal representation. Once this technique is justified, it can be used to show that *every repeating decimal represents a rational number.* The technique depends on five basic properties of limits of sequences which we will list and use but will not prove.

THEOREM 1

Let $\{a_n\}$ and $\{b_n\}$ be sequences, L and P be numbers such that $\{a_n\} \to L$ and $\{b_n\} \to P$, and let c be any number. Then:

(i) $\{ca_n\} \to cL$
(ii) $\{a_n + b_n\} \to L + P$
(iii) $\{a_n - b_n\} \to L - P$
(iv) $\{a_nb_n\} \to LP$
(v) $\left\{\dfrac{a_n}{b_n}\right\} \to \dfrac{L}{P}$ providing each $b_n \neq 0$ and $P \neq 0$

The technique illustrated previously is justified by parts (i) and (iii) of the theorem.

Example 18

If $x = .4\overline{57}$, then

$\{.4, .45, .457, .4575, .45757, \ldots\} \to x$

By (i)

$\{4, 4.5, 4.57, 4.575, 4.5757, \ldots\} \to 10x$

and

$\{457, 457.5, 457.57, 457.575, \ldots\} \to 1000x$

By (iii)

$\{453, 453, 453, \ldots\} \to 1000x - 10x = 990x$

Thus

$453 = 990x$

and

$$\frac{453}{990} = x$$

that is,

$$\frac{151}{330} = x \quad \square$$

Thus far, we have discussed only infinite *repeating* decimals. We could also consider sequences of finite decimals that have no repeating pattern.

Example 19

For the infinite decimal

$.1011011101111011111011111110\ldots$

the sequence of finite decimals is

$\{.1, .10, .101, .1011, .10110, .101101, \ldots\}$

but no "block of digits" repeats as in the cases we studied before. $\square$

Example 20

In the sequence

$\{1, 1.4, 1.41, 1.414, 1.4142, \ldots\}$

no "block" repeats. Notice that if we square each term we obtain the sequence

$\{1, 1.96, 1.9881, 1.999396, 1.99996164, \ldots\}$

which appears to be converging to 2, and hence if the first sequence has a limit, we expect it to be $\sqrt{2}$. $\square$

The sequence of finite decimals determined by a positive infinite decimal is always increasing. It is also always *bounded above*; that is, there is always a number which is larger than every term in the sequence. (We can obtain such a number by adding 1 to the first term in the sequence.)

If we look at a number line and view the sequence of terms given by an infinite decimal we feel, intuitively, that this sequence should have a limit. (See Figure 8–1.)

Figure 8–1

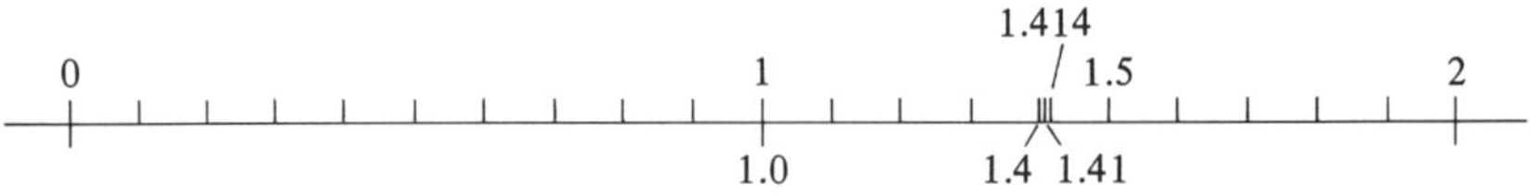

In order to show that such a sequence has a limit, we need Axiom 12 for the real numbers. Recall that Axiom 12 states:

> If S is a nonempty subset of R and a is a real number such that every element of S is less than a, then there is a smallest real number b which is greater than or equal to every element of S.

Example 21

If we look at the sequence

$\{1, 1.4, 1.41, 1.414, 1.4142, \ldots\}$

we see that every term in this sequence is less than 1.5. Thus, by Axiom 12, there is a smallest real number which is greater than or equal to each number in this set. This real number then turns out to be $\sqrt{2}$ which is the limit of this sequence. □

The above cases are generalized by the following theorem.

THEOREM 2

If $\{a_n\}$ is an increasing sequence of real numbers and there is some real number r such that $a_n < r$ for every positive integer n, then $\{a_n\}$ converges.

**Proof*:

Since every term in the set $\{a_1, a_2, \ldots\}$ is less than r, there is a smallest real number L such that every term a_n is less than or equal to L. We will now show that L is the limit of $\{a_n\}$. Let m be any positive integer and consider the real number $L - 1/m$. Since $L - 1/m < L$ and L is the smallest real number greater than or equal to each a_n, then for some positive integer w, a_w must be greater than $L - 1/m$, that is, $a_w > L - 1/m$

or $L - 1/m < a_w$. Since $\{a_n\}$ is increasing, for $n > w$, $a_w \leq a_n$. But we also know that $a_n \leq L < L + 1/m$ for every positive integer n. Thus we have, for $n > w$,

$$L - \frac{1}{m} < a_w \leq a_n \leq L < L + \frac{1}{m}$$

which implies

$$L - \frac{1}{m} < a_n < L + \frac{1}{m}$$

Thus for each positive integer m, there is a positive integer w such that whenever $n > w$,

$$L - \frac{1}{m} < a_n < L + \frac{1}{m}$$

By definition, L is the limit of $\{a_n\}$.

Since every positive infinite decimal determines an increasing sequence that is bounded above, every infinite decimal determines a convergent sequence. Thus *every positive infinite decimal represents a real number*, the number being the limit of this sequence. Since a negative infinite decimal is just the negative of a positive infinite decimal, it also represents a real number.

The converse of the above assertion can be shown, that is:

THEOREM 3

Every real number can be represented by an infinite decimal.

The proof of this assertion is quite sophisticated and we will not attempt it in this text.

By our earlier results we know that repeating decimals represent rational numbers and therefore the irrational numbers are represented by nonrepeating decimals. These infinite decimals provide arbitrarily close approximations for these numbers and in most computations we use a finite decimal approximation for an irrational number. For example, the logarithms of most rational numbers are irrational but we approximate them by finite decimals.

EXERCISES

1. Determine infinite decimal representations for the following rational numbers.
   (a) 5/6 (b) 1/7 (c) 4/13 (d) 7/18
   (e) 8/11 (f) 8/17 (g) 1/16 (h) 5/64

2. Determine the fractional form of each of the rational numbers represented by the following infinite decimals.
   (a) $.\overline{47}$ (b) $.1\overline{93}$ (c) $2.\overline{67}$ (d) $.\overline{1122}$ (e) $.0\overline{14}$
   (f) $.\overline{5234}$ (g) $.67\overline{32}$ (h) $.946\overline{88}$ (i) $.\overline{707}$ (j) $8.4\overline{59}$

*3. Can you describe, in terms of numerators and denominators, those rational numbers which can be represented as finite decimals?

4. Sequences of Partial Sums

With each sequence $\{a_n\}$ there is associated a *sequence of partial sums* $\{A_n\}$ where, for each positive integer n, A_n is defined by

$$A_n = a_1 + a_2 + a_3 + \cdots + a_n$$

Example 1

If

$$\{a_n\} = \left\{\frac{n^2 + 1}{3}\right\}$$

then

$$A_4 = a_1 + a_2 + a_3 + a_4$$
$$= \frac{2}{3} + \frac{5}{3} + \frac{10}{3} + \frac{17}{3} = \frac{34}{3}$$

and

$$A_5 = A_4 + a_5$$
$$= \frac{34}{3} + \frac{26}{3} = \frac{60}{3} = 20 \ \square$$

Notice that for any positive integer k, $A_{k+1} = A_k + a_{k+1}$.

It is frequently convenient to use summation notation to express partial sums. By summation notation we mean a capital Greek sigma "Σ" with variables and indices to indicate the terms to be added.

Example 2

$$\sum_{i=1}^{7} a_i$$

means

$$a_1 + a_2 + a_3 + a_4 + a_5 + a_6 + a_7$$

The 1 and 7 are the *bounds* of the summation and i is the *index*. □

For each sequence $\{b_n\}$ the partial sum B_n can be denoted by

$$B_n = \sum_{i=1}^{n} b_i$$

If $\{a_n\}$ is an arithmetic sequence, the partial sums can be expressed rather simply in terms of a_1 and the difference d. In Exercise 4 of Section 2 you were asked to determine a formula for a_n in terms of a_1 and d. If we write out the first few terms of the sequence we see that we have

$$\{a_1, a_1 + d, a_1 + d + d, a_1 + d + d + d, a_1 + d + d + d + d, \ldots\}$$

that is,

$$\{a_1, a_1 + d, a_1 + 2d, a_1 + 3d, a_1 + 4d, \ldots\}$$

In general, the nth term is given by

$$a_n = a_1 + (n - 1)d$$

If we form the sequence of partial sums A_n we have

$$\begin{aligned} A_n &= a_1 + a_2 + \cdots + a_n \\ &= (a_1 + d) + (a_1 + 2d) + (a_1 + 3d) + \cdots + (a_1 + (n - 1)d) \\ &= na_1 + (d + 2d + 3d + \cdots + (n - 1)d) \\ &= na_1 + d(1 + 2 + 3 + \cdots + (n - 1)) \end{aligned}$$

We need a simpler formula for the sum

$$1 + 2 + 3 + \cdots + (n - 1)$$

If we let

$$x = 1 + 2 + 3 + \cdots + (n - 1)$$

then

$$x = (n - 1) + (n - 2) + \cdots + 1$$

and

$$2x = n + n + \cdots + n$$

(This sum has $n - 1$ terms.)

Thus

$$2x = n(n - 1)$$

and

$$x = \frac{n(n - 1)}{2}$$

Since

$$A_n = na_1 + d(1 + 2 + \cdots + (n - 1))$$

then

$$A_n = na_1 + d\left(\frac{n(n-1)}{2}\right)$$

Example 3

What is the sum of the first 25 terms of the sequence

$\{7, 9, 11, 13, 15, 17, \ldots\}$?

In this case

$$\begin{aligned} a_1 &= 7 \\ n &= 25 \\ d &= 2 \end{aligned}$$

Thus

$$\begin{aligned} A_{25} &= 25 \cdot 7 + 2\left(\frac{25 \cdot 24}{2}\right) \\ &= 175 + 600 \\ &= 775 \quad \square \end{aligned}$$

If $\{b_n\}$ is a geometric sequence, the partial sums can be expressed rather simply in terms of b_1 and the ratio r. In Exercise 5 of Section 2 you were asked to determine a formula for b_n in terms of b_1 and r. If we write the first few terms of the sequence we see that we have

$$\{b_1, b_1 r, b_1 r^2, b_1 r^3, b_1 r^4, \ldots\}$$

Thus, in general,

$$b_n = b_1 r^{n-1}$$

If we form the sequence of partial sums B_n we have

$$\begin{aligned} B_n &= b_1 + b_1 r + b_1 r^2 + b_1 r^3 + \cdots + b_1 r^{n-1} \\ &= b_1(1 + r + r^2 + r^3 + \cdots + r^{n-1}) \end{aligned}$$

But

$$(1 + r + r^2 + r^3 + \cdots + r^{n-1})(1 - r) = 1 - r^n$$

Thus for $r \neq 1$,

$$1 + r + r^2 + r^3 + \cdots + r^{n-1} = \frac{1 - r^n}{1 - r}$$

Hence for $r \neq 1$,

$$B_n = b_1\left(\frac{1 - r^n}{1 - r}\right)$$

Example 4

What is the sum of the first six terms of

$$\left\{5, \frac{5}{2}, \frac{5}{4}, \frac{5}{8}, \frac{5}{16}, \frac{5}{32}, \frac{5}{64}, \ldots\right\}?$$

$$b_1 = 5$$
$$r = \tfrac{1}{2}$$
$$n = 6$$

Thus

$$\begin{aligned} B_6 &= 5\left(\frac{1-(\frac{1}{2})^6}{1-\frac{1}{2}}\right) \\ &= 5\left(\frac{1-\frac{1}{64}}{\frac{1}{2}}\right) \\ &= 5\left(\frac{\frac{63}{64}}{\frac{1}{2}}\right) \\ &= 5\left(\frac{63}{32}\right) = \frac{315}{32} \quad \square \end{aligned}$$

Under what conditions does the sequence $\{A_n\}$ of partial sums of $\{a_n\}$ converge? In general, this is a very difficult problem and techniques do not exist that will handle all questions easily. We will limit ourselves to some special results.

First of all notice that if $\{A_n\} \to L$, then $\{a_n\} \to 0$. This idea blends well with our intuition in that if all the A_n's are getting close to L, they are getting close to each other. Thus A_n and A_{n+1} must be close together and their difference must be small. We can prove that $\{a_n\} \to 0$ if $\{A_n\} \to L$ by first noticing that if $\{A_n\} \to L$, then $\{A_{n+1}\} \to L$ and

$$\{A_{n+1} - A_n\} \to L - L = 0$$

Since

$$A_{n+1} - A_n = a_n$$
$$\{a_n\} \to 0$$

Since $\{A_n\} \to L$ implies $\{a_n\} \to 0$, if $\{a_n\} \to K \neq 0$ or $\{a_n\}$ does not converge, then $\{A_n\}$ does not converge. There are, however, sequences $\{a_n\}$ in which $\{a_n\} \to 0$ but $\{A_n\}$ does not converge.

Example 5

$\left\{\dfrac{1}{n}\right\} \to 0$ but $\{A_n\}$ where

$$A_n = \sum_{i=1}^{n} \frac{1}{i}$$

does not converge. We will not present a proof but will indicate numerically what happens.

$$
\begin{aligned}
A_1 &= 1 \\
A_2 &= 1 + \tfrac{1}{2} \\
A_4 &= 1 + \tfrac{1}{2} + \tfrac{1}{3} + \tfrac{1}{4} > 2 \\
A_{30} &= 1 + \tfrac{1}{2} + \tfrac{1}{3} + \tfrac{1}{4} + (\tfrac{1}{5} + \tfrac{1}{7} + \tfrac{1}{8} + \cdots + \tfrac{1}{30}) > 3
\end{aligned}
$$

In fact, for any positive integer n, there is some A_k with $A_k > n$. □

If $\{a_n\}$ is an arithmetic sequence with $d \neq 0$, then $\{A_n\}$ does not converge since

$$\{a_n\} = \{a_1 + (n - 1)\,d\}$$

does not converge.

If $\{b_n\}$ is a geometric sequence with ratio $r \neq 1$, then

$$B_n = b_1\left(\frac{1 - r^n}{1 - r}\right)$$

In order for $\{B_n\}$ to converge, $\{b_n\} \to 0$ and since

$$\{b_n\} = \{b_1 r^{n-1}\}, \qquad \{b_n\} \to 0, \qquad \text{only if } \{b_1 r^{n-1}\} \to 0$$

If $b_1 = 0$, the sequence is the constant sequence 0 and converges. If $b_1 \neq 0$ and $|r| > 1$, then $\{|r^n|\}$ gets larger as n gets larger and $\{b_1 r^{n-1}\}$ does not converge to 0. On the other hand, if $|r| < 1$, then $\{r^{n-1}\} \to 0$ and therefore $\{b_1 r^{n-1}\} \to 0$. Thus if $|r| < 1$, there is a possibility that $\{B_n\}$ converges. For $|r| < 1$, $\{1 - r^n\} \to 1$ and therefore

$$\left(\frac{1}{1 - r}\right)\{1 - r^n\} \to \frac{1}{1 - r}$$

Thus

$$\left\{\frac{1 - r^n}{1 - r}\right\} \to \frac{1}{1 - r}$$

and

$$\left\{b_1\left(\frac{1 - r^n}{1 - r}\right)\right\} \to b_1\left(\frac{1}{1 - r}\right) = \frac{b_1}{1 - r}$$

Thus if $\{b_n\}$ is a geometric sequence with ratio r with $|r| < 1$,

$$\{B_n\} \to \frac{b_1}{1 - r}$$

Example 6

Find the limit of the sequence of partial sums of

$\{3, \frac{3}{2}, \frac{3}{4}, \frac{3}{8}, \ldots\}$

In this case $b_1 = 3$ and $r = \frac{1}{2}$

$$B_n = 3\left(\frac{1 - (\frac{1}{2})^n}{1 - \frac{1}{2}}\right) \quad \text{and} \quad B_n \to 3\left(\frac{1}{1 - \frac{1}{2}}\right) = 3(2) = 6$$

An infinite decimal such as

$.666\ldots$

can be viewed as the sequence of partial sums of the geometric sequence

$\{.6, .06, .006, .0006, \ldots\}$

with ratio .1. The limit of the first sequence is the limit of the sequence of partial sums and is

$$.6\left(\frac{1}{1 - .1}\right) = .6\left(\frac{1}{.9}\right) = \frac{.6}{.9} = \frac{2}{3} \quad \square$$

EXERCISES

1. Compute A_4, A_5, and A_{10} for the following sequences.
   (a) $\{3 + (n - 1)(.7)\}$ (b) $\{1/n\}$ (c) $\{1 - 1/n\}$
   (d) $\{1 + 1/n\}$ (e) $\{4(\frac{2}{3})^{n-1}\}$ (f) $\{(.96)(.01)^{n-1}\}$
2. Compute limit $\{A_n\}$ for those sequences in Exercise 1 in which the limit exists.
3. Suppose $0 \le a_n \le b_n$ for every positive integer n,

   $$A_n = \sum_{i=1}^{n} a_n \quad \text{and} \quad B_n = \sum_{i=1}^{n} b_n$$

   Explain why $\{A_n\}$ converges if $\{B_n\}$ converges.
4. Use Exercise 3 and the fact that $\{1/2^n\}$ is a geometric sequence to explain why

   $$\left\{\sum_{i=1}^{n} \frac{1}{2^{i^2}}\right\}$$

   converges.
5. Use Exercise 3 to explain why

   $$\left\{\sum_{i=1}^{n} \frac{n+1}{n^2}\right\}$$

   does not converge.

Appendix

Table 1 Four-place logarithms of numbers

N	0	1	2	3	4	5	6	7	8	9
10	0000	0043	0086	0128	0170	0212	0253	0294	0334	0374
11	0414	0453	0492	0531	0569	0607	0645	0682	0719	0755
12	0792	0828	0864	0899	0934	0969	1004	1038	1072	1106
13	1139	1173	1206	1239	1271	1303	1335	1367	1399	1430
14	1461	1492	1523	1553	1584	1614	1644	1673	1703	1732
15	1761	1790	1818	1847	1875	1903	1931	1959	1987	2014
16	2041	2068	2095	2122	2148	2175	2201	2227	2253	2279
17	2304	2330	2355	2380	2405	2430	2455	2480	2504	2529
18	2553	2577	2601	2625	2648	2672	2695	2718	2742	2765
19	2788	2810	2833	2856	2878	2900	2923	2945	2967	2989
20	3010	3032	3054	3075	3096	3118	3139	3160	3181	3201
21	3222	3243	3263	3284	3304	3324	3345	3365	3385	3404
22	3424	3444	3464	3483	3502	3522	3541	3560	3579	3598
23	3617	3636	3655	3674	3692	3711	3729	3747	3766	3784
24	3802	3820	3838	3856	3874	3892	3909	3927	3945	3962
25	3979	3997	4014	4031	4048	4065	4082	4099	4116	4133
26	4150	4166	4183	4200	4216	4232	4249	4265	4281	4298
27	4314	4330	4346	4362	4378	4393	4409	4425	4440	4456
28	4472	4487	4502	4518	4533	4548	4564	4579	4594	4609
29	4624	4639	4654	4669	4683	4698	4713	4728	4742	4757
30	4771	4786	4800	4814	4829	4843	4857	4871	4886	4900
31	4914	4928	4942	4955	4969	4983	4997	5011	5024	5038
32	5051	5065	5079	5092	5105	5119	5132	5145	5159	5172
33	5185	5198	5211	5224	5237	5250	5263	5276	5289	5302
34	5315	5328	5340	5353	5366	5378	5391	5403	5416	5428
35	5441	5453	5465	5478	5490	5502	5514	5527	5539	5551
36	5563	5575	5587	5599	5611	5623	5635	5647	5658	5670
37	5682	5694	5705	5717	5729	5740	5752	5763	5775	5786
38	5798	5809	5821	5832	5843	5855	5866	5877	5888	5899
39	5911	5922	5933	5944	5955	5966	5977	5988	5999	6010
40	6021	6031	6042	6053	6064	6075	6085	6096	6107	6117
41	6128	6138	6149	6160	6170	6180	6191	6201	6212	6222
42	6232	6243	6253	6263	6274	6284	6294	6304	6314	6325
43	6335	6345	6355	6365	6375	6385	6395	6405	6415	6425
44	6435	6444	6454	6464	6474	6484	6493	6503	6513	6522
45	6532	6542	6551	6561	6571	6580	6590	6599	6609	6618
46	6628	6637	6646	6656	6665	6675	6684	6693	6702	6712
47	6721	6730	6739	6749	6758	6767	6776	6785	6794	6803
48	6812	6821	6830	6839	6848	6857	6866	6875	6884	6893
49	6902	6911	6920	6928	6937	6946	6955	6964	6972	6981
50	6990	6998	7007	7016	7024	7033	7042	7050	7059	7067
51	7076	7084	7093	7101	7110	7118	7126	7135	7143	7152
52	7160	7168	7177	7185	7193	7202	7210	7218	7226	7235
53	7243	7251	7259	7267	7275	7284	7292	7300	7308	7316
54	7324	7332	7340	7348	7356	7364	7372	7380	7388	7396

Table 1 (*Continued*)

N	0	1	2	3	4	5	6	7	8	9
55	7404	7412	7419	7427	7435	7443	7451	7459	7466	7474
56	7482	7490	7497	7505	7513	7520	7528	7536	7543	7551
57	7559	7566	7574	7582	7589	7597	7604	7612	7619	7627
58	7634	7642	7649	7657	7664	7672	7679	7686	7694	7701
59	7709	7716	7723	7731	7738	7745	7752	7760	7767	7774
60	7782	7789	7796	7803	7810	7818	7825	7832	7839	7846
61	7853	7860	7868	7875	7882	7889	7896	7903	7910	7917
62	7924	7931	7938	7945	7952	7959	7966	7973	7980	7987
63	7993	8000	8007	8014	8021	8028	8035	8041	8048	8055
64	8062	8069	8075	8082	8089	8096	8102	8109	8116	8122
65	8129	8136	8142	8149	8156	8162	8169	8176	8182	8189
66	8195	8202	8209	8215	8222	8228	8235	8241	8248	8254
67	8261	8267	8274	8280	8287	8293	8299	8306	8312	8319
68	8325	8331	8338	8344	8351	8357	8363	8370	8376	8382
69	8388	8395	8401	8407	8414	8420	8426	8432	8439	8445
70	8451	8457	8463	8470	8476	8482	8488	8494	8500	8506
71	8513	8519	8525	8531	8537	8543	8549	8555	8561	8567
72	8573	8579	8585	8591	8597	8603	8609	8615	8621	8627
73	8633	8639	8645	8651	8657	8663	8669	8675	8681	8686
74	8692	8698	8704	8710	8716	8722	8727	8733	8739	8745
75	8751	8756	8762	8768	8774	8779	8785	8791	8797	8802
76	8808	8814	8820	8825	8831	8837	8842	8848	8854	8859
77	8865	8871	8876	8882	8887	8893	8899	8904	8910	8915
78	8921	8927	8932	8938	8943	8949	8954	8960	8965	8971
79	8976	8982	8987	8993	8998	9004	9009	9015	9020	9025
80	9031	9036	9042	9047	9053	9058	9063	9069	9074	9079
81	9085	9090	9096	9101	9106	9112	9117	9122	9128	9133
82	9138	9143	9149	9154	9159	9165	9170	9175	9180	9186
83	9191	9196	9201	9206	9212	9217	9222	9227	9232	9238
84	9243	9248	9253	9258	9263	9269	9274	9279	9284	9289
85	9294	9299	9304	9309	9315	9320	9325	9330	9335	9340
86	9345	9350	9355	9360	9365	9370	9375	9380	9385	9390
87	9395	9400	9405	9410	9415	9420	9425	9430	9435	9440
88	9445	9450	9455	9460	9465	9469	9474	9479	9484	9489
89	9494	9499	9504	9509	9513	9518	9523	9528	9533	9538
90	9542	9547	9552	9557	9562	9566	9571	9576	9581	9586
91	9590	9595	9600	9605	9609	9614	9619	9624	9628	9633
92	9638	9643	9647	9652	9657	9661	9666	9671	9675	9680
93	9685	9689	9694	9699	9703	9708	9713	9717	9722	9727
94	9731	9736	9741	9745	9750	9754	9759	9763	9768	9773
95	9777	9782	9786	9791	9795	9800	9805	9809	9814	9818
96	9823	9827	9832	9836	9841	9845	9850	9854	9859	9863
97	9868	9872	9877	9881	9886	9890	9894	9899	9903	9908
98	9912	9917	9921	9926	9930	9934	9939	9943	9948	9952
99	9956	9961	9965	9969	9974	9978	9983	9987	9991	9996

Table 2 Trigonometric functions

Degrees	Radians	Sin	Cos	Tan	Cot		
0	0.0000	0.0000	1.0000	0.0000		1.5708	90
1	0.0175	0.0175	0.9998	0.0175	57.290	1.5533	89
2	0.0349	0.0349	0.9994	0.0349	28.636	1.5359	88
3	0.0524	0.0523	0.9986	0.0524	19.081	1.5184	87
4	0.0698	0.0698	0.9976	0.0699	14.301	1.5010	86
5	0.0873	0.0872	0.9962	0.0875	11.430	1.4835	85
6	0.1047	0.1045	0.9945	0.1051	9.5144	1.4661	84
7	0.1222	0.1219	0.9925	0.1228	8.1443	1.4486	83
8	0.1396	0.1392	0.9903	0.1405	7.1154	1.4312	82
9	0.1571	0.1564	0.9877	0.1584	6.3138	1.4137	81
10	0.1745	0.1736	0.9848	0.1763	5.6713	1.3963	80
11	0.1920	0.1908	0.9816	0.1944	5.1446	1.3788	79
12	0.2094	0.2079	0.9781	0.2126	4.7046	1.3614	78
13	0.2269	0.2250	0.9744	0.2309	4.3315	1.3439	77
14	0.2443	0.2419	0.9703	0.2493	4.0108	1.3265	76
15	0.2618	0.2588	0.9659	0.2679	3.7321	1.3090	75
16	0.2793	0.2756	0.9613	0.2867	3.4874	1.2915	74
17	0.2967	0.2924	0.9563	0.3057	3.2709	1.2741	73
18	0.3142	0.3090	0.9511	0.3249	3.0777	1.2566	72
19	0.3316	0.3256	0.9455	0.3443	2.9042	1.2392	71
20	0.3491	0.3420	0.9397	0.3640	2.7475	1.2217	70
21	0.3665	0.3584	0.9336	0.3839	2.6051	1.2043	69
22	0.3840	0.3746	0.9272	0.4040	2.4751	1.1868	68
23	0.4014	0.3907	0.9205	0.4245	2.3559	1.1694	67
24	0.4189	0.4067	0.9135	0.4452	2.2460	1.1519	66
25	0.4363	0.4226	0.9063	0.4663	2.1445	1.1345	65
26	0.4538	0.4384	0.8988	0.4877	2.0503	1.1170	64
27	0.4712	0.4540	0.8910	0.5095	1.9626	1.0996	63
28	0.4887	0.4695	0.8829	0.5317	1.8807	1.0821	62
29	0.5061	0.4848	0.8746	0.5543	1.8040	1.0647	61
30	0.5236	0.5000	0.8660	0.5774	1.7321	1.0472	60
31	0.5411	0.5150	0.8572	0.6009	1.6643	1.0297	59
32	0.5585	0.5299	0.8480	0.6249	1.6003	1.0123	58
33	0.5760	0.5446	0.8387	0.6494	1.5399	0.9948	57
34	0.5934	0.5592	0.8290	0.6745	1.4826	0.9774	56
35	0.6109	0.5736	0.8192	0.7002	1.4281	0.9599	55
36	0.6283	0.5878	0.8090	0.7265	1.3764	0.9425	54
37	0.6458	0.6018	0.7986	0.7536	1.3270	0.9250	53
38	0.6632	0.6157	0.7880	0.7813	1.2799	0.9076	52
39	0.6807	0.6293	0.7771	0.8098	1.2349	0.8901	51
40	0.6981	0.6428	0.7660	0.8391	1.1918	0.8727	50
41	0.7156	0.6561	0.7547	0.8693	1.1504	0.8552	49
42	0.7330	0.6691	0.7431	0.9004	1.1106	0.8378	48
43	0.7505	0.6820	0.7314	0.9325	1.0724	0.8203	47
44	0.7679	0.6947	0.7193	0.9657	1.0355	0.8029	46
45	0.7854	0.7071	0.7071	1.0000	1.0000	0.7854	45
		Cos	Sin	Cot	Tan	Radians	Degrees

Table 3 Four-place values of trigonometric functions, angle θ in degrees and radians

Angle θ									
Degrees	**Radians**	**sin θ**	**csc θ**	**tan θ**	**cot θ**	**sec θ**	**cos θ**		
0° 00′	.0000	.0000	No value	.0000	No value	1.000	1.0000	1.5708	**90° 00′**
10	029	029	343.8	029	343.8	000	000	679	50
20	058	058	171.9	058	171.9	000	000	650	40
30	.0087	.0087	114.6	.0087	114.6	1.000	1.0000	1.5621	30
40	116	116	85.95	116	85.94	000	.9999	592	20
50	145	145	68.76	145	68.75	000	999	563	10
1° 00′	.0175	.0175	57.30	.0175	57.29	1.000	.9998	1.5533	**89° 00′**
10	204	204	49.11	204	49.10	000	998	504	50
20	233	233	42.98	233	42.96	000	997	475	40
30	.0262	.0262	38.20	.0262	38.19	1.000	.9997	1.5446	30
40	291	291	34.38	291	34.37	000	996	417	20
50	320	320	31.26	320	31.24	001	995	388	10
2° 00′	.0349	.0349	28.65	.0349	28.64	1.001	.9994	1.5359	**88° 00′**
10	378	378	26.45	378	26.43	001	993	330	50
20	407	407	24.56	407	24.54	001	992	301	40
30	.0436	.0436	22.93	.0437	22.90	1.001	.9990	1.5272	30
40	465	465	21.49	466	21.47	001	989	243	20
50	495	494	20.23	495	20.21	001	988	213	10
3° 00′	.0524	.0523	19.11	.0524	19.08	1.001	.9986	1.5184	**87° 00′**
10	553	552	18.10	553	18.07	002	985	155	50
20	582	581	17.20	582	17.17	002	983	126	40
30	.0611	.0610	16.38	.0612	16.35	1.002	.9981	1.5097	30
40	640	640	15.64	641	15.60	002	980	068	20
50	669	669	14.96	670	14.92	002	978	039	10
4° 00′	.0698	.0698	14.34	.0699	14.30	1.002	.9976	1.5010	**86° 00′**
10	727	727	13.76	729	13.73	003	974	981	50
20	756	756	13.23	758	13.20	003	971	952	40
30	.0785	.0785	12.75	.0787	12.71	1.003	.9969	1.4923	30
40	814	814	12.29	816	12.25	003	967	893	20
50	844	843	11.87	846	11.83	004	964	864	10
5° 00′	.0873	.0872	11.47	.0875	11.43	1.004	.9962	1.4835	**85° 00′**
10	902	901	11.10	904	11.06	004	959	806	50
20	931	929	10.76	934	10.71	004	957	777	40
30	.0960	.0958	10.43	.0963	10.39	1.005	.9954	1.4748	30
40	989	987	10.13	992	10.08	005	951	719	20
50	.1018	.1016	9.839	.1022	9.788	005	948	690	10
6° 00′	.1047	.1045	9.567	.1051	9.514	1.006	.9945	1.4661	**84° 00′**
		cos θ	**sec θ**	**cot θ**	**tan θ**	**csc θ**	**sin θ**	**Radians**	**Degrees**
								Angle θ	

Table 3 *(Continued)*

Angle θ									
Degrees	**Radians**	**sin θ**	**csc θ**	**tan θ**	**cot θ**	**sec θ**	**cos θ**		
6° 00′	.1047	.1045	9.567	.1051	9.514	1.006	.9945	1.4661	**84° 00′**
10	076	074	9.309	080	9.255	006	942	632	50
20	105	103	9.065	110	9.010	006	939	603	40
30	.1134	.1132	8.834	.1139	8.777	1.006	.9936	1.4573	30
40	164	161	8.614	169	8.556	007	932	544	20
50	193	190	8.405	198	8.345	007	929	515	10
7° 00′	.1222	.1219	8.206	.1228	8.144	1.008	.9925	1.4486	**83° 00′**
10	251	248	8.016	257	7.953	008	922	457	50
20	280	276	7.834	287	7.770	008	918	428	40
30	.1309	.1305	7.661	.1317	7.596	1.009	.9914	1.4399	30
40	338	334	7.496	346	7.429	009	911	370	20
50	367	363	7.337	376	7.269	009	907	341	10
8° 00′	.1396	.1392	7.185	.1405	7.115	1.010	.9903	1.4312	**82° 00′**
10	425	421	7.040	435	6.968	010	899	283	50
20	454	449	6.900	465	6.827	011	894	254	40
30	.1484	.1478	6.765	.1495	6.691	1.011	.9890	1.4224	30
40	513	507	6.636	524	6.561	012	886	195	20
50	542	536	6.512	554	6.435	012	881	166	10
9° 00′	.1571	.1564	6.392	.1584	6.314	1.012	.9877	1.4137	**81° 00′**
10	600	593	277	614	197	013	872	108	50
20	629	622	166	644	084	013	868	079	40
30	.1658	.1650	6.059	.1673	5.976	1.014	.9863	1.4050	30
40	687	679	5.955	703	871	014	858	1.4021	20
50	716	708	855	733	769	015	853	992	10
10° 00′	.1745	.1736	5.759	.1763	5.671	1.015	.9848	1.3963	**80° 00′**
10	774	765	665	793	576	016	843	934	50
20	804	794	575	823	485	016	838	904	40
30	.1833	.1822	5.487	.1853	5.396	1.017	.9833	1.3875	30
40	862	851	403	883	309	018	827	846	20
50	891	880	320	914	226	018	822	817	10
11° 00′	.1920	.1908	5.241	.1944	5.145	1.019	.9816	1.3788	**79° 00′**
10	949	937	164	974	066	019	811	759	50
20	978	965	089	.2004	4.989	020	805	730	40
30	.2007	.1994	5.016	.2035	4.915	1.020	.9799	1.3701	30
40	036	.2022	4.945	065	843	021	793	672	20
50	065	051	876	095	773	022	787	643	10
12° 00′	.2094	.2079	4.810	.2126	4.705	1.022	.9781	1.3614	**78° 00′**
10	123	108	745	156	638	023	775	584	50
20	153	136	682	186	574	024	769	555	40
30	.2182	.2164	4.620	.2217	4.511	1.024	.9763	1.3526	30
40	211	193	560	247	449	025	757	497	20
50	240	221	502	278	390	026	750	468	10
13° 00′	.2269	.2250	4.445	.2309	4.331	1.026	.9744	1.3439	**77° 00′**
		cos θ	**sec θ**	**cot θ**	**tan θ**	**csc θ**	**sin θ**	**Radians**	**Degrees**
								Angle θ	

Table 3 *(Continued)*

Angle θ									
Degrees	**Radians**	**sin θ**	**csc θ**	**tan θ**	**cot θ**	**sec θ**	**cos θ**		
13° 00′	.2269	.2250	4.445	.2309	4.331	1.026	.9744	1.3439	**77° 00′**
10	298	278	390	339	275	027	737	410	50
20	327	306	336	370	219	028	730	381	40
30	.2356	.2334	4.284	.2401	4.165	1.028	.9724	1.3352	30
40	385	363	232	432	113	029	717	323	20
50	414	391	182	462	061	030	710	294	10
14° 00′	.2443	.2419	4.134	.2493	4.011	1.031	.9703	1.3265	**76° 00′**
10	473	447	086	524	3.962	031	696	235	50
20	502	476	039	555	914	032	689	206	40
30	.2531	.2504	3.994	.2586	3.867	1.033	.9681	1.3177	30
40	560	532	950	617	821	034	674	148	20
50	589	560	906	648	776	034	667	119	10
15° 00′	.2618	.2588	3.864	.2679	3.732	1.035	.9659	1.3090	**75° 00′**
10	647	616	822	711	689	036	652	061	50
20	676	644	782	742	647	037	644	032	40
30	.2705	.2672	3.742	.2773	3.606	1.038	.9636	1.3003	30
40	734	700	703	805	566	039	628	974	20
50	763	728	665	836	526	039	621	945	10
16° 00′	.2793	.2756	3.628	.2867	3.487	1.040	.9613	1.2915	**74° 00′**
10	822	784	592	899	450	041	605	886	50
20	851	812	556	931	412	042	596	857	40
30	.2880	.2840	3.521	.2962	3.376	1.043	.9588	1.2828	30
40	909	868	487	994	340	044	580	799	20
50	938	896	453	.3026	305	045	572	770	10
17° 00′	.2967	.2924	3.420	.3057	3.271	1.046	.9563	1.2741	**73° 00′**
10	996	952	388	089	237	047	555	712	50
20	.3025	979	357	121	204	048	546	683	40
30	.3054	.3007	3.326	.3153	3.172	1.048	.9537	1.2654	30
40	083	035	295	185	140	049	528	625	20
50	113	062	265	217	108	050	520	595	10
18° 00′	.3142	.3090	3.236	.3249	3.078	1.051	.9511	1.2566	**72° 00′**
10	171	118	207	281	047	052	502	537	50
20	200	145	179	314	018	053	492	508	40
30	.3229	.3173	3.152	.3346	2.989	1.054	.9483	1.2479	30
40	258	201	124	378	960	056	474	450	20
50	287	228	098	411	932	057	465	421	10
19° 00′	.3316	.3256	3.072	.3443	2.904	1.058	.9455	1.2392	**71° 00′**
10	345	283	046	476	877	059	446	363	50
20	374	311	021	508	850	060	436	334	40
30	.3403	.3338	2.996	.3541	2.824	1.061	.9426	1.2305	30
40	432	365	971	574	798	062	417	275	20
50	462	393	947	607	773	063	407	246	10
20° 00′	.3491	.3420	2.924	.3640	2.747	1.064	.9397	1.2217	**70° 00′**
		cos θ	**sec θ**	**cot θ**	**tan θ**	**csc θ**	**sin θ**	**Radians**	**Degrees**
								Angle θ	

Table 3 *(Continued)*

Angle θ									
Degrees	Radians	sin θ	csc θ	tan θ	cot θ	sec θ	cos θ		
20° 00′	.3491	.3420	2.924	.3640	2.747	1.064	.9397	1.2217	**70° 00′**
10	520	448	901	673	723	065	387	188	50
20	549	475	878	706	699	066	377	159	40
30	.3578	.3502	2.855	.3739	2.675	1.068	.9367	1.2130	30
40	607	529	833	772	651	069	356	101	20
50	636	557	812	805	628	070	346	072	10
21° 00′	.3665	.3584	2.790	.3839	2.605	1.071	.9336	1.2043	**69° 00′**
10	694	611	769	872	583	072	325	1.2014	50
20	723	638	749	906	560	074	315	985	40
30	.3752	.3665	2.729	.3939	2.539	1.075	.9304	1.1956	30
40	782	692	709	973	517	076	293	926	20
50	811	719	689	.4006	496	077	283	897	10
22° 00′	.3840	.3746	2.669	.4040	2.475	1.079	.9272	1.1868	**68° 00′**
10	869	773	650	074	455	080	261	839	50
20	898	800	632	108	434	081	250	810	40
30	.3927	.3827	2.613	.4142	2.414	1.082	.9239	1.1781	30
40	956	854	595	176	394	084	228	752	20
50	985	881	577	210	375	085	216	723	10
23° 00′	.4014	.3907	2.559	.4245	2.356	1.086	.9205	1.1694	**67° 00′**
10	043	934	542	279	337	088	194	665	50
20	072	961	525	314	318	089	182	636	40
30	.4102	.3987	2.508	.4348	2.300	1.090	.9171	1.1606	30
40	131	.4014	491	383	282	092	159	577	20
50	160	041	475	417	264	093	147	548	10
24° 00′	.4189	.4067	2.459	.4452	2.246	1.095	.9135	1.1519	**66° 00′**
10	218	094	443	487	229	096	124	490	50
20	247	120	427	522	211	097	112	461	40
30	.4276	.4147	2.411	.4557	2.194	1.099	.9100	1.1432	30
40	305	173	396	592	177	100	088	403	20
50	334	200	381	628	161	102	075	374	10
25° 00′	.4363	.4226	2.366	.4663	2.145	1.103	.9063	1.1345	**65° 00′**
10	392	253	352	699	128	105	051	316	50
20	422	279	337	734	112	106	038	286	40
30	.4451	.4305	2.323	.4770	2.097	1.108	.9026	1.1257	30
40	480	331	309	806	081	109	013	228	20
50	509	358	295	841	066	111	001	199	10
26° 00′	.4538	.4384	2.281	.4877	2.050	1.113	.8988	1.1170	**64° 00′**
10	567	410	268	913	035	114	975	141	50
20	596	436	254	950	020	116	962	112	40
30	.4625	.4462	2.241	.4986	2.006	1.117	.8949	1.1083	30
40	654	488	228	.5022	1.991	119	936	054	20
50	683	514	215	059	977	121	923	1.1025	10
27° 00′	.4712	.4540	2.203	.5095	1.963	1.122	.8910	1.0996	**63° 00′**
		cos θ	sec θ	cot θ	tan θ	csc θ	sin θ	Radians	Degrees
								Angle θ	

Table 3 *(Continued)*

Angle θ									
Degrees	**Radians**	**sin θ**	**csc θ**	**tan θ**	**cot θ**	**sec θ**	**cos θ**		
27° 00′	.4712	.4540	2.203	.5095	1.963	1.122	.8910	1.0996	**63° 00′**
10	741	566	190	132	949	124	897	966	50
20	771	592	178	169	935	126	884	937	40
30	.4800	.4617	2.166	.5206	1.921	1.127	.8870	1.0908	30
40	829	643	154	243	907	129	857	879	20
50	858	669	142	280	894	131	843	850	10
28° 00′	.4887	.4695	2.130	.5317	1.881	1.133	.8829	1.0821	**62° 00′**
10	916	720	118	354	868	134	816	792	50
20	945	746	107	392	855	136	802	763	40
30	.4974	.4772	2.096	.5430	1.842	1.138	.8788	1.0734	30
40	.5003	797	085	467	829	140	774	705	20
50	032	823	074	505	816	142	760	676	10
29° 00′	.5061	.4848	2.063	.5543	1.804	1.143	.8746	1.0647	**61° 00′**
10	091	874	052	581	792	145	732	617	50
20	120	899	041	619	780	147	718	588	40
30	.5149	.4924	2.031	.5658	1.767	1.149	.8704	1.0559	30
40	178	950	020	696	756	151	689	530	20
50	207	975	010	735	744	153	675	501	10
30° 00′	.5236	.5000	2.000	.5774	1.732	1.155	.8660	1.0472	**60° 00′**
10	265	025	1.990	812	720	157	646	443	50
20	294	050	980	851	709	159	631	414	40
30	.5323	.5075	1.970	.5890	1.698	1.161	.8616	1.0385	30
40	352	100	961	930	686	163	601	356	20
50	381	125	951	969	675	165	587	327	10
31° 00′	.5411	.5150	1.942	.6009	1.664	1.167	.8572	1.0297	**59° 00′**
10	440	175	932	048	653	169	557	268	50
20	469	200	923	088	643	171	542	239	40
30	.5498	.5225	1.914	.6128	1.632	1.173	.8526	1.0210	30
40	527	250	905	168	621	175	511	181	20
50	556	275	896	208	611	177	496	152	10
32° 00′	.5585	.5299	1.887	.6249	1.600	1.179	.8480	1.0123	**58° 00′**
10	614	324	878	289	590	181	465	094	50
20	643	348	870	330	580	184	450	065	40
30	.5672	.5373	1.861	.6371	1.570	1.186	.8434	1.0036	30
40	701	398	853	412	560	188	418	1.0007	20
50	730	422	844	453	550	190	403	977	10
33° 00′	.5760	.5446	1.836	.6494	1.540	1.192	.8387	.9948	**57° 00′**
10	789	471	828	536	530	195	371	919	50
20	818	495	820	577	520	197	355	890	40
30	.5847	.5519	1.812	.6619	1.511	1.199	.8339	.9861	30
40	876	544	804	661	501	202	323	832	20
50	905	568	796	703	1.492	204	307	803	10
34° 00′	.5934	.5592	1.788	.6745	1.483	1.206	.8290	.9774	**56° 00′**
		cos θ	**sec θ**	**cot θ**	**tan θ**	**csc θ**	**sin θ**	**Radians**	**Degrees**
								Angle θ	

Table 3 *(Continued)*

Angle θ									
Degrees	Radians	sin θ	csc θ	tan θ	cot θ	sec θ	cos θ		
34° 00′	.5934	.5592	1.788	.6745	1.483	1.206	.8290	.9774	**56° 00′**
10	963	616	781	787	473	209	274	745	50
20	992	640	773	830	464	211	258	716	40
30	.6021	.5664	1.766	.6873	1.455	1.213	.8241	.9687	30
40	050	688	758	916	446	216	225	657	20
50	080	712	751	959	437	218	208	628	10
35° 00′	.6109	.5736	1.743	.7002	1.428	1.221	.8192	.9599	**55° 00′**
10	138	760	736	046	419	223	175	570	50
20	167	783	729	089	411	226	158	541	40
30	.6196	.5807	1.722	.7133	1.402	1.228	.8141	.9512	30
40	225	831	715	177	393	231	124	483	20
50	254	854	708	221	385	233	107	454	10
36° 00′	.6283	.5878	1.701	.7265	1.376	1.236	.8090	.9425	**54° 00′**
10	312	901	695	310	368	239	073	396	50
20	341	925	688	355	360	241	056	367	40
30	.6370	.5948	1.681	.7400	1.351	1.244	.8039	.9338	30
40	400	972	675	445	343	247	021	308	20
50	429	995	668	490	335	249	004	279	10
37° 00′	.6458	.6018	1.662	.7536	1.327	1.252	.7986	.9250	**53° 00′**
10	487	041	655	581	319	255	969	221	50
20	516	065	649	627	311	258	951	192	40
30	.6545	.6088	1.643	.7673	1.303	1.260	.7934	.9163	30
40	574	111	636	720	295	263	916	134	20
50	603	134	630	766	288	266	898	105	10
38° 00′	.6632	.6157	1.624	.7813	1.280	1.269	.7880	.9076	**52° 00′**
10	661	180	618	860	272	272	862	047	50
20	690	202	612	907	265	275	844	.9018	40
30	.6720	.6225	1.606	.7954	1.257	1.278	.7826	.8988	30
40	749	248	601	.8002	250	281	808	959	20
50	778	271	595	050	242	284	790	930	10
39° 00′	.6807	.6293	1.589	.8098	1.235	1.287	.7771	.8901	**51° 00′**
10	836	316	583	146	228	290	753	872	50
20	865	338	578	195	220	293	735	843	40
30	.6894	.6361	1.572	.8243	1.213	1.296	.7716	.8814	30
40	923	383	567	292	206	299	698	785	20
50	952	406	561	342	199	302	679	756	10
40° 00′	.6981	.6428	1.556	.8391	1.192	1.305	.7660	.8727	**50° 00′**
10	.7010	450	550	441	185	309	642	698	50
20	039	472	545	491	178	312	623	668	40
30	.7069	.6494	1.540	.8541	1.171	1.315	.7604	.8639	30
40	098	517	535	591	164	318	585	610	20
50	127	539	529	642	157	322	566	581	10
41° 00′	.7156	.6561	1.524	.8693	1.150	1.325	.7547	.8552	**49° 00′**
		cos θ	sec θ	cot θ	tan θ	csc θ	sin θ	Radians	Degrees
								Angle θ	

Table 3 *(Continued)*

Angle θ									
Degrees	Radians	sin θ	csc θ	tan θ	cot θ	sec θ	cos θ		
41° 00′	.7156	.6561	1.524	.8693	1.150	1.325	.7547	.8552	**49° 00′**
10	185	583	519	744	144	328	528	523	50
20	214	604	514	796	137	332	509	494	40
30	.7243	.6626	1.509	.8847	1.130	1.335	.7490	.8465	30
40	272	648	504	899	124	339	470	436	20
50	301	670	499	952	117	342	451	407	10
42° 00′	.7330	.6691	1.494	.9004	1.111	1.346	.7431	.8378	**48° 00′**
10	359	713	490	057	104	349	412	348	50
20	389	734	485	110	098	353	392	319	40
30	.7418	.6756	1.480	.9163	1.091	1.356	.7373	.8290	30
40	447	777	476	217	085	360	353	261	20
50	476	799	471	271	079	364	333	232	10
43° 00′	.7505	.6820	1.466	.9325	1.072	1.367	.7314	.8203	**47° 00′**
10	534	841	462	380	066	371	294	174	50
20	563	862	457	435	060	375	274	145	40
30	.7592	.6884	1.453	.9490	1.054	1.379	.7254	.8116	30
40	621	905	448	545	048	382	234	087	20
50	650	926	444	601	042	386	214	058	10
44° 00′	.7679	.6947	1.440	.9657	1.036	1.390	.7193	.8029	**46° 00′**
10	709	967	435	713	030	394	173	.7999	50
20	738	988	431	770	024	398	153	970	40
30	.7767	.7009	1.427	.9827	1.018	1.402	.7133	.7941	30
40	796	030	423	884	012	406	112	912	20
50	825	050	418	942	006	410	092	883	10
45° 00′	.7854	.7071	1.414	1.000	1.000	1.414	.7071	.7854	**45° 00′**
		cos θ	sec θ	cot θ	tan θ	csc θ	sin θ	Radians	Degrees
								Angle θ	

Table 4 Four-place logarithms of trigonometric functions, angle θ in degrees

Attach −10 to Logarithms Obtained from This Table

Angle θ	L sin θ	L csc θ	L tan θ	L cot θ	L sec θ	L cos θ	
0° 00′	No value	No value	No value	No value	10.0000	10.0000	**90° 00′**
10′	7.4637	12.5363	7.4637	12.5363	.0000	.0000	50′
20′	.7648	.2352	.7648	.2352	.0000	.0000	40′
30′	7.9408	12.0592	7.9409	12.0591	.0000	.0000	30′
40′	8.0658	11.9342	8.0658	11.9342	.0000	.0000	20′
50′	.1627	.8373	.1627	.8373	.0000	10.0000	10′
1° 00′	8.2419	11.7581	8.2419	11.7581	10.0001	9.9999	**89° 00′**
10′	.3088	.6912	.3089	.6911	.0001	.9999	50′
20′	.3668	.6332	.3669	.6331	.0001	.9999	40′
30′	.4179	.5821	.4181	.5819	.0001	.9999	30′
40′	.4637	.5363	.4638	.5362	.0002	.9998	20′
50′	.5050	.4950	.5053	.4947	.0002	.9998	10′
2° 00′	8.5428	11.4572	8.5431	11.4569	10.0003	9.9997	**88° 00′**
10′	.5776	.4224	.5779	.4221	.0003	.9997	50′
20′	.6097	.3903	.6101	.3899	.0004	.9996	40′
30′	.6397	.3603	.6401	.3599	.0004	.9996	30′
40′	.6677	.3323	.6682	.3318	.0005	.9995	20′
50′	.6940	.3060	.6945	.3055	.0005	.9995	10′
3° 00′	8.7188	11.2812	8.7194	11.2806	10.0006	9.9994	**87° 00′**
10′	.7423	.2577	.7429	.2571	.0007	.9993	50′
20′	.7645	.2355	.7652	.2348	.0007	.9993	40′
30′	.7857	.2143	.7865	.2135	.0008	.9992	30′
40′	.8059	.1941	.8067	.1933	.0009	.9991	20′
50′	.8251	.1749	.8261	.1739	.0010	.9990	10′
4° 00′	8.8436	11.1564	8.8446	11.1554	10.0011	9.9989	**86° 00′**
10′	.8613	.1387	.8624	.1376	.0011	.9989	50′
20′	.8783	.1217	.8795	.1205	.0012	.9988	40′
30′	.8946	.1054	.8960	.1040	.0013	.9987	30′
40′	.9104	.0896	.9118	.0882	.0014	.9986	20′
50′	.9256	.0744	.9272	.0728	.0015	.9985	10′
5° 00′	8.9403	11.0597	8.9420	11.0580	10.0017	9.9983	**85° 00′**
10′	.9545	.0455	.9563	.0437	.0018	.9982	50′
20′	.9682	.0318	.9701	.0299	.0019	.9981	40′
30′	.9816	.0184	.9836	.0164	.0020	.9980	30′
40′	8.9945	11.0055	8.9966	11.0034	.0021	.9979	20′
50′	9.0070	10.9930	9.0093	10.9907	.0023	.9977	10′
6° 00′	9.0192	10.9808	9.0216	10.9784	10.0024	9.9976	**84° 00′**
	L cos θ	L sec θ	L cot θ	L tan θ	L csc θ	L sin θ	Angle θ

Table 4 *(Continued)*

Attach −10 to Logarithms Obtained from This Table

Angle θ	L sin θ	L csc θ	L tan θ	L cot θ	L sec θ	L cos θ	
6° 00′	9.0192	10.9808	9.0216	10.9784	10.0024	9.9976	**84° 00′**
10′	.0311	.9689	.0336	.9664	.0025	.9975	50′
20′	.0426	.9574	.0453	.9547	.0027	.9973	40′
30′	.0539	.9461	.0567	.9433	.0028	.9972	30′
40′	.0648	.9352	.0678	.9322	.0029	.9971	20′
50′	.0755	.9245	.0786	.9214	.0031	.9969	10′
7° 00′	9.0859	10.9141	9.0891	10.9109	10.0032	9.9968	**83° 00′**
10′	.0961	.9039	.0995	.9005	.0034	.9966	50′
20′	.1060	.8940	.1096	.8904	.0036	.9964	40′
30′	.1157	.8843	.1194	.8806	.0037	.9963	30′
40′	.1252	.8748	.1291	.8709	.0039	.9961	20′
50′	.1345	.8655	.1385	.8615	.0041	.9959	10′
8° 00′	9.1436	10.8564	9.1478	10.8522	10.0042	9.9958	**82° 00′**
10′	.1525	.8475	.1569	.8431	.0044	.9956	50′
20′	.1612	.8388	.1658	.8342	.0046	.9954	40′
30′	.1697	.8303	.1745	.8255	.0048	.9952	30′
40′	.1781	.8219	.1831	.8169	.0050	.9950	20′
50′	.1863	.8137	.1915	.8085	.0052	.9948	10′
9° 00′	9.1943	10.8057	9.1997	10.8003	10.0054	9.9946	**81° 00′**
10′	.2022	.7978	.2078	.7922	.0056	.9944	50′
20′	.2100	.7900	.2158	.7842	.0058	.9942	40′
30′	.2176	.7824	.2236	.7764	.0060	.9940	30′
40′	.2251	.7749	.2313	.7687	.0062	.9938	20′
50′	.2324	.7676	.2389	.7611	.0064	.9936	10′
10° 00′	9.2397	10.7603	9.2463	10.7537	10.0066	9.9934	**80° 00′**
10′	.2468	.7532	.2536	.7464	.0069	.9931	50′
20′	.2538	.7462	.2609	.7391	.0071	.9929	40′
30′	.2606	.7394	.2680	.7320	.0073	.9927	30′
40′	.2674	.7326	.2750	.7250	.0076	.9924	20′
50′	.2740	.7260	.2819	.7181	.0078	.9922	10′
11° 00′	9.2806	10.7194	9.2887	10.7113	10.0081	9.9919	**79° 00′**
10′	.2870	.7130	.2953	.7047	.0083	.9917	50′
20′	.2934	.7066	.3020	.6980	.0086	.9914	40′
30′	.2997	.7003	.3085	.6915	.0088	.9912	30′
40′	.3058	.6942	.3149	.6851	.0091	.9909	20′
50′	.3119	.6881	.3212	.6788	.0093	.9907	10′
12° 00′	9.3179	10.6821	9.3275	10.6725	10.0096	9.9904	**78° 00′**
10′	.3238	.6762	.3336	.6664	.0099	.9901	50′
20′	.3296	.6704	.3397	.6603	.0101	.9899	40′
30′	.3353	.6647	.3458	.6542	.0104	.9896	30′
40′	.3410	.6590	.3517	.6483	.0107	.9893	20′
50′	.3466	.6534	.3576	.6424	.0110	.9890	10′
13° 00′	9.3521	10.6479	9.3634	10.6366	10.0113	9.9887	**77° 00′**
	L cos θ	**L sec θ**	**L cot θ**	**L tan θ**	**L csc θ**	**L sin θ**	**Angle θ**

Table 4 *(Continued)*

Attach −10 to Logarithms Obtained from This Table

Angle θ	L sin θ	L csc θ	L tan θ	L cot θ	L sec θ	L cos θ	
13° 00′	9.3521	10.6479	9.3634	10.6366	10.0113	9.9887	**77° 00′**
10′	.3575	.6425	.3691	.6309	.0116	.9884	50′
20′	.3629	.6371	.3748	.6252	.0119	.9881	40′
30′	.3682	.6318	.3804	.6196	.0122	.9878	30′
40′	.3734	.6266	.3859	.6141	.0125	.9875	20′
50′	.3786	.6214	.3914	.6086	.0128	.9872	10′
14° 00′	9.3837	10.6163	9.3968	10.6032	10.0131	9.9869	**76° 00′**
10′	.3887	.6113	.4021	.5979	.0134	.9866	50′
20′	.3937	.6063	.4074	.5926	.0137	.9863	40′
30′	.3986	.6014	.4127	.5873	.0141	.9859	30′
40′	.4035	.5965	.4178	.5822	.0144	.9856	20′
50′	.4083	.5917	.4230	.5770	.0147	.9853	10′
15° 00′	9.4130	10.5870	9.4281	10.5719	10.0151	9.9849	**75° 00′**
10′	.4177	.5823	.4331	.5669	.0154	.9846	50′
20′	.4223	.5777	.4381	.5619	.0157	.9843	40′
30′	.4269	.5731	.4430	.5570	.0161	.9839	30′
40′	.4314	.5686	.4479	.5521	.0164	.9836	20′
50′	.4359	.5641	.4527	.5473	.0168	.9832	10′
16° 00′	9.4403	10.5597	9.4575	10.5425	10.0172	9.9828	**74° 00′**
10′	.4447	.5553	.4622	.5378	.0175	.9825	50′
20′	.4491	.5509	.4669	.5331	.0179	.9821	40′
30′	.4533	.5467	.4716	.5284	.0183	.9817	30′
40′	.4576	.5424	.4762	.5238	.0186	.9814	20′
50′	.4618	.5382	.4808	.5192	.0190	.9810	10′
17° 00′	9.4659	10.5341	9.4853	10.5147	10.0194	9.9806	**73° 00′**
10′	.4700	.5300	.4898	.5102	.0198	.9802	50′
20′	.4741	.5259	.4943	.5057	.0202	.9798	40′
30′	.4781	.5219	.4987	.5013	.0206	.9794	30′
40′	.4821	.5179	.5031	.4969	.0210	.9790	20′
50′	.4861	.5139	.5075	.4925	.0214	.9786	10′
18° 00′	9.4900	10.5100	9.5118	10.4882	10.0218	9.9782	**72° 00′**
10′	.4939	.5061	.5161	.4839	.0222	.9778	50′
20′	.4977	.5023	.5203	.4797	.0226	.9774	40′
30′	.5015	.4985	.5245	.4755	.0230	.9770	30′
40′	.5052	.4948	.5287	.4713	.0235	.9765	20′
50′	.5090	.4910	.5329	.4671	.0239	.9761	10′
19° 00′	9.5126	10.4874	9.5370	10.4630	10.0243	9.9757	**71° 00′**
10′	.5163	.4837	.5411	.4589	.0248	.9752	50′
20′	.5199	.4801	.5451	.4549	.0252	.9748	40′
30′	.5235	.4765	.5491	.4509	.0257	.9743	30′
40′	.5270	.4730	.5531	.4469	.0261	.9739	20′
50′	.5306	.4694	.5571	.4429	.0266	.9734	10′
20° 00′	9.5341	10.4659	9.5611	10.4389	10.0270	9.9730	**70° 00′**
	L cos θ	**L sec θ**	**L cot θ**	**L tan θ**	**L csc θ**	**L sin θ**	**Angle θ**

Table 4 *(Continued)*

Attach −10 to Logarithms Obtained from This Table

Angle θ	L sin θ	L csc θ	L tan θ	L cot θ	L sec θ	L cos θ	
20° 00′	9.5341	10.4659	9.5611	10.4389	10.0270	9.9730	**70° 00′**
10′	.5375	.4625	.5650	.4350	.0275	.9725	50′
20′	.5409	.4591	.5689	.4311	.0279	.9721	40′
30′	.5443	.4557	.5727	.4273	.0284	.9716	30′
40′	.5477	.4523	.5766	.4234	.0289	.9711	20′
50′	.5510	.4490	.5804	.4196	.0294	.9706	10′
21° 00′	9.5543	10.4457	9.5842	10.4158	10.0298	9.9702	**69° 00′**
10′	.5576	.4424	.5879	.4121	.0303	.9797	50′
20′	.5609	.4391	.5917	.4083	.0308	.9692	40′
30′	.5641	.4359	.5954	.4046	.0313	.9687	30′
40′	.5673	.4327	.5991	.4009	.0318	.9682	20′
50′	.5704	.4296	.6028	.3972	.0323	.9677	10′
22° 00′	9.5736	10.4264	9.6064	10.3936	10.0328	9.9672	**68° 00′**
10′	.5767	.4233	.6100	.3900	.0333	.9667	50′
20′	.5798	.4202	.6136	.3864	.0339	.9661	40′
30′	.5828	.4172	.6172	.3828	.0344	.9656	30′
40′	.5859	.4141	.6208	.3792	.0349	.9651	20′
50′	.5889	.4111	.6243	.3757	.0354	.9646	10′
23° 00′	9.5919	10.4081	9.6279	10.3721	10.0360	9.9640	**67° 00′**
10′	.5948	.4052	.6314	.3686	.0365	.9635	50′
20′	.5978	.4022	.6348	.3652	.0371	.9629	40′
30′	.6007	.3993	.6383	.3617	.0376	.9624	30′
40′	.6036	.3964	.6417	.3583	.0382	.9618	20′
50′	.6065	.3935	.6452	.3548	.0387	.9613	10′
24° 00′	9.6093	10.3907	9.6486	10.3514	10.0393	9.9607	**66° 00′**
10′	.6121	.3879	.6520	.3480	.0398	.9602	50′
20′	.6149	.3851	.6553	.3447	.0404	.9596	40′
30′	.6177	.3823	.6587	.3413	.0410	.9590	30′
40′	.6205	.3795	.6620	.3380	.0416	.9584	20′
50′	.6232	.3768	.6654	.3346	.0421	.9579	10′
25° 00′	9.6259	10.3741	9.6687	10.3313	10.0427	9.9573	**65° 00′**
10′	.6286	.3714	.6720	.3280	.0433	.9567	50′
20′	.6313	.3687	.6752	.3248	.0439	.9561	40′
30′	.6340	.3660	.6785	.3215	.0445	.9555	30′
40′	.6366	.3634	.6817	.3183	.0451	.9549	20′
50′	.6392	.3608	.6850	.3150	.0457	.9543	10′
26° 00′	9.6418	10.3582	9.6882	10.3118	10.0463	9.9537	**64° 00′**
10′	.6444	.3556	.6914	.3086	.0470	.9530	50′
20′	.6470	.3530	.6946	.3054	.0476	.9524	40′
30′	.6495	.3505	.6977	.3023	.0482	.9518	30′
40′	.6521	.3479	.7009	.2991	.0488	.9512	20′
50′	.6546	.3454	.7040	.2960	.0495	.9505	10′
27° 00′	9.6570	10.3430	9.7072	10.2928	10.0501	9.9499	**63° 00′**
	L cos θ	**L sec θ**	**L cot θ**	**L tan θ**	**L csc θ**	**L sin θ**	**Angle θ**

Table 4 *(Continued)*

Attach −10 to Logarithms Obtained from This Table

Angle θ	L sin θ	L csc θ	L tan θ	L cot θ	L sec θ	L cos θ	
27° 00′	9.6570	10.3430	9.7072	10.2928	10.0501	9.9499	**63° 00′**
10′	.6595	.3405	.7103	.2897	.0508	.9492	50′
20′	.6620	.3380	.7134	.2866	.0514	.9486	40′
30′	.6644	.3356	.7165	.2835	.0521	.9479	30′
40′	.6668	.3332	.7196	.2804	.0527	.9473	20′
50′	.6692	.3308	.7226	.2774	.0534	.9466	10′
28° 00′	9.6716	10.3284	9.7257	10.2743	10.0541	9.9459	**62° 00′**
10′	.6740	.3260	.7287	.2713	.0547	.9453	50′
20′	.6763	.3237	.7317	.2683	.0554	.9446	40′
30′	.6787	.3213	.7348	.2652	.0561	.9439	30′
40′	.6810	.3190	.7378	.2622	.0568	.9432	20′
50′	.6833	.3167	.7408	.2592	.0575	.9425	10′
29° 00′	9.6856	10.3144	9.7438	10.2562	10.0582	9.9418	**61° 00′**
10′	.6878	.3122	.7467	.2533	.0589	.9411	50′
20′	.6901	.3099	.7497	.2503	.0596	.9404	40′
30′	.6923	.3077	.7526	.2474	.0603	.9397	30′
40′	.6946	.3054	.7556	.2444	.0610	.9390	20′
50′	.6968	.3032	.7585	.2415	.0617	.9383	10′
30° 00′	9.6990	10.3010	9.7614	10.2386	10.0625	9.9375	**60° 00′**
10′	.7012	.2988	.7644	.2356	.0632	.9368	50′
20′	.7033	.2967	.7673	.2327	.0639	.9361	40′
30′	.7055	.2945	.7701	.2299	.0647	.9353	30′
40′	.7076	.2924	.7730	.2270	.0654	.9346	20′
50′	.7097	.2903	.7759	.2241	.0662	.9338	10′
31° 00′	9.7118	10.2882	9.7788	10.2212	10.0669	9.9331	**59° 00′**
10′	.7139	.2861	.7816	.2184	.0677	.9323	50′
20′	.7160	.2840	.7845	.2155	.0685	.9315	40′
30′	.7181	.2819	.7873	.2127	.0692	.9308	30′
40′	.7201	.2799	.7902	.2098	.0700	.9300	20′
50′	.7222	.2778	.7930	.2070	.0708	.9292	10′
32° 00′	9.7242	10.2758	9.7958	10.2042	10.0716	9.9284	**58° 00′**
10′	.7262	.2738	.7986	.2014	.0724	.9276	50′
20′	.7282	.2718	.8014	.1986	.0732	.9268	40′
30′	.7302	.2698	.8042	.1958	.0740	.9260	30′
40′	.7322	.2678	.8070	.1930	.0748	.9252	20′
50′	.7342	.2658	.8097	.1903	.0756	.9244	10′
33° 00′	9.7361	10.2639	9.8125	10.1875	10.0764	9.9236	**57° 00′**
10′	.7380	.2620	.8153	.1847	.0772	.9228	50′
20′	.7400	.2600	.8180	.1820	.0781	.9219	40′
30′	.7419	.2581	.8208	.1792	.0789	.9211	30′
40′	.7438	.2562	.8235	.1765	.0797	.9203	20′
50′	.7457	.2543	.8263	.1737	.0806	.9194	10′
34° 00′	9.7476	10.2524	9.8290	10.1710	10.0814	9.9186	**56° 00′**
	L cos θ	**L sec θ**	**L cot θ**	**L tan θ**	**L csc θ**	**L sin θ**	**Angle θ**

Table 4 (*Continued*)

Attach −10 to Logarithms Obtained from This Table

Angle θ	L sin θ	L csc θ	L tan θ	L cot θ	L sec θ	L cos θ	
34° 00′	9.7476	10.2524	9.8290	10.1710	10.0814	9.9186	**56° 00′**
10′	.7494	.2506	.8317	.1683	.0823	.9177	50′
20′	.7513	.2487	.8344	.1656	.0831	.9169	40′
30′	.7531	.2469	.8371	.1629	.0840	.9160	30′
40′	.7550	.2450	.8398	.1602	.0849	.9151	20′
50′	.7568	.2432	.8425	.1575	.0858	.9142	10′
35° 00′	9.7586	10.2414	9.8452	10.1548	10.0866	9.9134	**55° 00′**
10′	.7604	.2396	.8479	.1521	.0875	.9125	50′
20′	.7622	.2378	.8506	.1494	.0884	.9116	40′
30′	.7640	.2360	.8533	.1467	.0893	.9107	30′
40′	.7657	.2343	.8559	.1441	.0902	.9098	20′
50′	.7675	.2325	.8586	.1414	.0911	.9089	10′
36° 00′	9.7692	10.2308	9.8613	10.1387	10.0920	9.9080	**54° 00′**
10′	.7710	.2290	.8639	.1361	.0930	.9070	50′
20′	.7727	.2273	.8666	.1334	.0939	.9061	40′
30′	.7744	.2256	.8692	.1308	.0948	.9052	30′
40′	.7761	.2239	.8718	.1282	.0958	.9042	20′
50′	.7778	.2222	.8745	.1255	.0967	.9033	10′
37° 00′	9.7795	10.2205	9.8771	10.1229	10.0977	9.9023	**53° 00′**
10′	.7811	.2189	.8797	.1203	.0986	.9014	50′
20′	.7828	.2172	.8824	.1176	.0996	.9004	40′
30′	.7844	.2156	.8850	.1150	.1005	.8995	30′
40′	.7861	.2139	.8876	.1124	.1015	.8985	20′
50′	.7877	.2123	.8902	.1098	.1025	.8975	10′
38° 00′	9.7893	10.2107	9.8928	10.1072	10.1035	9.8965	**52° 00′**
10′	.7910	.2090	.8954	.1046	.1045	.8955	50′
20′	.7926	.2074	.8980	.1020	.1055	.8945	40′
30′	.7941	.2059	.9006	.0994	.1065	.8935	30′
40′	.7957	.2043	.9032	.0968	.1075	.8925	20′
50′	.7973	.2027	.9058	.0942	.1085	.8915	10′
39° 00′	9.7989	10.2011	9.9084	10.0916	10.1095	9.8905	**51° 00′**
10′	.8004	.1996	.9110	.0890	.1105	.8895	50′
20′	.8020	.1980	.9135	.0865	.1116	.8884	40′
30′	.8035	.1965	.9161	.0839	.1126	.8874	30′
40′	.8050	.1950	.9187	.0813	.1136	.8864	20′
50′	.8066	.1934	.9212	.0788	.1147	.8853	10′
40° 00′	9.8081	10.1919	9.9238	10.0762	10.1157	9.8843	**50° 00′**
10′	.8096	.1904	.9264	.0736	.1168	.8832	50′
20′	.8111	.1889	.9289	.0711	.1179	.8821	40′
30′	.8125	.1875	.9315	.0685	.1190	.8810	30′
40′	.8140	.1860	.9341	.0659	.1200	.8800	20′
50′	.8155	.1845	.9366	.0634	.1211	.8789	10′
41° 00′	9.8169	10.1831	9.9392	10.0608	10.1222	9.8778	**49° 00′**
	L cos θ	L sec θ	L cot θ	L tan θ	L csc θ	L sin θ	Angle θ

Table 4 (*Continued*)

Attach −10 to Logarithms Obtained from This Table

Angle θ	L sin θ	L csc θ	L tan θ	L cot θ	L sec θ	L cos θ	
41° 00′	9.8169	10.1831	9.9392	10.0608	10.1222	9.8778	**49° 00′**
10′	.8184	.1816	.9417	.0583	.1233	.8767	50′
20′	.8198	.1802	.9443	.0557	.1244	.8756	40′
30′	.8213	.1787	.9468	.0532	.1255	.8745	30′
40′	.8227	.1773	.9494	.0506	.1267	.8733	20′
50′	.8241	.1759	.9519	.0481	.1278	.8722	10′
42° 00′	9.8255	10.1745	9.9544	10.0456	10.1289	9.8711	**48° 00′**
10′	.8269	.1731	.9570	.0430	.1301	.8699	50′
20′	.8283	.1717	.9595	.0405	.1312	.8688	40′
30′	.8297	.1703	.9621	.0379	.1324	.8676	30′
40′	.8311	.1689	.9646	.0354	.1335	.8665	20′
50′	.8324	.1676	.9671	.0329	.1347	.8653	10′
43° 00′	9.8338	10.1662	9.9697	10.0303	10.1359	9.8641	**47° 00′**
10′	.8351	.1649	.9722	.0278	.1371	.8629	50′
20′	.8365	.1635	.9747	.0253	.1382	.8618	40′
30′	.8378	.1622	.9772	.0228	.1394	.8606	30′
40′	.8391	.1609	.9798	.0202	.1406	.8594	20′
50′	.8405	.1595	.9823	.0177	.1418	.8582	10′
44° 00′	9.8418	10.1582	9.9848	10.0152	10.1431	9.8569	**46° 00′**
10′	.8431	.1569	.9874	.0126	.1443	.8557	50′
20′	.8444	.1556	.9899	.0101	.1455	.8545	40′
30′	.8457	.1543	.9924	.0076	.1468	.8532	30′
40′	.8469	.1531	.9949	.0051	.1480	.8520	20′
50′	.8482	.1518	9.9975	.0025	.1493	.8507	10′
45° 00′	9.8495	10.1505	10.0000	10.0000	10.1505	9.8495	**45° 00′**
	L cos θ	L sec θ	L cot θ	L tan θ	L csc θ	L sin θ	Angle θ

Answers to Odd-Numbered Exercises

CHAPTER 1

Section 1

1. (a) $\{a, b, c, d, f, g, h, k\}$ (b) $\{a.c.e\}$ (c) $\{b.d.f\}$
 (d) $\{g, h, k\}$ (e) $\{b, d, f\}$ (f) $\{b, d, f, g, h, k\}$
5. (a) True (b) True (c) True (d) True
 (e) True (f) True (g) False (h) True
 (i) True (j) False (k) True (l) True

Section 2

1. (a) True (b) True (c) False
 (d) False (e) True (f) False

Section 3

1. (a) False (b) False (c) False (d) True (e) False
 (f) True (g) True (h) False (i) True (j) True
 (k) True (l) True (m) True (n) True

Section 4

1. Yes
3. 1021 is prime.

5. (a) $\frac{3}{7}$ (b) $\frac{4}{15}$ (c) $\frac{11}{7}$ (d) $\frac{4}{25}$
(e) $\frac{49}{27}$ (f) $\frac{4}{3}$ (g) $\frac{5}{7}$ (h) $\frac{10}{13}$

7. You do not need to try any divisor larger than $\sqrt{n}$.

Section 5

1. (a) 729 (b) $\frac{1}{729}$ (c) 3 (d) $\frac{1}{3}$
(e) 27 (f) $\frac{1}{27}$ (g) 9 (h) $\frac{1}{9}$

3. (a) 25 (b) $5^{1/2}$ (c) $\frac{1}{25}$ (d) 25
(e) 54.78 (f) $5^{.21}$ (g) $5^{22/21}$ (h) $5^{7/8}$

5. (a) $5\sqrt{2}$ (b) $4\sqrt{6}$ (c) $2x^4y^2\sqrt{15x}$ (d) $6x^2y^4\sqrt{2xy}$
(e) $4x^3y^4z^6\sqrt{2xz}$ (f) $2a^4b^3c^4\sqrt{17a}$ (g) $5x^5y$ (h) $11a^5b^6\sqrt{2ab}$

7. (a) $3\sqrt[3]{2}$ (b) $2x^3y\sqrt[4]{3x^2y^2}$ (c) $.1\sqrt[5]{.1}$ (d) $-2\sqrt[3]{9}$
(e) $-\sqrt[3]{17}$ (f) $\sqrt[3]{15}/3$ (g) $\sqrt[4]{6}/2$ (h) $\sqrt[3]{5y^2}/y$
(i) $2x^2y^3\sqrt[3]{2x^2y^2}$ (j) $2a^3b^2\sqrt[7]{8a^4b^2}$

9. (a) $2\sqrt{3}$ (b) $2\sqrt[3]{15}$ (c) 10 (d) $23\sqrt{61}$
(e) .3 (f) $\sqrt{2}$ (g) 10 (h) $3\sqrt[5]{143}$

11. (a) $\sqrt{21}/3$ (b) $\sqrt{14}/7$ (c) $\sqrt{3}/3$ (d) $\sqrt[3]{20}/2$ (e) $\sqrt{ab}/b$
(f) $\sqrt[3]{ab^2}/b$ (g) $\sqrt[4]{20}/2$ (h) $\sqrt[3]{.02}/.1$ (i) $\sqrt{2\sqrt[3]{9}}/3$ (j) $\sqrt{2\sqrt[3]{6}}/2$

13. (a) $3x^2y$ (b) $\sqrt{30}x^{5/2}y^{7/2}$ (c) $3a^{13/3}b^{8/3}$
(d) $8a^3b^6$ (e) $\sqrt[3]{2}\sqrt{3}a^{8/3}b^{19/6}$ (f) $(5x^2y/3)^{1/3}$ or $(\sqrt[3]{45}x^{2/3}y^{1/3})/3$
(g) $(2a^{1/15}b^{-8/15})/3$ or $2a^{1/15}/3b^{8/15}$ (h) $x^{7/6}y^{1/4}$

Section 6

1. (a) $8x^2 - 22x - 21$ (b) $16x^2 + 66x + 35$ (c) $6x^2 - 71x + 55$
(d) $6x^2 + 35x + 50$ (e) $20a^2 + ab - 63b^2$ (f) $12x^2 + 46xy - 77y^2$
(g) $16x^2 + 22xz - 45z^2$ (h) $30x^2 - 113xy + 91y^2$

3. (a) $x^4 + 2x^3 - 2x^2 - 6x - 3$ (b) $x^3 + y^3$
(c) $4x^4 + 5x^3 - 5x^2 + 13x - 5$ (d) $a^4 - a^3b + 2a^2b^2 + ab^3 + 3b^4$
(e) $x^5 + 4x^4 - 6x^3 - 15x^2 - 11x - 35$
(f) $a^6 + 2a^5b - 23a^4b^2 - 54a^3b^3 - 82a^2b^4 + 53ab^5 - 35b^6$

5. (a) $(x - 2y)^2$
(b) $(b^2 - 2a)(b^4 + 2b^2a + 4a^2)(b^2 + 2a)(b^4 - 2b^2a + 4a^2)$
(c) $a(a + 3b)(a + 2b)$ (d) $2x(x - 8)(x + 1)$
(e) $(a^2 + 4)(a + 2)(a - 2)$ (f) $3x(x - 3)(x + 1)$
(g) $(x^2 + x + 1)(x - 1)$ (h) $(y - 2)(y^2 + 2y + 4)$
(i) $(x + 1)(x^2 - x + 1)$ (j) $(x^2 + 1)(x - 1)(x + 1)$
(k) $(x - \sqrt[3]{2})(x^2 + \sqrt[3]{2}x + \sqrt[3]{4})$ (l) $(y - 1)(y^4 + y^3 + y^2 + y + 1)$
(m) $(a + 1)(a^4 - a^3 + a^2 - a + 1)$ (n) $(a + 2)(a^4 - 2a^3 + 4a^2 - 8a + 16)$
(o) $(x - 2)(x^6 + 2x^5 + 4x^4 + 8x^3 + 16x^2 + 32x + 64)$

7. (a) $(5x - y)(4x + 3y)$ (b) $(x + 4y)(x - 2y)$
(c) $(2a + 1)(3a - 4)$ (d) $(3a + 2)(2a - 3)$
(e) $(5x + 6y)(x - y)$ (f) $(5x + 3y)(2x - 3y)$
(g) $4x^3(x - 1)^2$ (h) $2xy(3x - 7y)(2x + 5y)$

(i) $3x^2y^2(x^2 + y^2)(x + y)(x - y)$
(j) $(x - 1)(x - 3)(x^2 + 3x + 9)(x^2 + x + 1)$
(k) $(x - \sqrt{7})(x + \sqrt{7})$ (l) $(y - \sqrt{3})(y + \sqrt{3})$
(m) $(x - \frac{1}{2})(x + \frac{1}{3})$ (n) $(x + \frac{4}{5})(x + \frac{2}{3})$

9. (a) $\dfrac{5(\sqrt{7} + \sqrt{5})}{2}$ (b) $\dfrac{\sqrt{11} - \sqrt{7}}{4}$ (c) $\dfrac{13 + 2\sqrt{30}}{7}$

(d) $\dfrac{17 - \sqrt{285}}{2}$ (e) $\dfrac{a - 2\sqrt{ab} + b}{a - b}$ (f) $\dfrac{\sqrt{42} + 3\sqrt{2}}{4}$

(g) $(\sqrt[3]{25} + \sqrt[3]{10} + \sqrt[3]{4})/2$ (h) $(\sqrt[3]{x^2} + \sqrt[3]{xy} + \sqrt[3]{y^2})/(x - y)$

11. (a) $(x - 2)(x - 3)(x - 7)$ (b) $(x + 3)(x^2 + 3x + 12)$
(c) $(x - 4)(x^2 + x + 1)$ (d) $(x - 2)(x - 3)(x + 3)(x + 2x + 4)$
(e) $(x - 1)(x - 3)(x^3 + 9x^2 + 27x + 54)$

15. 2^n

17. $(x - y)^n = x^n - nx^{n-1}y + \dfrac{n(n-1)}{2}x^{n-2}y^2 - \dfrac{n(n-1)(n-2)}{2 \cdot 3}x^{n-3}y^3 + \cdots + (-1)^n y^n$

The $(k + 1)$st term is

$$(-1)^{k+1}\frac{n!}{(n-k)!k!}x^{n-k}y^k$$

Section 7

1. (a) $\dfrac{x + 3y}{x + y}$ (b) $\dfrac{x + y}{x^2 + xy + y^2}$ (c) $\dfrac{x - 5y}{x - 3y}$

(d) $\dfrac{x + 2y}{5(x - 2y)}$ (e) $\dfrac{(x^2 + 1)(x + 1)}{x^2 + x + 1}$ (f) $\dfrac{x^4 + x^2 + 1}{x^2 + 1}$

(g) $\dfrac{x - 4}{2x - 5}$ (h) $\dfrac{(5x - 4)(2x + 7)}{2x(5x^3 + x^2 - 4)}$

3. (a) $\dfrac{7x^2 - 3x}{(x + 1)(x - 1)}$ (b) $\dfrac{1}{x + 1}$ (c) $\dfrac{x + 3y}{3x + y}$

(d) $\dfrac{1}{3}$ (e) $\dfrac{5x - 1}{6x}$ (f) $\dfrac{(x + 2y)[(x - 2y)^2 + 1]}{x - 2y}$

(g) $\dfrac{2(x - 1)}{3(x - 5)}$ (h) $\dfrac{5(x^2 + x + 1)}{3(x + 1)}$

5. (a) $\dfrac{6x^3 - 16x^2 + 74x + 8}{(x - 5)(x - 2)(x + 3)(x + 2)}$ (b) $\dfrac{15x^2 - 2x - 5}{x(x - 1)(x + 1)}$

(c) $\dfrac{12x^2 + 2x + 2}{(x + 1)^2(x - 1)^2}$ (d) $\dfrac{x^3 + 5x^2 + 6x - 9}{(x - 4)(x + 4)(x + 3)}$

(e) $\dfrac{5x^3 - 11x^2 + 5x - 5}{(x - 2)^2(x + 2)}$ (f) $\dfrac{-(x^3 + 2x^2 - 3x - 3)}{(x + 2)(x - 2)}$

(g) $\dfrac{2x^3 + 7x^2 - 9x - 36}{(x + 1)(x - 4)(x + 3)}$ (h) $\dfrac{2x^4 + x^3 + 9x^2 - 206x + 20}{(x + 5)(x - 5)(x^2 + 5x + 25)}$

7. (a) $\dfrac{(2x+5)(x+1)}{(x-1)(3x-1)}$ (b) $\dfrac{(x-3)(x+4)}{(x+6)(x-2)}$ (c) $\dfrac{(x+1)(x-2)}{(x+2)(x-1)}$

(d) $\dfrac{(x-7)(x^2+4x+5)}{(x+1)(x+2)(2x+1)}$ (e) $\dfrac{x^2-1}{x^2+1}$ (f) $\dfrac{y}{x-y}$

(g) $\dfrac{(x^2+2x+4)(x^2+x+1)}{(x+1)(x+2)}$ (h) $\dfrac{(y-1)^5}{(y-2)^3(y^2+2y+4)}$

CHAPTER 2

Section 1

1. (a) $x = \frac{7}{3}$ (b) $x = 4\sqrt{2}$ (c) $y = \frac{2}{3}$
(d) $z = 15(4+\sqrt{2})/14$ (e) $10/(\sqrt[3]{4}-\sqrt[3]{2})$ or $10\sqrt[3]{2}+5\sqrt[3]{4}+10$

3. (a) $x = \frac{1713}{118}$ (b) $x = 1$ (c) $y = \frac{21}{22}$
(d) $x = -35(\sqrt{5}+3\sqrt{2})/153$ (e) $x = -\frac{2146}{481}$

5. (a) $r = d/t$ (b) $\frac{220}{19}$ yds/sec (c) 500 mph (d) $t = d/r$
(e) 1 sec (f) 36 sec (g) 1 sec

7. (a) 18 in (b) 10.5 in (c) 2.5 in

9. (a) 63.48 sq. cm. (b) $h = 2A/(b_1+b_2)$, $b_1 = (2A/h) - b_2$, $b_2 = (2A/h) - b_1$
(c) 25.21 in (d) 11.33 cm.

11. $x = \frac{4}{3} - 2y$, $x = -\frac{14}{3}$

Section 2

1. (a) $x > -\frac{3}{2}$ (b) $x < 4$ (c) $x < \frac{40}{3}$ (d) $x \le 9$
(e) $x \ge -\frac{11}{8}$ (f) $x < \frac{235}{98}$ (g) $y < 2$ (h) $x < 26$
(i) $x \le \frac{38}{47}$ (j) $x < \frac{59}{60}$

3. (a) $-\frac{3}{2} < x < 4$ (b) $-\frac{11}{8} \le x \le 9$ (c) $\sqrt{3}/3 < x < 2\sqrt{2}$
(d) $\frac{16}{7} \le x < \frac{5}{3}$ (e) $\frac{1}{3} < x < \frac{4}{3}$ (f) all numbers
(g) $x \ge \frac{24}{13}$ or $x \le \frac{14}{13}$ (h) all numbers

Section 3

1. $x = \frac{13}{5}$, $y = \frac{7}{5}$

3. $x = \frac{43}{39}$, $y = -\frac{145}{78}$

5. $x = \frac{17}{3}$, $y = -\frac{23}{3}$

7. $x = \frac{156}{55}$, $y = \frac{111}{220}$

9. $x = \frac{14826}{18117}$, $y = -\frac{504}{18117}$

11. $x = (\sqrt{35}+\sqrt{10}-\sqrt{21}-\sqrt{6})/2$
$y = (\sqrt{105}+\sqrt{30}-3\sqrt{7}-5\sqrt{2})/2$

13. $x = (pd-qb)/(ad-bc)$, $y = (aq-cp)/(ad-bc)$

15. $y = \frac{937}{904}$, $x = \frac{1177}{452}$

17. $x = 4$, $y = -2$, $z = -3$

19. $x = 2.9$, $y = -.2$, $z = .5$

21. $x = (pck - qbk + rbf)/(ack + gbf)$
$y = (aqk - arf + gpf)/(ack + gbf)$
$z = (acr + gbq - gcp)/(ack + gbf)$

23. $$x = \frac{4\sqrt{22} + 12\sqrt{7} - 8\sqrt{33} - 3\sqrt{3} + 3\sqrt{2}}{\sqrt{21} - 2\sqrt{3} + \sqrt{35} - 3\sqrt{2} + \sqrt{110} + 3\sqrt{11}}$$
$$y = \frac{-4\sqrt{33} - 12\sqrt{6} + 8\sqrt{55} + 3\sqrt{5} + 3\sqrt{3}}{\sqrt{21} - 2\sqrt{3} + \sqrt{35} - 3\sqrt{2} + \sqrt{119} + 3\sqrt{11}}$$
$$z = \frac{4\sqrt{21} - 8\sqrt{3} - 8\sqrt{35} + 16\sqrt{2} - 3\sqrt{10} - 9}{\sqrt{21} - 2\sqrt{3} + \sqrt{35} - 3\sqrt{2} + \sqrt{110} + 3\sqrt{11}}$$

25. $m = (q - k)/(p - h)$, $b = (pk - qh)/(p - h)$

Section 4

1. (a) $x = \frac{1}{4}, x = -\frac{3}{2}$ (b) $x = -\frac{1}{5}, x = \frac{7}{2}$
(c) $x = 3$ (d) $x = \frac{1}{6}, x = 16$
(e) $x = -\frac{2}{5}$ (f) $x = -\frac{1}{6}, x = \frac{5}{7}$
(g) $x = 0, x = -7, x = -\frac{3}{2}$ (h) $x = 0, x = 1, x = -\frac{1}{6}$
(i) $t = 4, t = -4$ (j) $w = 2, w = -2$

3. (a) $x = 3, x = -3$ (b) $y = 5, y = -5$
(c) $y = 7, x = -7$ (d) $x = 4, x = -6$
(e) $x = 6, x = -8$ (f) $x = 5, x = -9$
(g) $x = 4, x = -10$ (h) $x = 1, x = -5$
(i) $x = 1, x = -9$ (j) $x = 2, x = -8$
(k) $x = 2, x = -4$ (l) $x = -p + q, x = -p - q$
(m) $x = p + q, x = p - q$ (n) $x = -p + q, x = -p - q$

Section 5

1. (a) $x = -2 + 2\sqrt{3}, x = -2 - 2\sqrt{3}$
(b) $x = (7 + \sqrt{61})/2, x = (7 - \sqrt{61})/2$
(c) $x = 1, x = -2$
(d) $x = -6 + \sqrt{41}, x = -6 - \sqrt{41}$
(e) $x = 1, x = -\frac{5}{2}$
(f) $x = 1, x = -\frac{5}{3}$
(g) $x = (-1 + \sqrt{17})/8, x = (-1 - \sqrt{17})/8$
(h) $x = -1, x = \frac{1}{3}$

3. (a) $x = (-3 - \sqrt{41})/4, x = (-3 + \sqrt{41})/4$
(b) $x = (-5 - \sqrt{33})/2, x = (-5 + \sqrt{33})/2$
(c) $x = (-1 + \sqrt{253})/14, x = (-1 - \sqrt{253})/14$
(d) $x = 5, x = -1$
(e) $x = (-7 + 2\sqrt{7})/10, x = (-7 - 2\sqrt{7})/10$
(f) $x = (-\sqrt{35} + \sqrt{35 + 28\sqrt{14}})/14, x = (-\sqrt{35} - \sqrt{35 + 28\sqrt{14}})/14$
(g) $x = (-17 - \sqrt{2081})/28, x = (-17 + \sqrt{2081})/28$
(h) $x = (15 - \sqrt{33})/32, x = (15 + \sqrt{33})/32$

(i) no solutions
(j) no solutions

5. (a) $\left(x + \frac{5 - \sqrt{65}}{2}\right)\left(x + \frac{5 + \sqrt{65}}{2}\right)$

(b) $\left(x + \frac{2 - \sqrt{22}}{3}\right)\left(x + \frac{2 + \sqrt{22}}{3}\right)$

(c) $\left(x - \frac{3 + \sqrt{29}}{10}\right)\left(x - \frac{3 - \sqrt{29}}{10}\right)$

(d) $\left(x - \frac{17 + \sqrt{161}}{32}\right)\left(x - \frac{17 - \sqrt{161}}{32}\right)$

(e) $\left(x - \frac{3 + \sqrt{541}}{14}\right)\left(x - \frac{3 - \sqrt{541}}{14}\right)$

(f) $x\left(x - \frac{7 + \sqrt{33}}{8}\right)\left(x - \frac{7 - \sqrt{33}}{8}\right)$

(g) no solution

(h) $\left(x - \frac{5 + 2\sqrt{6.73}}{.8}\right)\left(x - \frac{5 - 2\sqrt{6.73}}{.8}\right)$

7. $-B/2A$
9. $-B/A$

Section 6

1. (a) $-5 < x < 4$ (b) $x < \frac{1}{2}$ or $x > \frac{7}{5}$
(c) $4 < x < 5$ (d) $-1 < x < 0$ or $x > 4$
(e) $\frac{1}{6} < x < \frac{2}{5}$ (f) $0 < x < \frac{2}{15}$ or $1 < x$
(g) $x > 2$ or $x < -2$ (h) $-3 < x < 3$
(i) $x < \frac{2}{3}$ or $x > \frac{3}{4}$ (j) $x > -1$

3. (a) $x > 3$ or $x < -3$ (b) $-\sqrt{5} < x < \sqrt{5}$
(c) $x > 1$ or $x < -5$ (d) $-7 < x < 13$
(e) $x > 2$ or $x < -3$ (f) $-\frac{3}{2} < x < \frac{5}{2}$
(g) $x < -4 - \sqrt{.5}$ or $x > \sqrt{.5} - 4$ (h) $x < \sqrt{7} + 3$ or $x > 3 - \sqrt{7}$
(i) $-\frac{1}{6} < x < \frac{7}{6}$ (j) $-10 - .2 < x < 10 - .2$

5. $Ax^2 + Bx + C < 0$ if x is between the two roots and $Ax^2 + Bx + C \geq 0$ otherwise.

Section 7

1. $x = 5$
3. $x = 8, x = -\frac{3}{2}$
5. $x = (5 + \sqrt{41})/4, x = (5 - \sqrt{41})/4$
7. $x = -3 + \sqrt{5}, x = -3 - \sqrt{5}$
9. $x = 1, y = 2,$ or $x = -1, y = -2$
11. $x = -8\sqrt{5}/15, y = (10 + 2\sqrt{5})/5$ or $x = 8\sqrt{5}/5, y = (10 - 2\sqrt{5})/5$
13. $x = \pm\sqrt{\frac{8}{25}}, y = \pm\sqrt{\frac{3}{35}}$
15. $x = 4 + 2\sqrt{3}, y = 8 + 3\sqrt{17}$ or $x = 4 - 2\sqrt{3}, y = 17 - 8\sqrt{3}$

Section 8

1. (a) 32 (b) 0 (c) -20 (d) $-\frac{128}{3}$ (e) $\frac{179}{42}$ (f) -7

3. (a) $x = -\frac{625548}{804804}$ or approximately $-.777$
   $y = \frac{4014}{156780}$ or approximately .026
   (b) $x = \frac{313636}{88920}$ or approximately 3.527
   $y = -\frac{7500}{8892}$ or approximately $-.843$
   (c) $x = (\sqrt{21} - 4\sqrt{15})/(2\sqrt{5} + \sqrt{15})$
   $y = (\sqrt{30} - \sqrt{14})/(2\sqrt{5} + \sqrt{15})$
   (d) $x = 1.875$, $y = -.05$

Section 9

1. $y = 5/2x$, $y = 25$
3. $y = 55/x$, $y = \frac{11}{2}$
5. $2y + x^2 = 9$, $y = -\frac{91}{2}$
7. $x^2y^2 = 196$, $y = \pm\frac{7}{5}$
9. $6y^3 - 625x^2 = 1500$, $y = 80\sqrt{3}/3$

Section 10

1. $6 + \sqrt{6}, 6 - \sqrt{6}$
3. .23 qt vermouth and .77 qt the 10:1 mixture
5. \$1250
7. $x/(1 - x)$
9. $\sqrt{165550}$ or about 129.03 miles
11. $\pm\sqrt{5}$ and $\pm\sqrt{3}$
13. 9 parts peanuts to 2 parts each of the others
15. 3200 lbs
17. at least 42¢
19. \$142.40
21. about 11%
23. 83%
25. \$60.48
27. at least 2.90
29. about 16 (15.95)
31. (a) 4.18 (b) 2.34 (c) 1.79 (d) 60.44 (e) no
33. 6.4 gallons
35. less than \$24.80
37. about .36 qt of oil
39. $2\sqrt[3]{2}$, multiplies it by 8
41. \$2.89, \$11.10
43. sales less than \$133,333 or greater than \$250,000
45. .46 gal of first and 5.4 gal of second
47. 12 months
49. $\frac{64}{3}$ or 21.33; $\frac{16}{3}$ or 5.33
51. 64
53. \$1280 a year
55. \$13,200; \$9,360, \$7,100
57. (a) $P = -100p^2 + 325p - 105$ (b) \$145
    (c) \$1.62 yields a profit of \$159.06

CHAPTER 3

Section 1

1. (a) 13 (b) $\sqrt{185}$ (c) $\sqrt{680}$
   (d) $\sqrt{7.402}$ (e) $\sqrt{14} - \sqrt{6}$ (f) $\sqrt{43 - 2\sqrt{209} - 4\sqrt{15}}$
   (g) $2\sqrt{a^2 + b^2}$ (h) $|a - b|\sqrt{2}$ (i) $\sqrt{a^2 + b^2}$
   (j) $|a|\sqrt{2}$

5. (a) $(5, 0), (3, 4), (3, -4), (-3, 4), (2, \sqrt{21})$

9. $\sqrt{(x - 5)^2 + y^2} + \sqrt{(x + 5)^2 + y^2} = 20$

11. $\sqrt{(x - 2)^2 + (y - 7)^2} = |y - 1|$

12. $|\sqrt{(x - 4)^2 + y^2} - \sqrt{(x + 4)^2 + y^2}| = 1$

Section 3

3. (a) $y = 5x + 3$ (b) $y = -3x - 5$
(c) $y - 2 = -\frac{2}{3}(x - 5)$ (d) $y - 5 = 10(x - 1)$
(e) $y - 1 = -\frac{2}{5}(x - 7)$ (f) $y - 1 = 2(x - 6)$
(g) $y = 2x + 4$ (h) $x = 4$
(i) $y - 5 = 5(x - 1)$ (j) $y - 5 = -\frac{1}{5}(x - 1)$
(k) $y = 3$ (l) $x = 7$

Section 4

1. $y = -2x + 16$
3. yes; direct
5. 1450; 700
7. $78.67
9. (a) $y = .1x - 6000$ (b) $7,000 (c) $60,000

Section 5

7. (a) b decreases and the ellipse becomes "thinner"; b increases and the ellipse becomes "thicker."
(b) b increases and the ellipse gets "thicker"; b decreases and the ellipse becomes "thinner."
(c) Decreases.

Section 6

7. It does not meet the X-axis.
If $A > 0$, it is above the X-axis.
If $A < 0$, it is below.

9. $\left(\frac{11 + \sqrt{145}}{2}, \frac{310 + 26\sqrt{145}}{4}\right), \left(\frac{11 - \sqrt{145}}{2}, \frac{310 - 26\sqrt{145}}{4}\right)$

11. no solutions

Section 7

13. yes
17. The graph of the first is the set of asymptotes of the second.

Section 8

1. (a) $-5 \leq x \leq 5; 0 \leq f(x) \leq 5$
(b) $x \neq 1; g(x) \neq 1$
(c) all real numbers; $0 < h(x) \leq 9$
(d) x and $k(x)$ can be any real number.
(e) $x \neq -1$ and $x \neq 4$; $p(x)$ can be any number.
5. All have the same shape.
7. The larger n, the "flatter" the curve at $x = 0$.

Section 9

1. (a) min at (2, 19) (b) min at $(-1.25, -10.125)$
(c) max at $(\frac{2}{3}, \frac{47}{3})$ (d) max at $(-.25, 24.25)$
(e) max at $(-3, 54)$; min at $(3, -54)$ (f) max at $(-1, 5)$; min at $(1, -3)$
(g) max at (0, 17); min at $(-2, 0)$ and (2, 0)
(h) max at (0, 64); min at $(-\sqrt{15}, -11)$ and $(\sqrt{15}, -11)$

3. $64\sqrt{2}$ or about 90.51 ft/sec

5. (a) 500 (b) 58 (c) Part a is the same; part b becomes 68

Section 10

1. (a) $x^3 + x^2 + 1$; all real numbers (b) $x^5 + x^3$; all real numbers
(c) $x^3/\sqrt{x-2}$; $x > 2$ (d) $\sqrt{x-2}/(x+1)$; $x > 2$
(e) $(x^2+1)/(x+1)$; $x \neq -1$ (f) $x^3 - x^2 - 1$; all real numbers
(g) $x^3/(x+1)$; $x \neq -1$ (h) $1/(x+1)(x^2+1)$; $x \neq -1$

3. (a) $x^6 + 1$; all real numbers (b) $(x^2+1)^3$; all real numbers
(c) $x - 1$; $x \geq 2$ (d) $\sqrt{\dfrac{-2x-1}{x+1}}$; $-1 < x \leq -\dfrac{1}{2}$
(e) $\dfrac{x^2+2x+2}{(x+1)^2}$; $x \neq -1$ (f) $(x-2)^{3/2}$; $x \geq 2$
(g) $\sqrt{\dfrac{-x^2-2x}{(x+1)^2}}$; $-2 \leq x < -1$; $-1 < x \leq 0$
(h) $\dfrac{(x+1)^3}{1+(x+1)^3}$; $x \neq (-1, -2)$

CHAPTER 4

Section 1

1. 1.3222 **3.** 1.0791 **5.** 1.1461 **7.** 6.4746
9. -1.699 **11.** 2.2922 **13.** 0.6232 **15.** 0.9542
17. 0.5441 **19.** -3.097 **21.** -0.0458 **23.** -0.3010
25. 0.1761 **27.** $1 + p, 2 + p, 3 + p, p - 2$

Section 2

1. (a) 1.6928 (b) 8.9694 − 10 (c) 4.8686 (d) 3.0334
(e) 9.8825 − 10 (f) 1.3909 (g) 9.9279 − 10 (h) 3.9754
(i) 1.7952 (j) 8.8915 − 10

3. (a) 1.7734 (b) 2.7963 (c) 0.9653 (d) 8.8536 − 10
(e) 1.8857 (f) 9.9787 − 10 (g) 1.3921 (h) 1.2963
(i) 7.599 − 10 (j) 4.6625

Section 3

1. (a) 18,680 (b) 0.5420 (c) 0.05177 (d) 1.471
(e) 0.2208 (f) 11.01 (g) 0.3232 (h) 0.3804

3. (a) 2.124 (b) 2.720 (c) 0.9530 (d) 6.106 (e) 1.162

Section 4

1. \$957.40
3. \$6,218
5. 35 quarters, $8\frac{3}{4}$ years or 9 years depending on bank policy
7. 1.5%/quarter or 6%/year compounded quarterly

Section 5

1. (a) 3 (b) 4 (c) -2 (d) $\frac{1}{2}$ (e) $\frac{1}{3}$
   (f) 4 (g) $\frac{1}{4}$ (h) -2 (i) $\frac{1}{3}$ (j) $\frac{4}{3}$
3. (a) 1.4651 (b) 1.0686 (c) 0.9040 (d) 0.4777
   (e) -0.9767 (f) 1.5972 (g) 1.4728 (h) 1.7918
   (i) 1.4350 (j) -2.6450

Section 7

1. $y = 4(3)^x$
3. $y = 5 + 7(1.4639)^x$
5. 40.8 million, 83.3 million
7. 12.99 gm

CHAPTER 5

Section 1

1.

	0°	90°	180°	270°
sin	0	1	0	−1
cos	1	0	−1	0
tan	0	*	0	*
cot	*	0	*	0
csc	*	1	*	−1
sec	1	*	−1	*

* Does not exist.

3. (a) 1/2 (b) $-\sqrt{2}/2$ (c) $-\sqrt{3}/3$ (d) $-\sqrt{3}/3$
   (e) $-\sqrt{3}/2$ (f) $-2\sqrt{3}/3$ (g) $-\sqrt{2}$ (h) $-\sqrt{3}/2$
   (i) $\sqrt{3}$ (j) 2
7. Range of csc is $(-\infty, -1) \cup (1, \infty)$.
   Range of sec is $(-\infty, -1) \cup (1, \infty)$.

Section 2

1. (a) $\pi/6$ (b) $2\pi/3$ (c) $-\pi/4$ (d) $3\pi/4$
   (e) $25\pi/6$ (f) $5\pi/6$ (g) $5\pi/4$ (h) $5\pi/12$

3.

	0	$\frac{\pi}{6}$	$\frac{\pi}{4}$	$\frac{\pi}{3}$	$\frac{\pi}{2}$	$\frac{2\pi}{3}$	$\frac{3\pi}{4}$	$\frac{5\pi}{6}$	π	$\frac{7\pi}{6}$	$\frac{5\pi}{4}$	$\frac{4\pi}{3}$	$\frac{3\pi}{2}$	$\frac{5\pi}{3}$	$\frac{7\pi}{4}$	$\frac{11\pi}{6}$
sin	0	$1/2$	$\frac{\sqrt{2}}{2}$	$\frac{\sqrt{3}}{2}$	1	$\frac{\sqrt{3}}{2}$	$\frac{\sqrt{2}}{2}$	$1/2$	0	$-1/2$	$\frac{-\sqrt{2}}{2}$	$\frac{-\sqrt{3}}{2}$	-1	$\frac{-\sqrt{3}}{2}$	$\frac{-\sqrt{2}}{2}$	$-1/2$
cos	1	$\frac{\sqrt{3}}{2}$	$\frac{\sqrt{2}}{2}$	$1/2$	0	$-1/2$	$\frac{-\sqrt{2}}{2}$	$\frac{-\sqrt{3}}{2}$	-1	$\frac{-\sqrt{3}}{2}$	$\frac{-\sqrt{2}}{2}$	$-1/2$	0	$1/2$	$\frac{\sqrt{2}}{2}$	$\frac{\sqrt{3}}{2}$
tan	0	$\frac{\sqrt{3}}{3}$	1	$\sqrt{3}$	*	$-\sqrt{3}$	-1	$\frac{-\sqrt{3}}{3}$	0	$\frac{\sqrt{3}}{2}$	1	$\sqrt{3}$	*	$-\sqrt{3}$	-1	$\frac{-\sqrt{3}}{2}$
cot	*	$\sqrt{3}$	1	$\frac{\sqrt{3}}{3}$	0	$\frac{-\sqrt{3}}{3}$	-1	$-\sqrt{3}$	*	$\sqrt{3}$	1	$\frac{\sqrt{3}}{3}$	0	$\frac{-\sqrt{3}}{3}$	-1	$-\sqrt{3}$
sec	1	$\frac{2\sqrt{3}}{3}$	$\sqrt{2}$	2	*	-2	$-\sqrt{2}$	$\frac{-2\sqrt{3}}{3}$	-1	$\frac{-2\sqrt{3}}{3}$	$-\sqrt{2}$	-2	*	2	$-\sqrt{2}$	$\frac{2\sqrt{3}}{3}$
csc	*	2	$\sqrt{2}$	$\frac{2\sqrt{3}}{3}$	1	$\frac{2\sqrt{3}}{3}$	$\sqrt{2}$	2	*	-2	$-\sqrt{2}$	$\frac{-2\sqrt{3}}{3}$	-1	$\frac{-2\sqrt{3}}{3}$	$-\sqrt{2}$	-2

* Does not exist.

Section 3

11. $\cos x \doteq .9165$
$\tan x \doteq .4364$
$\cot x \doteq 2.291$
$\sec x \doteq 1.091$
$\csc x \doteq 2.5$

13. $\sin x \doteq .9539$
$\tan x \doteq 3.17$
$\cot x \doteq .3145$
$\sec x \doteq \frac{10}{3}$
$\csc x \doteq 1.048$

15. $\sin x = s\sqrt{1+s^2}/(1+s^2)$
$\cos x = \sqrt{1+s^2}/(1+s^2)$
$\cot x = 1/s$
$\sec x = \sqrt{1+s^2}$
$\csc x = \sqrt{1+s^2}/s$

17. $\sin x \doteq -0.9798$
$\tan x \doteq 4.899$
$\cot x \doteq 0.2041$
$\sec x \doteq -5$
$\csc x \doteq -1.0206$

19. $\cos x = -\sqrt{1-t^2}$
$\tan x = -t\sqrt{1-t^2}/(1-t^2)$

21. $\cos x = -\sqrt{1+t^2}/(t^2+1)$
$\sin x = -t\sqrt{1+t^2}/(1+t^2)$

Section 5

1. (a) 0.4067 (b) 0.4848 (c) 1.0724 (d) 2.904
(e) 3.42 (f) 1.589

Section 7

7. (a) $(\sqrt{2}/4)(1+\sqrt{3})$ (b) $(\sqrt{2}/4)(1+\sqrt{3})$ (c) $(\sqrt{2}/4)(\sqrt{3}-1)$
(d) $2-\sqrt{3}$ (e) $2+\sqrt{3}$ (f) $\sqrt{2-\sqrt{2}}/2$
(g) $\sqrt{2+\sqrt{2}}/2$ (h) $\sqrt{2}-1$

Section 8

1. (a) $\pi/3$ (b) $-\pi/2$ (c) $\pi/6$ (d) $\pi/4$
(e) $7\pi/18$ (f) $-.4108\pi$ (g) $\frac{53}{180}\pi$ (h) $\frac{113}{180}\pi$

3. (a) π (b) $-\pi/4$ (c) $\frac{11}{180}\pi$ (d) $\pi/3$
(e) $-\frac{2}{3}\pi$ (f) .3681

5. (a) 0.7071 (b) $\frac{4}{3}$ (c) -1.732 (d) 0.4359
(e) 0.7526 (f) $\frac{13}{10}$

7. (a) $\pi/2$ (b) $\pi/2$ (c) $\pi/2$ (d) $\pi/2$
(e) $\pi/2$ (f) $\pi/2$

CHAPTER 6

Section 1

1. (a) 0.2639 (b) 0.6494 (c) 2.391 (d) 2.778
(e) 0.7955 (f) 0.9799 (g) 0.5025 (h) 0.7221

3. (a) 0.5193 (b) 0.7210 (c) 0.2611 (d) 0.3959
(e) 0.8467 (f) 0.3239 (g) 0.2330 (h) 0.0953

5. (a) 0.2124 (b) 1.062 (c) 0.4305 (d) 0.3596
(e) 0.9649 (f) 0.9734

7. (a) 15.61°	(b) 82.36°	(c) 63.43°	(d) 16.4°
(e) 52.5°	(f) 22.75°		
9. (a) 10.82	(b) 0.5492	(c) 0.0038	(d) 2.025
(e) −3.239	(f) 8.627	(g) 0.8518	(h) 0.9365

Section 2

1. 25.37 in
33.47 in
3. 39.83°, 50.17°
5. 38.59 in, 35.65 in
7. 172.08 ft (plus height of end of string above ground)
9. 2.448 ft/sec
11. 348.7 ft

Section 3

1. $c = 28.64$, $A = 19.8$, $B = 106.2$
3. $c = 28.37$, $A = 8141'$, $B = 4459'$
5. $c = 75$, $a = 10.39$, $b = 13.84$
7. $A = 46.7$, $B = 96.7$, $C = 36.6$
9. $B = 37.3$, $C = 107.7$, $c = 74.7$ or $B = 127.3$, $C = 17.7$, $c = 14.43$
11. $A = 12.75$, $C = 136.25$, $c = 18.66$
19. 95.43 *ft*
21. Approx. 28 min
23. 73.62, 117.72
57.6°, 113.2°, 121.2°

Section 4

1. 112.7
3. 4983
5. 96110
9. 7545 ft^2
11. 17,468 ft^2
13.

Sides	Angle°
63, 73	90
52, 70	103.45
52, 73	85.7
63, 70	80.85

Section 5

1. $x = \frac{\pi}{3} + \frac{n\pi}{2}$ n an integer

3. $x = \frac{\pi}{3} + n\pi$ or $\frac{2\pi}{3} + n\pi$ $n \in Z$

5. $[68.2° + n(180°)]/5 \qquad n \in Z$

7. $x = n\pi$ or $\dfrac{3\pi}{4} + 2n\pi$ or $\dfrac{5\pi}{4} + 2n\pi \qquad n \in Z$

9. $x = \dfrac{3\pi}{2} + 2n\pi$ or $x = \dfrac{\pi}{4} + 2n\pi$ or $x = \dfrac{3\pi}{4} + 2n\pi \qquad n \in Z$

11. $x = \dfrac{\pi}{15} + \dfrac{n\pi}{5}$ or $\dfrac{2\pi}{15} + \dfrac{n\pi}{5} \qquad n \in Z$

15. $x = \dfrac{\pi}{6} + \dfrac{n\pi}{3} \qquad n \in Z$

CHAPTER 7

Section 1

1. (a) $10 + 16i$ (b) $62i$ (c) $\frac{53}{30} + \frac{4}{15}i$
(d) $-\frac{1}{6} + \frac{38}{15}i$ (e) $2\sqrt{3} + \sqrt{6} + (3\sqrt{2} - 2)i$ (f) $8\sqrt{2} - 2\sqrt{30}i$

3. (a) $-8i$ (b) -16.81 (c) $-7\sqrt{7}i$ (d) $4i$
(e) $2i\sqrt{3}$ (f) $9i\sqrt{2}$ (g) $i\sqrt[4]{2}$ (h) $\pi i\sqrt{\pi}$

5. (a) $-i, -1, i, 1, -i, -1, i, 1, -i$
(b) Divide $-n$ by 4 and compute i^{-r} where r is the remainder. It will be 1 if $r = 0$, -1 if r is even, $-i$ if $r = 1$, and i if $r = 3$.

11. (a) $z = -2 \pm 2i$ (b) $z = \dfrac{-1 \pm i\sqrt{3}}{2}$
(c) $z = \dfrac{-3 \pm 3i\sqrt{3}}{2}$ (d) $z = \dfrac{1 \pm i\sqrt{3}}{2}$
(e) $1, \dfrac{1 \pm i\sqrt{3}}{2}$ (f) $b, b\left(\dfrac{-1 \pm i\sqrt{3}}{2}\right)$
(g) $\pm 1, \pm i$ (h) $\pm b, \pm bi$

13. (a) $\dfrac{4i \pm \sqrt{-8 + 12i}}{2}$ (b) $\dfrac{2 + 5i \pm \sqrt{-37 + 4i}}{2i}$
(c) $\dfrac{-6 - 5i \pm \sqrt{-18 + 112i}}{8 - 10i}$ (d) $\dfrac{-3\sqrt{10}i \pm \sqrt{222 + 276i}}{2\sqrt{5} - 4\sqrt{5}i}$

Section 2

1. (a) $\sqrt{29}$ (b) $\sqrt{58}$ (c) $\sqrt{317}$ (d) $\sqrt{145/8}$
(e) $\sqrt{2545/14}$ (f) $\sqrt{5}$ (g) $.1\sqrt{65}$ (h) $.1\sqrt{653}$
(i) $\sqrt{5}$ (j) $\sqrt{a}$

3. (a) 68° (b) −23° (c) −52° (d) 48° (e) −14°
(f) 51° (g) 60° (h) 121° (i) 90° (j) 90°

5. (a) $30°; 2; 2(\cos 30° + i \sin 30°)$
(b) $45°; 5\sqrt{2}; 5\sqrt{2}(\cos 45° + i \sin 45°)$

(c) $-60°$; 2; $2(\cos -60° + i \sin -60°)$
(d) $-90°$; 7; $7(\cos -90° + i \sin -90°)$
(e) $90°$; $\sqrt{18}$; $\sqrt{18}(\cos 90° + i \sin 90°)$
(f) Approximately $53.13°$; 5; $5(\cos 53.13° + i \sin 53.13°)$

7. (a) $28(\cos 45° + i \sin 45°)$ (b) $.48(\cos 98° + i \sin 98°)$
(c) $15.4(\cos 163° + i \sin 163°)$ (d) $\frac{8}{21}(\cos 49 + i \sin 49)$

9. (a) $27\sqrt{3}/2 + \frac{27}{2}i$ (b) $8 + 8\sqrt{3i}$ (c) $-1.44i$
(d) $-16\sqrt{2}/243 - (16\sqrt{2}/243)i$ (e) -1 (f) i

11. (a) $\pm[\sqrt{2}/2 + (\sqrt{2}/2)i]$; $\pm[\sqrt{2}/2 - (\sqrt{2}/2)i]$
(b) $\pm(.92 + .38i)$; $\pm(.38 - .92i)$
(c) $\pm\sqrt{3}$; $\pm i\sqrt{3}$
(d) $\pm(.99 + .3i)$; $\pm(.13 - .99i)$
(e) $\pm(.48 - .13i)$; $\pm(.13 + .48i)$
(f) $\pm(3.84 - 1.37i)$; $\pm(1.37 + 3.84i)$

CHAPTER 8

Section 1

1. (a) $\frac{8}{15}; \frac{13}{17}; \frac{103}{107}$ (b) $\frac{5}{26}; \frac{10}{101}; \frac{100}{10001}$ (c) $\frac{3}{7}; \frac{8}{12}; \frac{98}{102}$ (d) $\frac{32}{25}; \frac{117}{100}; \frac{10197}{10000}$
3. d, f, g are geometric; h is arithmetic.
5. (a) $b, br, br^2, br^3, br^4, br^5$ (b) br^{n-1}
7. d positive; d negative

Section 2

1. In exercise 1: (a) converges to 1; (b) converges to 0; (c) converges to 1; (d) converges to 1.

In exercise 2: (a) converges to 0; (b) converges to 0; (d) converges to 0; (e) converges to 1; (i) converges to 1; (j) converges to 1.

Section 3

1. (a) $.8\overline{3}$ (b) $.\overline{142857}$ (c) $.\overline{307692}$ (d) $.3\overline{8}$
(e) $.\overline{72}$ (f) $.\overline{47058823529}$ (g) $.0625\overline{0}$ (h) $.078125\overline{0}$
3. Those, which when expressed in lowest terms, have denominators that are divisible by no primes other than 2 and 5.

Section 4

1. (a) 17.2; 19.9; 61.5
(b) $\frac{25}{12}; \frac{137}{60}; \frac{7381}{2520}$
(c) $\frac{23}{12}; \frac{263}{60}; \frac{17819}{2520}$
(d) $\frac{73}{12}; \frac{377}{60}; \frac{32581}{2520}$
(e) $\frac{260}{27}; \frac{844}{81}; 4[(3^{10} - 2^{10})/3^9]$ or approximately 11.79
(f) .97, .97, .97 to two decimal places.

Index

Index of Mathematical Symbols

76 77 78 79 80 9 8 7 6 5 4 3 2 1